Synthese Library

Studies in Epistemology, Logic, Methodology, and Philosophy of Science

Volume 530

The aim of *Synthese Library* is to provide a forum for the best current work in the methodology and philosophy of science and in epistemology, all broadly understood. A wide variety of different approaches have traditionally been represented in the Library, and every effort is made to maintain this variety, not for its own sake, but because we believe that there are many fruitful and illuminating approaches to the philosophy of science and related disciplines.

Special attention is paid to methodological studies which illustrate the interplay of empirical and philosophical viewpoints and to contributions to the formal (logical, set-theoretical, mathematical, information-theoretical, decision-theoretical, etc.) methodology of empirical sciences. Likewise, the applications of logical methods to epistemology as well as philosophically and methodologically relevant studies in logic are strongly encouraged. The emphasis on logic will be tempered by interest in the psychological, historical, and sociological aspects of science. In addition to monographs *Synthese Library* publishes thematically unified anthologies and edited volumes with a well-defined topical focus inside the aim and scope of the book series. The contributions in the volumes are expected to be focused and structurally organized in accordance with the central theme(s), and should be tied together by an extensive editorial introduction or set of introductions if the volume is divided into parts. An extensive bibliography and index are mandatory.

Adolfo García de la Sienra

The Logical Foundations of Nonstandard Analysis

With Applications to Economic Theory

Springer

Adolfo García de la Sienra
Universidad Veracruzana
Xalapa, Veracruz, Mexico

ISSN 0166-6991 ISSN 2542-8292 (electronic)
Synthese Library
ISBN 978-3-032-05676-4 ISBN 978-3-032-05677-1 (eBook)
https://doi.org/10.1007/978-3-032-05677-1

This work was supported by Research leading to this book was made possible by the Consejo Nacional de Ciencia y Tecnología (CONACYT) of Mexico, now Secretaría de Ciencia, Humanidades, Tecnología e Innovación (SECIHTI) through the project *Lógica, metodología y filosofía de la ciencia* CF-2023-I-161.

This Springer imprint is published by the registered company Springer Nature Switzerland AG
The registered company address is: Gewerbestrasse 11, 6330 Cham, Switzerland

This book is dedicated to my sons,
Rodrigo and Adolfo

Preface

It has been my intention in writing this book to achieve two goals: on the one hand, to provide a complete chain of demonstrations that lead from the solid foundation of class theory to the main results of nonstandard analysis along with some of its applications to economic theory; on the other hand, to minimize the length of the chain by restricting it to the strictly necessary links. I have used a unified language that is standard in logical theory and more digestible for philosophers, yet also attractive for mathematicians and economists. The novelty of the book is primarily in the organization and minimalist completeness of the subject, as well as in a logical foundation for nonstandard economics, based on Debreu's [1] concept of social system. I have aimed to provide my own demonstrations of well-known theorems, simpler and more consistent with the adopted notation.

Ackermann [2] provided a very elegant and austere foundation for class theory that I have taken as point of departure. The original system had to be complemented with additional axioms that I introduce in order to get a complete system for class and set theory. The present work leads the reader from this system, step by step, to a universe rich and complex enough to provide the foundations for nonstandard analysis and develop it sufficiently for its applications in economic theory. The idea has been to introduce the minimum theory and results required to lead the reader (particularly one with philosophical training) from Cantor's Paradise to the Garden of Delights of nonstandard analysis and its applications in economic theory. Almost all of the theorems presented are well known, but the sequence in which they are ordered and some of the demonstrations are not. In particular, finding the minimal sequence leading to saturation results turned out to be an exacting task.

Eager to develop or apply nonstandard analysis, mathematicians generally dedicate little space to the complete and systematic development of its foundations. On the other hand, logicians barely sketch the construction of the nonstandard universe (not even [3] fully develops the demonstrations) and fall short of showing that such a construction satisfies the requirements demanded by analysis (as, e.g., Goldblatt formulates them in 1988, pp. 158–162, who, by the way, also does not show it). This work aims to introduce the necessary apparatus to present a complete and integral foundation of nonstandard analysis.

After reviewing the necessary set theory for the task, I present a fairly complete account of Boolean algebra theory, including demonstrations of Tarski's theorems on fixed points of lattices and a detailed proof of Stone's Representation Theorem. The constructions in this chapter will be used in Chap. 3 to present a somewhat different proof of Gödel's Completeness Theorem. I develop topics—aimed at completeness—that are not strictly necessary for the development of nonstandard analysis but were within reach of developments that are. They may be omitted to stay strictly within the minimalist chain, but they may be useful for a course in Boolean algebras.

In Chap. 3, I review first-order logic and prove the existence of "non-Archimedean" spaces using Anatoli Maltsev's Lemma (Compactness Theorem). Later, in Chap. 4, using the ultraproduct construction, I proceed to develop the theory of ultraproducts up to the natural embedding. I place particular emphasis on the construction of saturated models, as saturated universes have nice properties that allow for the adequate development of the topological spaces required by nonstandard economics.

In Chap. 5, I construct the required nonstandard space and prove that this space satisfies the requirements demanded by analysis, as Goldblatt formulates them. I include a discussion of internal entities with some methods to deal with them.

In Chap. 6, I develop in some detail the theory of nonstandard numbers, also called "hyperreal numbers." The development is sufficient to provide, in Chap. 7, an introduction to infinitesimal calculus that is sufficient for a host of applications in economics.

Chapter 8 is dedicated to an exposition of the nonstandard topologies required by economic theory, but not before providing an introduction to the standard topology needed to transition to the nonstandard topological notions.

In Chap. 9, I introduce nonstandard versions of the classical theorems of the Separating Hyperplane and the Fixed Point of Fan-Glicksberg, as well as Kakutani's, which turns out to be a corollary of the latter. I also address the use of the Lyapunov-Loeb Theorem in economic theory.

I conclude the last chapter, Chap. 10, by developing in detail Richter's [4] Theorem on the existence of nonstandard representations of utility. Then, I introduce a nonstandard formulation of Debreu's [1] concept of a social system, prove the existence of equilibria for social systems, and conclude the book with an application of the concept to economic theory, specifically to the theory of pure exchange.

Although relatively self-contained, and dealing with almost all topics in textbooks, due to its monographic nature, this work does not intend to be a textbook on set theory and logic, although it can be used as such in a graduate course. I have aimed to leave to the reader some easy demonstrations as exercises, or the completion of others.

Xalapa, Veracruz, Mexico
Spring 2025

Adolfo García de la Sienra

References

1. Debreu, G. (1952). A social equilibrium existence theorem. *Proceedings of the National Academy of Sciences, 38*(10), 886–893.
2. Ackermann, W. (1956). Zur axiomatik der mengenlehre. *Mathematische Annalen, 131*(4), 336–345.
3. Chang, C. C., & Keisler, H. J. (1990). *Model theory*. Elsevier.
4. Richter, M. K. (1971). Rational choice. In J. S. Chipman, L. Hurwicz, M. K. Richter, & H. F. Sonnenschein (Eds.), *Preferences, utility, and demand* (pp. 29–58). Harcourt Brace Jovanovich.

Contents

Chapter 1
Classes and Sets

Abstract This chapter presents a comprehensive formulation of Wilhelm Ackermann's class theory. After discussing his initial formulation, which lacked the Axiom of Regularity, the axiom is introduced, and from the resulting complete system—referred to as system A—the set theory needed for the purposes of the book is developed. In fact, the treatment is quite detailed and can serve as an introduction to set theory. Moreover, it provides the axiomatic foundations for the development of category theory, as it allows for the existence of classes that are not sets. The power of system A is demonstrated by deriving the Zermelo–Fraenkel system with the Axiom of Choice (ZFC). Starting with the usual basic notions, the chapter develops in detail the theory of ordinal numbers, the rudiments of cardinal arithmetic, and includes a proof of Zorn's Lemma.

Keywords Ackermann's class theory · Set theory

1.1 Preamble

I introduce in this chapter the theory of classes required for the purposes of this treatise. Starting from a modification of Wilhelm Ackermann's sober and elegant theory of classes (1956)—a modification that I will call A—, I derive the axioms of Zermelo-Fraenkel set theory with the axiom of choice (ZFC) and all the set-theoretic apparatus that the treatise needs, with particular emphasis on ordinal numbers. This derivation is possible because Ackermann's theory is logically more powerful than ZFC, and it is convenient because both philosophers and mathematicians are more accustomed to operating within the confines of the latter theory. Although the treatment is not complete, it is more than sufficient as an introduction to the discipline.

Ackermann (1956: 337) stated the following:

> We assume a certain domain of things that can be left indeterminate in and of itself; but it must be broad enough for the totalities and sets that are to be formed to belong to it.

A. García de la Sienra, *The Logical Foundations of Nonstandard Analysis*,
Synthese Library 530, https://doi.org/10.1007/978-3-032-05677-1_1

However, the starting point of Ackermann's system cannot be "a certain domain of things that can be left indeterminate in and of itself", but rather a predicate in the vernacular (German for Ackermann; English for us) that cannot be interpreted as a totality (*Gesamtheit*), and that does not determine any totality: the predicate "*ist ein Klasse*" or "*is a class*", when it comes to providing the foundations of pure set theory; or "*ist ein Klasse oder ein Grundobjekt*", that is, "*it is a class or it is an urelement*", when it comes to providing the foundations of a set theory that admits urelements, which Ackermann calls *Grundobjekte*. This means that the foundation of class and set theory presupposes the conception of logic as language—as *lingua characteristica*—and, not as calculus—as *calculus raciocinator*.[1] In other words, A does not have an interpretation domain that can be thought of or conceived as a whole. This means that the universal quantifier of A does not refer to a previously determined domain.

Like A, the quantifier of ZF also does not have a predetermined domain, but unlike A, the variables of ZF take exclusively sets as values, while those of A take classes that may not be sets—and, in some interpretations, they also take urelements as values. Moreover, the language of A contains the predicate "*is a set*" ("*ist ein Menge*"), which refers to the elements of a well-defined class: the class of sets. Clearly, there is no such "totality" in ZF, and this should be sufficient to establish the difference between the two systems.

The language $\mathcal{L}_{\mathsf{A}}$ of A consists of two categories of symbols: logical symbols and parameters, presented in separate lists:

A. Logical symbols

0. Parentheses: (,).
1. Connectives: $\neg$, $\wedge$, $\vee$, $\rightarrow$, $\leftrightarrow$.
2. Variables: a countable sequence x, y, z, u, v, with or without subscripts.
3. The equality symbol: $=$.
4. The symbol ε for the epsilon operator of Hilbert.
5. The symbol λ for the lambda operator of Church.

The language $\mathcal{L}_{\mathsf{A}}$ of A consists of two categories of symbols: logical symbols and parameters, presented in separate lists:

B. Parameters

0. The universal and existential quantifiers: $\forall$ and $\exists$, respectively.
1. The symbol $\in$ for class membership.
2. The predicate I, 'is an urelement'; and the predicate M, 'is a set'.

The basic terms of the language are the variables. However, through the epsilon and lambda operators, it is possible to construct additional terms, as stipulated below.

[1] For this distinction see van Heijenoort (1967) and Hintikka (1988).

(1) If τ, τ_1, τ_2 are terms and ϕ is an expression of the form $\tau_1 = \tau_2$ or $\tau_1 \in \tau_2$, then ϕ is a well-formed formula (wff). $M\tau$ is also considered a well-formed formula, which will be equivalent to $\tau \in V$. Likewise, $I\tau$ is a wff.
(2) If ϕ is a wff, then $(\neg\phi)$ is a wff.
(3) If ϕ and ψ are wffs, then $(\phi \wedge \psi)$, $(\phi \vee \psi)$, $(\phi \rightarrow \psi)$, and $(\phi \leftrightarrow \psi)$ are wffs.
(4) If ϕ is a wff and x is an individual variable, then $\forall x\, \phi$ is a wff and $\exists x\, \phi$ is a wff.
(5) Only formulas constructed using the four previous rules are wffs.

I write $\phi(x_1, \ldots, x_n)$ to indicate that the free variables of ϕ, if there are any, are among $x_1, \ldots, x_n$. If $\phi(x)$ is a formula in which x is the only variable that appears free, and there is an object that satisfies $\phi(x)$,

$$\varepsilon_x \phi(x) \tag{1.1}$$

is a term that denotes an arbitrarily chosen object that satisfies the formula $\phi(x)$. Therefore, the formula

$$\exists x \phi(x) \rightarrow \phi[\varepsilon_x \phi(x)] \tag{1.2}$$

is always true. Similarly, if $\phi(x)$ is a formula in which only x appears free,

$$\lambda_x \phi(x) \tag{1.3}$$

is a term that denotes the class of all x that satisfy the formula $\phi(x)$.

A central notion in A is that of a condition whose definition does not refer to the property "is a set", although it may refer to properties that have already formed certain sets. Other objects may enter into the definition of the condition as parameters, and the associated collection will only represent a set if these objects are sets that we can consider as already defined. We thus define the concept of a *bounded condition* $\phi(x)$ as one that contains only parameters of sets or urelements, and in which M or V do not occur.

Let us observe that if the axioms allow proving the existence of a set or a urelement, then it is legitimate to introduce a name for it. We define a *set term* as the name of a set whose existence is deduced from the axioms and that is introduced as such in some extension of $\mathcal{L}_{\mathsf{A}}$; a term of urelement is obviously a designator of some urelement. In a bounded condition, all terms that appear are terms of sets or urelements. It is typical to consider real numbers as urelements that constitute a designated set $\mathbb{R}$. In the constructive structures or universes we will consider later, the set $\lambda_x I(x)$ of urelements will precisely be $\mathbb{R}$.

A does not include an axiom of choice because Ackermann believed that such an axiom expresses a logical truth. This thesis has not been widely accepted, and some authors—such as Patrick Suppes—have even been skeptical of the axiom of choice and have sought to avoid it whenever possible. Of course, Gödel (1938) demonstrated that the axiom is consistent with the other axioms of set theory, but this should not be surprising if, in fact, the axiom is a logical truth, as Hintikka

(2014) also argued. The thesis is that the axiom of choice is a self-evident principle, akin to what Aristotle and the medieval philosophers referred to as a principle of *scientia*. This is why Ackermann does not include it as an *axiom of set theory* but rather as a *logical axiom* forming part of the logical presuppositions of his system.

According to Hintikka,

> the axiom of choice is a completely obvious truth of first-order logic. It can be derived from the usual first-order logic by an innocuous liberalization of one of its rules. This liberalization does not make first-order logic semantically any less evident. (Hintikka 2014: 23)

Indeed, if the truth of a quantified sentence implies the existence of appropriate "witness individuals", then, if there is an object satisfying a condition $\phi(x)$, it should certainly be possible to designate that object as a "witness" to such truth. The operator that performs this action is Hilbert's epsilon operator. Once we have introduced the axioms and developed the theory sufficiently, we will show how the axiom of choice is derived using this operator.

Since I will add some axioms to Ackermann's original theory, as presented in his seminal 1956 text, as well as replace some of them, I will refer to the original theory as A^*.

1.2 The Axioms of A^*

The first axiom of A^*, Ackermann's original theory, states that if $\phi(x)$ is a condition in which the variable x appears free, then there must exist a class (not necessarily a set) y such that an object x belongs to y if and only if $\phi(x)$, provided that anything satisfying $\phi(x)$ is either a set or an urelement. In such a case, '$x \in y$' and '$\phi(x)$' are considered "sentences with the same meaning".[2] Indeed, by virtue of the first axiom, there exists a class V such that $x \in V$ if and only if x is a set; that is, V is the class of all sets. We will treat the expressions 'Mx' and '$x \in V$' as synonymous.

If we define the predicate Ox (read as "x is an object") as $Mx \vee Ix$, this axiom can be formulated as follows:

$$\forall x(\phi(x) \rightarrow Ox) \rightarrow \exists y \forall z(z \in y \leftrightarrow \phi(z)). \tag{1.4}$$

This axiom states that if the fact that any x has the property ϕ implies that x is an object, then there exists a class y whose elements are precisely the objects that have the property ϕ. The class y is denoted by the term '$\{x \mid \phi(x)\}$,' which is read as "the class of all x such that $\phi(x)$". Another way to denote this class is $\lambda_x \phi(x)$. The lambda operator should only be applied to properties that satisfy this axiom.

The following is the classical axiom of extensionality; it states that classes with the same elements are identical.

---

[2] See item 1 in §1 of Ackermann (1956: 336). See also α' in Ackermann (1956: 344).

$$\forall x \forall y \forall z((z \in x \leftrightarrow z \in y) \rightarrow x = y). \tag{1.5}$$

To formulate this axiom in the same terms as Ackermann, it is useful at this point to introduce a definition.

Definition 1.2.1 If x and y are classes, we say that x is a *subclass* of y, denoted by $x \subseteq y$, if and only if every element of x is an element of y. Symbolically:

$$x \subseteq y \text{ means } \forall z(z \in x \rightarrow z \in y). \tag{1.6}$$

If $x \subseteq y$ but $x \neq y$, we write $x \subset y$.

Using the symbol for inclusion, the extensionality axiom can also be formulated as follows:

$$\forall x \forall y((x \subseteq y \land y \subseteq x) \rightarrow x = y). \tag{1.7}$$

Note that the axiom of extensionality states a sufficient condition for the identity of objects that are not urelements; that is, objects that are classes.

The third axiom of A* specifies under what conditions a formula with free variables determines a set. If $\phi(y, x_1, \ldots, x_n)$ is a condition in which the terms M (or V) do not appear, and the variables $x_1, \ldots, x_n$ appear freely in $\phi(y, x_1, \ldots, x_n)$, it is required that they can only be instantiated by sets or urelements for the condition ϕ to determine a set:

$$(Ox_1 \land \cdots \land Ox_n) \rightarrow (\forall y(\phi(y, x_1, \ldots, x_n) \rightarrow Oy) \rightarrow \exists z(Mz \land \forall u(u \in z \leftrightarrow \phi(u, x_1, \ldots, x_n)))) \tag{1.8}$$

If $x_1, \ldots, x_n$ do not appear freely in ϕ, the antecedent of (1.8) is omitted.

The fourth axiom of the elegant and austere system A* asserts that anything that is an element of or a subset of a set is either a set or an urelement:

$$(Mx \land (y \in x \lor y \subseteq x)) \rightarrow Oy \tag{1.9}$$

It is useful to list these four axioms for easier future reference.

A*1 $\forall x(\phi(x) \rightarrow Ox) \rightarrow \exists x \forall z(z \in y \leftrightarrow \phi(z))$.
A*2 $\forall x \forall y((x \subseteq y \land y \subseteq x) \rightarrow x = y)$.
A*3 If the term M does not appear in ϕ,

$$(Ox_1 \land \cdots \land Ox_n) \rightarrow (\forall y(\phi(y, x_1, \ldots, x_n) \rightarrow Oy) \rightarrow \exists z(Mz \land \forall u(u \in z \leftrightarrow \phi(u, x_1, \ldots, x_n)))).$$

A*4 $(Mx \land (y \in x \lor y \subseteq x)) \rightarrow Oy$.

1.3 The Axioms of A

The fifth axiom, which does not originally appear in the system A^*, is the axiom of regularity, foundational axiom, or *Fundierungsaxiom*, which states that every nonempty set has an element that shares no elements with itself. Symbolically:

$$\forall x((Mx \wedge x \neq \emptyset) \rightarrow \exists y(y \in x \wedge \neg\exists z(z \in y \wedge z \in x))).$$

I will designate this axiom as A^*5.

Lévy (1959: 162, Theorem 5) demonstrated that the system A^*, consisting of the axioms A^*1–A^*5, implies a principle called the *principle of partial reflection*—denoted as R.[3] *Roughly speaking*, R asserts that if the class V of all sets possesses a property expressed by a certain condition, $\phi(x)$, then there exists a set $v \in V$ that also possesses the same property, $\phi(x)$. A more precise characterization is given in the following definition.

Definition 1.3.1 Let $\phi(y, x_1, \ldots, x_n)$ be a formula in which at most the variables $y, x_1, \ldots, x_n$ appear free. The *principle of partial reflection* (R) is the schema

$$(x_1, \ldots, x_n \in V \wedge \phi(V, x_1, \ldots, x_n)) \rightarrow \exists v(v \in V \wedge \phi(v, x_1, \ldots, x_n)).$$

Lévy and Vaught (1961) demonstrated, moreover, that the principle of partial reflection implies, together with the axioms A^*1, A^*2, A^*4, and A^*5, the axiom A^*3.[4] The proof is as follows.

Theorem 1.3.2 *Assume the axioms* A^*1, A^*2, *R, and* A^*4. *Then* A^*3 *holds.*

Proof Let $\phi(y, x_1, \ldots, x_n)$ be a formula in which the term V does not appear. Assume $x_1, \ldots, x_n \in V$ and that

$$\forall y\,(\phi(y, x_1, \ldots, x_n) \rightarrow y \in V).$$

By A^*1,

$$\exists z\, u \in z \leftrightarrow \phi(u, x_1, \ldots, x_n). \tag{1}$$

Then,

$$\begin{aligned} u \in z &\Rightarrow \phi(u, x_1, \ldots, x_n) \\ &\Rightarrow u \in V. \end{aligned} \tag{2}$$

[3] Cf. Lévy (1959), Theorem 5, and p. 163: "Theorem 5 was proved in A^*, i.e., we made use of the axiom of foundation in its proof".

[4] Cf. p. 1057: "In the axioms of A^*, A3 may be replaced by R".

From (2), it follows that $z \subseteq V$, and thus,

$$u \in z \to (\phi(u, x_1, \ldots, x_n) \wedge z \subseteq V). \tag{3}$$

Moreover, it is obvious that

$$(\phi(u, x_1, \ldots, x_n) \wedge z \subseteq V) \to u \in z. \tag{4}$$

From (3) and (4), it is concluded that V satisfies the condition $\psi(w)$:

$$\exists z\,(u \in z \leftrightarrow (\phi(u) \wedge z \subseteq w)). \tag{5}$$

That is, we have

$$\exists z\,(u \in z \leftrightarrow (\phi(u) \wedge z \subseteq V)).$$

Thus, by R, there exists $v \in V$ such that $\psi(v)$:

$$\exists z\,(u \in z \leftrightarrow (\phi(u) \wedge z \subseteq v)). \tag{6}$$

In other words, there exist a set v and a set z' such that $u \in z'$ syss $\phi(u)$ and $z' \subseteq v$. Since $u \in z'$ syss $\phi(u)$ syss $u \in z$, A*2 implies $z' = z$. Finally, since $v \in V$, A*4 implies $z \in V$. ⊣

The replacement of axiom A*3 with axiom R constitutes the modification of the extended Ackermann system, namely the system A*, which serves as the foundation for the theory of classes presented in this book. The new system will be denoted by the symbol A, where axiom A*3 is substituted with R. The system A is thus composed of the axioms A*1, A*2, R, A*4, and A*5, now renamed A1 through A5:

- A1 $\forall x(\phi(x) \to Ox) \to \exists x \forall z(z \in y \leftrightarrow \phi(z))$.
- A2 $\forall x \forall y((x \subseteq y \wedge y \subseteq x) \to x = y)$.
- A3 Let $\phi(y, x_1, \ldots, x_n)$ be a formula in which at most the variables $y, x_1, \ldots, x_n$ are free. Then,

$$(x_1, \ldots, x_n \in V \wedge \phi(V, x_1, \ldots, x_n)) \to$$
$$\to \exists v(v \in V \wedge \phi(v, x_1, \ldots, x_n)).$$

- A4 $(Mx \wedge (y \in x \vee y \subseteq x)) \to Oy$.
- A5 $\forall x((Mx \wedge x \neq \emptyset) \to \exists y(y \in x \wedge \neg \exists z(z \in y \wedge z \in x)))$.

1.4 Derivation of ZFC

We will now demonstrate that the system A is at least as powerful as ZFC. The existence of a nonempty set of urelements is an assumption exogenous to A and is of an "empirical" nature: A does not assert or deny the existence of any urelement. On the other hand, A can prove the existence of certain sets. The first theorem of A establishes the existence of an empty set.

Theorem 1.4.1 *There exists precisely one empty set, i.e., a set without any elements.*

Proof If $\phi(x)$ is the condition $x \neq x$, a substitution case of Theorem 1.3.2 (that is, of A*3) is

$$\forall x(x \neq x \rightarrow Ox) \rightarrow \exists y(My \wedge \forall z(z \in y \leftrightarrow z \neq z)) \tag{1}$$

Since $x \neq x$ is clearly false for any substitution of x,

$$x \neq x \rightarrow Ox \tag{2}$$

is satisfiable for any x. From (1) and (2), we derive

$$\exists y(My \wedge \forall z(z \in y \leftrightarrow z \neq z)). \tag{3}$$

To prove the uniqueness of this set, suppose there are more than one, and let y and y' both be empty sets. If y' were not identical to y, $A2$ would imply

$$\neg\forall z\,(z \in y' \leftrightarrow z \in y), \tag{4}$$

which leads to

$$\exists z\,(\neg(z \in y' \rightarrow z \in y) \vee \neg(z \in y \rightarrow z \in y')) \tag{5}$$

or

$$\exists z\,((z \in y' \wedge z \notin y) \vee (z \in y \wedge z \notin x)). \tag{6}$$

From this, it follows that either y' has an element or y does. Since both y' and y are empty by hypothesis, this proves that there cannot be more than one empty set. ⊣

This y is denoted by the symbol $\varnothing$. Thus, the class V of sets contains at least the element $\varnothing$. If the theory is applied to urelements, then V also contains as an element the set of urelements. In some applications below, we will assume that this set is nonempty and that it has no elements in common with V. Hence, urelements x are characterized by $x \neq \varnothing$ and $\forall y\ y \notin x$.

The following theorem might seem strange to some, but it is very natural since its negation would imply that there is an element in the set without elements that is not contained in some set.

Theorem 1.4.2 *The empty set is a subset of every set.*

$$\forall x \, \emptyset \subseteq x. \tag{1.10}$$

Proof For any object y, it is true that $y \notin \emptyset$. From this, it tautologically follows that

$$y \notin \emptyset \vee y \in x,$$

or equivalently,

$$y \in \emptyset \rightarrow y \in x. \dashv$$

Another set whose existence can be proven by A is the one that contains one or two given objects.

Theorem 1.4.3 *For any objects x and y, there exists a set that contains precisely both objects:*

$$\forall x \forall y ((Ox \wedge Oy) \rightarrow \\ \exists w (Mw \wedge \forall u (u \in w \leftrightarrow (u = x \vee u = y)))).$$

Proof Suppose Ox and Oy, and observe that $z = x \vee z = y$ implies Oz. Therefore, it holds that

$$(Ox \wedge Oy) \rightarrow \forall z ((z = x \vee z = y) \rightarrow Oz). \tag{1}$$

A substitution case of Theorem 1.3.2 is

$$((Ox \wedge Oy) \rightarrow \forall z ((z = x \vee z = y) \rightarrow Oz)) \rightarrow \\ \exists w (Mw \wedge \forall u (u \in w \leftrightarrow u = x \vee u = y)). \tag{2}$$

From (1) and (2), the desired conclusion follows. $\dashv$

The set w is known as "the unordered pair of x and y and is also denoted as $\{x, y\}$ or $\{y, x\}$. If $x = y$, the set is written as $\{x\}$ and is called the *singleton set of* x.

Definition 1.4.4 For any set x, we define the *intersection* of the elements of x, $\bigcap x$, as the set containing the objects that belong to all members of x:

$$\bigcap x \equiv_d \{y \mid \forall z (z \in x \rightarrow y \in z)\}.$$

If x consists of two elements u and v, $\bigcap x$ is denoted as $u \cap v$. If the elements of x are indexed, $\{z_i\}_{i \in I}$, their intersection is written as $\bigcap_{i \in I} z_i$. If the cardinality of I is α, it is also written as $\bigcap_{i=1}^{\alpha} z_i$. If α is a positive integer n, it can be written as $z_1 \cap \cdots \cap z_n$.

Since $\bigcap x$ consists of the objects that simultaneously belong to all the elements of x, $y \in \bigcap x$ implies that y is in each and every element of x. Thus, if any of these elements z is empty (because it is either the empty set or an urelement), y cannot belong to z, and therefore $y \notin \bigcap x$ for all y; that is, $\bigcap x = \varnothing$.

Theorem 1.4.5 *For any set x, $\bigcap x$ exists and is a set:*

$$Mx \rightarrow \exists w(Mw \wedge \forall y(y \in w \leftrightarrow \forall z(z \in x \rightarrow y \in z))).$$

Proof Let x be a set, and let $\phi(y)$ represent the condition

$$\forall z(z \in x \rightarrow y \in z). \tag{1}$$

If any element z of x is the empty set or a urelement, no y satisfies (1); in that case, $\bigcap x = \varnothing$. If all the elements z of x are sets and $\varnothing$ is not among them, A4 implies that any y satisfying condition (1) is an object, as it is an element of every z in x, and x is a set. That is,

$$\forall y(\phi(y) \rightarrow Oy). \tag{2}$$

However, a substitution case of Theorem 1.3.2 is

$$Ox \rightarrow (\forall y(\phi(y) \rightarrow Oy) \rightarrow \\ \exists w(Mw \wedge \forall v(v \in w \leftrightarrow \forall z(z \in x \rightarrow v \in z)))).\dashv$$

Another important set operation is the union of sets.

Definition 1.4.6 For any set x, we define the *union* of the elements of x, $\bigcup x$, as the class containing as elements those objects that belong to at least one of the members of x:

$$\bigcup x \equiv_d \{y \mid \exists z(z \in x \wedge y \in z)\}.$$

If x consists of two elements u and v, $\bigcup x$ is denoted as $u \cup v$. The notational observations made regarding the intersection apply, *mutatis mutandis*, to the symbols $\bigcup$ and $\cup$.

Theorem 1.4.7 *For any set x, $\bigcup x$ exists and is a set:*

$$Mx \rightarrow \exists w(Mw \wedge \forall y(y \in w \leftrightarrow \exists z(z \in x \wedge y \in z))).$$

Proof Let x be a set, and let $\phi(y)$ represent the condition

$$\exists z(z \in x \wedge y \in z). \tag{1}$$

If x is empty or x contains no nonempty sets, then $\bigcup x = \varnothing$. If x contains at least one nonempty set z, all its elements are objects, so we have again

$$\forall y(\phi(y) \rightarrow Oy). \tag{2}$$

A substitution case of $A3$ is

$$Ox \rightarrow (\forall y(\phi(y) \rightarrow Oy) \rightarrow$$
$$\exists w(Mw \wedge \forall v(v \in w \leftrightarrow \exists z(z \in x \wedge v \in z)))). \dashv$$

Another important concept is the class of all subsets of a given set.

Definition 1.4.8 If x is a set, the *power set* of x is defined as the collection of all subsets of x and is denoted as $\mathcal{P}(x)$:

$$\mathcal{P}(x) \equiv_d \{y \mid My \wedge y \subseteq x\}.$$

Theorem 1.4.9 *If x is a set, $\mathcal{P}(x)$ exists and is a set.*

Proof Let x be a set, and let $\phi(y)$ represent the condition

$$y \subseteq x. \tag{1}$$

By virtue of Axiom A4, we know that any y satisfying condition (1) is an object. Furthermore, a substitution case of Theorem 1.3.2 is

$$Ox \rightarrow (\forall y(\phi(y) \rightarrow Oy) \rightarrow$$
$$\exists z(Mz \wedge \forall u(u \in z \leftrightarrow u \subseteq x))). \dashv \tag{2}$$

We will now show that the Axiom of Separation (*Aussonderung*) from ZF is derivable in A.

Theorem 1.4.10 (Separation) *For any set x and bounded condition $\phi(y)$, there exists the set of all elements of x that satisfy $\phi(y)$:*

$$\forall x\,(Mx \rightarrow \exists z\,(Mz \wedge \forall y\,(y \in z \leftrightarrow (y \in x \wedge \phi(y))))). \tag{1.11}$$

Proof Let x be a set, and let $\phi(y)$ be a bounded condition on y. Define $\psi(y)$ as the condition '$y \in x \wedge \phi(y)$'. Then $\psi(y)$ is also a bounded condition that is only satisfied by objects, i.e.,

$$\forall y(\psi(y) \rightarrow Oy). \tag{1}$$

Now, the following formula is a substitution case of Theorem 1.3.2:

$$Ox \to (\forall y(\psi(y) \to Oy)) \to$$
$$\exists z\, (Mz \wedge \forall y\, (y \in z \leftrightarrow (y \in x \wedge \phi(y)))). \dashv \tag{1.12}$$

If the set I of urelements is infinite, this guarantees the existence of an infinite set. However, even within pure set theory, it is possible to prove the existence of an infinite set. For any set x, the set $x \cup \{x\}$, denoted x^+, is called the *successor* of x.

Definition 1.4.11 A *successor class* is a class y that satisfies the following two conditions:

(1) $\varnothing \in y$;
(2) if $x \in y$, then $x^+ \in y$.

We say that x is a *successor* of $\varnothing$ if x has the form $x^n \equiv x^{\overbrace{+\cdots+}^{n \text{ veces}}}$.

The following proposition holds in the A system.

Theorem 1.4.12 *There exists a set that contains as elements precisely* $\varnothing$ *and the successors of* $\varnothing$.

Proof By Theorem 1.4.1, we know that $\varnothing$ is a set. By Theorem 1.4.3, unit sets of sets are sets. Furthermore, by Theorem 1.4.7, the union of two sets is a set. Thus, we have:

$$\forall x\, (Mx \to M(x \cup \{x\})). \tag{1}$$

This implies that all successors of $\varnothing$ are sets.

Let $\phi(x))$ be the condition '$x = \varnothing$ or x is a successor of $\varnothing$'. Clearly, the following assertion is true:

$$\forall x\, (\phi(x) \to Mx). \tag{2}$$

Now, a substitution case of Theorem 1.3.2 is

$$\forall x\, (\phi(x) \to Mx) \to \exists x\, (My \wedge \forall z\, (z \in y \leftrightarrow \phi(z))). \tag{3}$$

From (2) and (3), it follows that

$$\exists x\, (My \wedge \forall z\, (z \in y \leftrightarrow \phi(z))). \dashv$$

It is useful to gather in a single definition all the notions defined so far, as well as a few additional ones.

Definition 1.4.13 Let x and y be sets or urelements. We define:

(1) *Unordered pair:* The *unordered pair* of x and y is the set $\{u \mid u = x \vee u = y\}$.
(2) *Singleton set:* The *singleton set* of x is the unordered pair $\{x, x\}$.
(3) *Power set:* The *power set* of the set x, $\mathcal{P}(x)$, is the family of all subsets of x.
(4) *Ordered pair:* The *ordered pair* of x and y is the set $\{\{x\}, \{x, y\}\}$, denoted by $\langle x, y\rangle$.
(5) *Symbol* $\langle x\rangle$: The symbol $\langle x\rangle$ denotes x.
(6) *Ordered n-tuple:* The *ordered n-tuple* of $x_1, \ldots, x_n$ is defined as the ordered pair:

$$\langle\langle x_1, \ldots, x_{n-1}\rangle, x_n\rangle,$$

and is denoted by $\langle x_1, \ldots, x_n\rangle$.
(7) *Subset and superset:* The set x is *included* in y (or is a *subset* of y) if every element of x is an element of y. In this case, y is called a *superset* of x, and this is denoted by $x \subseteq y$.
(8) *Proper subset:* The set x is *properly included* in y if $x \subseteq y$ but $x \neq y$. In this case, x is called a *proper subset* of y, and this is denoted by $x \subset y$.
(9) *Empty set:* The set $\emptyset$ is called the *empty set*.
(10) *Disjoint sets:* The sets x and y are *disjoint* if they have no elements in common.
(11) *Intersection of sets*: The *intersection* of the sets in x is the set of objects that belong to all elements of x.
(12) *Union of sets:* The *union* of the sets in x is the set of objects that belong to at least one element of x.
(13) *Univocal set:* The set x is *univocal* if, for every object y, there exists at most one z such that $\langle y, z\rangle \in x$. We express that x is univocal as **univ** (x).
(14) *Disjointed set:* The set x is *disjointed* if $x \neq \emptyset$, the elements of x are sets, $\emptyset \notin x$, and $y, z \in x$ implies $y \cap z = \emptyset$. We express that x is disjointed as **disy** (x).

An important concept is that of the Cartesian product of two sets, which is a certain set of ordered pairs.

Definition 1.4.14 If x and y are sets, the *Cartesian product* of x and y is the class

$$\{\langle u, v\rangle \mid u \in x \wedge v \in y\}.$$

The following theorem guarantees the existence of the Cartesian product of any two sets.

Theorem 1.4.15 *Given any two sets x and y, there always exists a set whose elements are the ordered pairs $\langle u, v\rangle$ with $u \in x$ and $v \in y$:*

$$\forall x\, \forall y\, \exists z\, \forall w\, (Mx \wedge My \wedge Mz \wedge Mw) \rightarrow$$
$$(w \in z \leftrightarrow \exists u\, \exists v\, (u \in x \wedge v \in y \wedge w = \langle u, v\rangle)). \qquad (1.13)$$

Proof Let $\phi(w)$ be the clause

$$\exists u \exists v(u \in x \wedge v \in y \wedge w = \langle u, v\rangle).$$

This clause is bounded and, since x and y are sets, u and v are urelements or sets, and *a fortiori*, w is also a set. By applying Theorem 1.3.2,

$$\exists z((Mz \wedge \forall w(w \in z \leftrightarrow \phi(w))).$$

Therefore, $z = \{\langle u, v\rangle \mid u \in x \wedge v \in y\}$ is a set. $\dashv$

The following theorem expresses one form of the axiom of choice and is equivalent to all others. Note that in the proof we make use of Hilbert's epsilon operator and a logical principle.

Theorem 1.4.16 *For every x, if x is a disjointed set, then there exists a set y that contains exactly one element from each of the sets in x:*

$$\forall x(\mathbf{disy}\, x \rightarrow \exists y \forall z\, (z \in x \rightarrow \exists u\, (u \in z \wedge z \cap y = \{u\}))).$$

Proof Let x be a disjointed set. Then $x \neq \emptyset$, $\emptyset \notin x$, all its elements are sets, and these elements are pairwise disjoint. For each $z \in x$, we can choose a *unique* arbitrary element u, which must exist as z is a nonempty set. Thus, for each $z \in x$, we select an element denoted as '$\varepsilon_u \varphi(z, u)$', where $\varphi(z, u)$ is the condition '$u \in z$'.

If $\psi(v)$ is the condition '$\exists z\, (z \in x \wedge v = \varepsilon_u \phi(z, u))$', then it holds that

$$\forall v\, (\psi(v) \rightarrow Ov).$$

However, the following statement is a substitution case of Theorem 1.3.2:

$$\forall v\, (\psi(v) \rightarrow Ov) \rightarrow \exists y\, (My \wedge \forall w\, (w \in y \leftrightarrow \psi(w))). \dashv$$

Condition $\phi(x, y)$ is called a *functional* in x if $\phi(x, y)$ and $\phi(x, z)$ imply $y = z$ for every x. For any set u, the following theorem—known in **ZF** as the schema of replacement—guarantees the existence of a function with domain u and image v, consisting of the y that satisfy the functional $\phi(x, y)$ for some $x \in u$.

Theorem 1.4.17 *For any set u, if the bounded condition $\phi(x, y)$ is a functional in x, then there exists a set v such that $y \in v$ if and only if $x \in u$ and $\phi(x, y)$:*

$$\forall u((Mu \wedge \forall x \forall y \forall z((x \in u \wedge \phi(x, y) \wedge \phi(x, z)) \rightarrow y = z) \\ \rightarrow \exists v(Mv \wedge \forall y(y \in v \leftrightarrow \exists x(x \in u \wedge \phi(x, y)))).$$

Proof Assume that u is an arbitrary set and that $\phi(x, y)$ is a functional in x. Let $\psi(y)$ be the condition "there exists x in u such that $\phi(x, y)$". It is easy to see (by the definition of functional) that

$$\forall y\,(\psi(y) \to y \in V),$$

so by A1 it follows that

$$\exists z\,\forall w\,(w \in z \leftrightarrow \psi(w)).$$

Since $\psi(w)$ implies $w \in V$, we have $z \subseteq V$. It remains to show that $z \in V$.

Let $\theta(v)$ be the condition "there exists z such that, for all w, $w \in z \leftrightarrow \psi(w)$ and $z \subseteq v$". Observe that $\theta(V)$ is indeed the case, as it implies

$$\exists z\,(\forall w\,(w \in z \leftrightarrow \psi(w)) \wedge z \subseteq V).$$

By the principle of reflection, it follows that

$$\exists v\,(v \in V \wedge \exists z'\,(\forall w\,(w \in z' \leftrightarrow \psi(w)) \wedge z' \subseteq v)).$$

Using A 2 and A 4, we conclude that $z \in V$. ⊣

Several useful and interesting propositions follow within the theoretical framework we have introduced so far.

Theorem 1.4.18 *The following sentences are true within system A.*

(1) $\forall x\, x \notin \emptyset$.
(2) *For any natural number n:*
$$\forall x_1 \cdots \forall x_n \forall y_1 \cdots \forall y_n\,(\langle x_1, \ldots, x_n\rangle = \langle y_1, \ldots, y_n\rangle \leftrightarrow (x_1 = y_1 \wedge \cdots \wedge x_n = y_n)).$$
(3) $\forall x(Mx \to \exists y \forall z(z \in y \leftrightarrow \neg z \in x))$.
(4) $\forall x(Mx \to \exists y \forall z(z \in y \leftrightarrow \exists u\langle z, u\rangle \in x))$.
(5) $\forall x \exists y(My \wedge \forall z \forall u(\langle u, z\rangle \in y \leftrightarrow \langle z, u\rangle \in x))$.
(6) $\forall x(\mathbf{univ}\, x \to \exists y \forall z\,(z \in y \leftrightarrow \exists u\,\langle z, u\rangle \in x)))$.

The proof is left as an exercise for the reader.

In Theorem 1.4.18, (1) states that (as expected) the set $\emptyset$ has no elements. (2) asserts the identity condition for n-tuples in terms of the identity of their components. (3) states that, for every set, there exists another set whose elements are precisely the sets that do not belong to the first set. The meaning of (4) is that, for every set x, there exists a set y whose elements are all the individuals that are the first components of pairs belonging to x; naturally, if there are no pairs in x, this set is empty. (5) states that for every set x, there exists a set y whose elements are precisely the pairs $\langle u, z\rangle$ where $\langle z, u\rangle$ belongs to x. Finally, (6) states that, for every set x and univocal set y, there exists a set whose elements are precisely the first elements of ordered pairs in x. Based on these results, we can introduce the following series of definitions.

Definition 1.4.19 The previous theorems entitle us to introduce the following definitions.

(1) The *complement* of a set x is the set $\{y \mid y \notin x\}$, whose existence is guaranteed by Theorem 1.4.18, (3); it is denoted as $\overline{x}$.
(2) The *inverse* of x is the set $\{\langle v, u\rangle \mid \langle u, v\rangle \in x\}$, whose existence is guaranteed by Theorem 1.4.18, (5); it is denoted as x^{-1}.
(3) The *domain* of x is the set $\{z \mid \langle z, u\rangle \in x\}$, whose existence is guaranteed by Theorem 1.4.18, (4); it is denoted as $\mathbf{dom}\, x$ and is empty if x contains no ordered pairs. The image of x, $\mathbf{im}\, x$, is precisely $\mathbf{dom}\, x^{-1}$.
(4) A set consisting solely of ordered pairs is called a *binary relation*.
(5) If y is univocal, the *restriction* of y to x is the set $\{z \mid z \in \mathbf{dom}\, y \cap x\}$, whose existence is asserted by Theorem 1.4.18, (6); it is denoted as $y \restriction x$.

We could continue developing set theory in a highly formal manner, but it is now more practical to simplify the notation and provide less formal proofs. From this point forward, we will use uppercase letters to denote sets and lowercase letters such as f or g to denote univocal relations. This shift is introduced with the following definition:

Definition 1.4.20 We define the following important set-theoretical notions.

(1) *Function*: f is a *function* from A to B, denoted $f : A \to B$, if and only if $f \subseteq A \times B$, $\mathbf{dom}\, f = A$, and f is univocal ($\mathbf{univ}\, f$). If $\langle x, y\rangle \in f$, we write $y = f(x)$. The *codomain* of f is B, and the *image* of f is $\{y \mid y = f(x) \text{ for some } x \in A\}$, denoted as $\mathbf{im}\, f$. If $X \subseteq A$, $\mathbf{im}\,(f\restriction X)$ is written as $f(X)$. For an integer $n \geq 0$ and any nonempty set A, an *n-ary closed operation* on A is a function $f : A \times \cdots \times A \to A$.
(2) *Injectivity*: A function f is *one-to-one* (or *injective*) if $f(x) = f(y)$ implies $x = y$ for all $x, y \in \mathbf{dom}\, f$.
(3) *Surjectivity*: A function f is *onto* (or *surjective*) if $\mathbf{im}\, f = \mathbf{cod}\, f$.
(4) *Bijectivity*: A function f is *bijective* if f is both injective and surjective.
(5) *Identity Function*: The *identity function* on a set A is $\iota : A \to A$, defined by $\iota(x) = x$ for all $x \in A$.
(6) *Function composition*: If $f : A \to B$ and $g : B \to C$, the *composition* of f and g, denoted $g \circ f$, is the function $h : A \to C$ such that $h(x) = g[f(x)]$ for all $x \in A$ (see the commutative diagram).

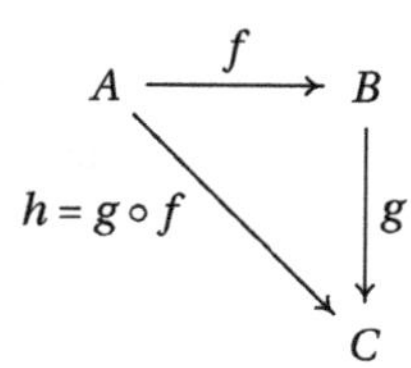

(7) *Equipotence*: Sets A and B are *equinumerous* or *equipotent*, written $A \sim B$, if $A = B = \emptyset$ or if there exists a bijection f from A to B.
(8) *Finite and infinite sets*: A set A is *finite* if A is equipotent to a successor of $\emptyset$. A set A is *infinite* if A is not finite.

(9) *Partial order*: A relation R is a *partial order* on A if R is reflexive, antisymmetric, and transitive.
(10) *Total order*: A relation R is a *total order* on A if R is a partial order and connected on A.
(11) *Well-ordering*: A relation R *well-orders* A if R is a total order and every nonempty subset of A has a least element under R.
(12) *Natural numbers*: A *natural number* is an element of the set ω, defined as the intersection of all sets of successors.
(13) *Ordinal numbers*: An *ordinal number* is a set of sets well-ordered by $\in$, where each initial segment is a member of the set.
(14) *Limit ordinals*: A *limit ordinal* is a nonempty ordinal with no immediate predecessor.
(15) *Predecessor and initial segment*: The *initial segment* of subset A of an ordinal number, determined by a, denoted $s(A, \in a)$, is the set of all predecessors of a in A.
(16) *Transitive set*: A set x is *transitive* if all elements of x are subsets of x.
(17) Let A be a set of successors (whose existence is guaranteed by Theorem 1.4.12), and let

$$S = \{B \mid B \subseteq A \text{ and } B \text{ is a set of successors}\}.$$

By Theorem 1.3.2, S is a set. Thus, we can define the set ω as the intersection $\bigcap S$.
(18) If $\langle A, R\rangle$ is a well-order and $a \in A$, the *initial segment of A determined by a* is the set

$$s(A, R, a) = \{x \mid x \in A \wedge x \neq a \wedge \langle x, a\rangle \in R\}.$$

$s(A, R, a)$ is the set of predecessors of a that belong to A.
(19) An *ordinal number* is a set of sets α that is well-ordered by $\in$, whose first element with respect to $\in$ is $\emptyset$, and such that the initial segment of α determined by a, $s(\alpha, \in, a)$, is equal to a itself for every $a \in \alpha$. We write **ord** x to express that x is an ordinal number.
(20) A *limit ordinal* is an ordinal different from $\emptyset$ that has no immediate predecessor; i.e., it is an ordinal α such that $\alpha \neq \emptyset$ and $\alpha \neq \beta^+$ for all $\beta \in \alpha$.
(21) If α and β are ordinals, we say that α *precedes* β if and only if $\alpha \in \beta$. We write $\alpha \underline{\in} \beta$ if α precedes β or $\alpha = \beta$. We also write $\alpha < \beta$ to express that α precedes β, and $\alpha \leq \beta$ if and only if $\alpha \underline{\in} \beta$.

Thus, we conclude the presentation of the most fundamental definitions of set theory. We need to establish some truths about ordinal numbers, in particular cardinal numbers, to develop first-order logic. If R is a relation, we will occasionally write Rxy to express the same as $\langle x, y\rangle \in R$.

The following is a useful lemma, whose proof is left to the reader.

Lemma 1.4.21 *The composition of functions has the following properties for any functions* $f : A \to B$, $g : B \to C$, *and* $h : C \to D$:

(1) $h \circ (g \circ f) = (h \circ g) \circ f$ *(function composition is associative).*
(2) *If* ι_A *is the identity on* A *and* ι_B *the identity on* B, *then* $f \circ \iota_A = f$ *and* $\iota_B \circ f = f$.
(3) *If* f *is a bijection, then* $f^{-1} \circ f = \iota_A$ *and* $f \circ f^{-1} = \iota_B$.

1.5 Relationship Between A and ZFC

An important question, given that the Zermelo-Fraenkel system is the one actively used by mathematicians, concerns the relationship between A and ZF. Utilizing previous results by Lévy (1959, 1960, 1965) and Lévy and Vaught (1961), Reinhardt (1970) demonstrated, using metamathematical methods, that A is equivalent to ZF; that is, the relativization to V of a statement ϕ is provable in A* if and only if ϕ is provable in ZF. We have provided the necessary tools here, particularly to show that A implies ZFC, using a non-metamathematical proof of the separation theorem.

The list of ZFC axioms correlated with the corresponding theorems in A is as follows.

(ZF1) Axiom of Extensionality (Axiom A2):
Sets with the same elements are equal.

$$\forall x \forall y \forall z\, ((x \in V \wedge y \in V) \to ((z \in x \leftrightarrow z \in y) \to x = y)).$$

(ZF2) Axiom of Union or Sum (Theorem 1.4.7):
For any set x, *there exists a set whose elements are precisely the elements of the elements of* x *that are sets.*

$$\forall x (x \in V \to \exists y (y \in V \wedge \forall z (z \in y \leftrightarrow \exists u (z \in u \wedge u \in x)))).$$

(ZF3) Axiom of Power Set (Theorem 1.4.8):
For any set x, *there exists the set of all subclasses of* x.

$$\forall x (x \in V \to \exists y (y \in V \wedge \forall z (z \in y \leftrightarrow z \subseteq x))).$$

(ZF4) Axiom of Regularity (Axiom A5):
Every nonempty set has an element that shares no elements in common with itself.

$$\forall x ((x \in V \wedge x \neq \emptyset) \to \exists y (y \in x \wedge \neg \exists z (z \in y \wedge z \in x))).$$

(ZF5) Axiom of Infinity (Theorem 1.4.12):
There exists at least one set of successors.

(ZF6) Replacement Schema Axiom (Theorem 1.4.17):

For any set u, if the bounded condition $\phi(x, y)$ is functional in x, then there exists a set v such that $y \in v$ if and only if $x \in u$ and $\phi(x, y)$:

$$\forall u\big((Mu \wedge \forall x \forall y \forall z((x \in u \wedge \phi(x, y) \wedge \phi(x, z)) \rightarrow y = z)) \\ \rightarrow \exists v(Mv \wedge \forall y(y \in v \leftrightarrow \exists x(x \in u \wedge \phi(x, y)))).$$

(C) AXIOM OF CHOICE (Theorem 1.4.16)
For any x, if x is a disjointed set, then there exists a unique set y such that, for each element z of x, there is precisely one element u of this same z such that the pair $\langle z, u\rangle$ is in y.

$$\forall x(\mathbf{disy}\, x \rightarrow \exists y(\mathbf{univ}\, y \wedge \forall z(z \in x \rightarrow \exists u(u \in z \wedge \langle z, u\rangle \in y)))).$$

However, A is more powerful than ZFC, as it provides the conceptual framework for category theory by admitting classes that are not sets.

1.6 Ordinal Numbers

The first remarkable properties of ordinal numbers are transitivity and well-ordering.

Theorem 1.6.1 **ord** x *if and only if x is transitive and well-ordered by* $\in$.

Proof Let x be an ordinal. By definition, x is well-ordered by $\in$. If $a \in x$,

$$a = s(x, \in, a) = \{y \mid y \in x \wedge y \neq a \wedge y \in a\} \subset x.$$

This proves that x is transitive.

Conversely, suppose x is transitive and well-ordered by $\in$, and let $a \in x$. Since x is transitive, $a \subset x$, so $y \in x$ for every $y \in a$, and by the regularity theorem, $y \neq a$. Hence, only the elements of a satisfy '$y \in x \wedge y \neq a \wedge y \in a$.' But this is the membership condition for $s(x, \in, a)$, so $s(x, \in, a) = a$. ⊣

Another notable property of ordinal numbers is that (except for the first) they are composed of ordinal numbers.

Theorem 1.6.2 *Every element of an ordinal is an ordinal.*

Proof Assume **ord** α and $a \in \alpha$. Then $a \subset \alpha$. Thus, $A \subseteq a$ implies $A \subset \alpha$, so A has a first element. Clearly, $\in$ is connected in a, since if x, y are distinct elements of a, then $x, y \in \alpha$, and thus $x \in y$ or $y \in x$. This establishes that a is well-ordered by $\in$.

If $b \in a$, then $b \in \alpha$, and thus $s(a, \in, b) = s(\alpha, \in, b) = b$. ⊣

While the membership relation orders the ordinals, so does the inclusion relation.

Theorem 1.6.3 *If α and β are ordinals, then $\alpha \subseteq \beta$ or $\beta \subseteq \alpha$.*

Proof Let α and β be arbitrary ordinals. If $\alpha = \beta$, the result is clear. We can therefore assume that $\alpha \neq \beta$. We will show that $\alpha \in \beta$ or $\beta \in \alpha$.

First, the intersection $\alpha \cap \beta$ is nonempty because it contains $\emptyset$. Indeed, if $\hat{\gamma} \in \alpha$ is the first element of α, then there is no element γ in α such that $\gamma \in \hat{\gamma}$. But

$$\emptyset = s(\alpha, \in, \hat{\gamma}) = \{\gamma \mid \gamma \in \alpha \wedge \gamma \in \hat{\gamma}\} \in \alpha.$$

This establishes that $\emptyset \in \alpha$, and the same holds for β.

Assume that there exist $\gamma_\alpha \in \alpha$ and $\gamma_\beta \in \beta$ such that $\gamma_\beta \notin \alpha$ and $\gamma_\alpha \notin \beta$. To reach a contradiction, note that $\gamma_\alpha \in \alpha \setminus (\alpha \cap \beta)$ and $\gamma_\beta \in \beta \setminus (\alpha \cap \beta)$, so the sets $\alpha \setminus (\alpha \cap \beta)$ and $\beta \setminus (\alpha \cap \beta)$ are nonempty. Since $\alpha \setminus (\alpha \cap \beta) \subseteq \alpha$ and $\beta \setminus (\alpha \cap \beta) \subseteq \beta$, $\alpha \setminus (\alpha \cap \beta)$ has a first element $\hat{\gamma}_\alpha$, and $\beta \setminus (\alpha \cap \beta)$ has a first element $\hat{\gamma}_\beta$.

Observe that if $\eta \in \hat{\gamma}_\alpha$ but $\eta \notin \alpha \cap \beta$, then $\hat{\gamma}_\alpha$ would not be the first element of $\alpha \setminus (\alpha \cap \beta)$, so $\hat{\gamma}_\alpha \subseteq \alpha \cap \beta$. Suppose now that η is an arbitrary element of $\alpha \cap \beta$. Then $\eta \in \hat{\gamma}_\alpha$ or $\hat{\gamma}_\alpha \in \eta$ because $\eta, \hat{\gamma}_\alpha \in \alpha$, which is well-ordered. However, $\hat{\gamma}_\alpha \notin \eta$ because the hypothesis $\eta \in \alpha \cap \beta$ implies $\eta \in \alpha$ and $\eta \in \beta$, so (by the transitivity of α and β) $\eta \subset \alpha \cap \beta$, which in turn implies that $\hat{\gamma}_\alpha \in \alpha \cap \beta$, contradicting the fact that $\hat{\gamma}_\alpha$ is the first element of $\alpha \setminus (\alpha \cap \beta)$. It follows that $\eta \in \hat{\gamma}_\alpha$, so $\alpha \cap \beta \subseteq \hat{\gamma}_\alpha$. In summary, $\alpha \cap \beta = \hat{\gamma}_\alpha$. An analogous argument shows that $\alpha \cap \beta = \hat{\gamma}_\beta$. This implies $\hat{\gamma}_\alpha = \hat{\gamma}_\beta$, which is impossible because $\alpha \setminus (\alpha \cap \beta)$ and $\beta \setminus (\alpha \cap \beta)$ have no common elements. Consequently, either all elements of α are in β, or vice versa. ⊣

The following is a trichotomy law for ordinals.

Theorem 1.6.4 *If α and β are ordinals, then exactly one of the following holds: $\alpha \in \beta$, $\beta \in \alpha$, or $\alpha = \beta$.*

Proof Let α and β be any ordinals such that $\alpha \neq \beta$. By Theorem 1.6.3, either $\alpha \subset \beta$ or $\beta \subset \alpha$. I will show that $\alpha \in \beta$ or $\beta \in \alpha$. Without loss of generality, assume that $\alpha \subset \beta$. Then it is immediate that the first element $\hat{\gamma}$ of $\beta \setminus \alpha$ is equal to α. Thus, $\alpha = \hat{\gamma} \in \beta \setminus \alpha \subseteq \beta$, and therefore $\alpha \in \beta$. ⊣

For any class of ordinals A, we define a *lower bound* of A as an ordinal γ_* such that $\gamma_* \leq \alpha$ for every $\alpha \in A$. Similarly, an *upper bound* for A is an ordinal γ^* such that $\alpha \leq \gamma^*$ for every $\alpha \in A$. The *infimum* of A, $\inf A$, is the greatest lower bound of A; the supremum of A, $\sup A$, is the least upper bound of A. Since every set of ordinals is always bounded below, we say that A is *bounded* if there exists an ordinal γ such that $\alpha \leq \gamma$ for every α in A.

Theorem 1.6.5 *If A is a set of ordinals, both $\bigcap A$ and $\bigcup A$ are ordinals. Furthermore, $\inf A = \bigcap A$ and $\sup A = \bigcup A$.*

Proof Without loss of generality, we can assume that $A \neq \emptyset$. All elements of A have only ordinals as elements (Theorem 1.6.2), so both $\bigcup A$ and $\bigcap A$ are nonempty sets of ordinals (since at least $\emptyset \in \alpha$ for all $\alpha \in A$).

By Theorem 1.6.4, the relation $\in$ is connected in $\bigcup A$. We will show that if $B \subseteq \bigcup A$, then B has a first element. If β is any element of B, the subset $\beta \cap B$ of β has a first element γ. If γ were not the first element of B, there would exist some $\gamma' \in B$ such that $\gamma' \in \gamma \leq \beta$, and thus $\gamma' \in \beta$. However, this contradicts the fact that γ is the first element of $\beta \cap B$. Hence, γ must also be the first element of B.

Finally, if $\alpha \in \bigcup A$, then $\alpha \in \beta$ for some ordinal $\beta \in A$. Therefore, $s(\bigcup A, \in, \alpha) = s(\beta, \in, \alpha) = \alpha$. It follows that $\alpha^* = \bigcup A$ is an ordinal.

The ordinal α^* is an upper bound of A because $\alpha \in \alpha^*$ for all $\alpha \in A$. Indeed, since $\alpha \neq \alpha^*$, either $\alpha \in \alpha^*$ or $\alpha^* \in \alpha$ (Theorem 1.6.4). However, $\alpha^* \in \alpha$ is not possible because it would imply $\alpha^* \subset \alpha \subset \alpha^*$. Moreover, α^* is the least upper bound of A. If it were not, there would exist an ordinal γ such that $\alpha \leq \gamma < \alpha^*$ for all $\alpha \in A$. In that case, $\gamma \in \bigcup A$, and thus $\gamma \in \alpha \leq \gamma$ for some $\alpha \in A$, which is impossible.

Similarly, $\alpha_* = \bigcap A$ is also an ordinal. Since it is a set of ordinals, α_* is totally ordered by $\in$. If $B \subseteq \alpha_*$, then $B \subseteq \bigcup A = \alpha^*$, so B has a first element. If $\alpha \in \bigcap A$, then $\alpha \in \beta$ for every ordinal $\beta \in A$. Hence, $s(\bigcap A, \in, \alpha) = s(\beta, \in, \alpha) = \alpha$. It follows that $\alpha_* = \bigcap A$ is an ordinal.

That α_* is the infimum of A follows from the fact that $\alpha_* \subseteq \alpha$ for every $\alpha \in A$, so $\alpha \leq \alpha_*$ is not possible; thus, $\alpha_* \in \alpha$ for all $\alpha \in A$ (Theorem 1.6.4). Moreover, if $\gamma \in \alpha$ for all $\alpha \in A$, then $\gamma \in \bigcap A = \alpha_*$. $\dashv$

Theorem 1.6.6 *If α, β, and γ are ordinals, then the following hold:*

(1) *Not $\alpha < \alpha$;*
(2) *If $\alpha < \beta$, then not $\beta < \alpha$;*
(3) *If $\alpha < \beta$ and $\beta < \gamma$, then $\alpha < \gamma$;*
(4) *Exactly one of the following holds: $\alpha < \beta$, $\beta < \alpha$, $\alpha = \beta$.*

Proof Let α, β, and γ be arbitrary ordinals.

(1) By axiom A5, $\alpha \notin \alpha$, which establishes that $\alpha < \alpha$ is not true.
(2) Suppose $\alpha \in \beta$. If $\beta \in \alpha$ as well, then $\alpha \subset \beta \subset \alpha$, which implies $\alpha \neq \alpha$, a contradiction.
(3) The assumption that $\alpha \in \beta$ and $\beta \in \gamma$ implies $\beta \subseteq \gamma$. Thus, $\alpha \in \gamma$.
(4) This follows immediately from Theorem 1.6.4. $\dashv$

Theorem 1.6.7 *If $\alpha^+ = \beta^+$, then $\alpha = \beta$.*

Proof Assume $\alpha \cup \{\alpha\} = \beta \cup \{\beta\}$. This implies that $\alpha \in \beta \cup \{\beta\}$ and $\beta \in \alpha \cup \{\alpha\}$. If $\alpha \neq \beta$, it follows that $\alpha \in \beta$ and $\beta \in \alpha$. However, this leads to $\alpha < \alpha$, which is impossible. Therefore, $\alpha = \beta$. $\dashv$

Theorem 1.6.8 *If* **ord** *α, then $\bigcup \alpha^+ = \alpha$.*

Proof Since $\bigcup_{\xi \in \alpha} \xi \subseteq \bigcup_{\xi \subset \alpha} \xi \subseteq \alpha$,

$$\bigcup \alpha^+ = \bigcup_{\xi \in \alpha^+} \xi = \left(\bigcup_{\xi \in \alpha} \xi \right) \cup \alpha = \alpha. \dashv$$

Theorem 1.6.9 *If α is an ordinal, then there is no ordinal β such that $\alpha < \beta < \alpha^+$.*

Proof Suppose $\alpha < \beta < \alpha^+$. Then $\alpha \subset \beta \subset \alpha \cup \{\alpha\}$, which immediately implies $\beta \not\subseteq \alpha$. Consequently, there exists some $x \in \beta$ that is not in α. Therefore, $x = \alpha$. However, this implies that $\alpha^+ = \alpha \cup \{\alpha\} \subseteq \beta$, which contradicts the assumption that $\beta < \alpha^+$.⊣

Theorem 1.6.10 *α^+ is an ordinal if and only if α is an ordinal.*

Proof If α^+ is an ordinal, then by Theorem 1.6.2, the elements of α^+ are ordinals, and $\alpha \in \alpha^+$.

Now assume α is an ordinal. If β and γ are distinct elements of α^+, at most one of them (say γ) is in $\{\alpha\}$, i.e., $\gamma = \alpha$. In this case, $\beta < \gamma$. If both are in α, then either $\beta < \gamma$ or $\gamma < \beta$. This proves that $\in$ is connected in α^+.

Let A be any subset of α^+. If $\alpha \notin A$, then $A \subseteq \alpha$, and A has a first element. If $\alpha \in A$, then $A \setminus \{\alpha\} \subseteq \alpha$ has a first element $\hat{\gamma}$. Since $\hat{\gamma} < \alpha$, $\hat{\gamma}$ is the first element of A. This establishes that α^+ is well-ordered by $\in$.

Finally, if $\beta \in \alpha^+$, then $\beta \in \alpha$ or $\beta = \alpha$. In the first case,

$$s(\alpha^+, <, \beta) = s(\alpha, <, \beta) = \beta.$$

In the second case,

$$s(\alpha^+, <, \beta) = s(\alpha^+, <, \alpha) = \{x \mid x \in \alpha^+ \wedge x \neq \alpha \wedge x \in \alpha\} = \alpha. \dashv$$

Theorem 1.6.11 *If $n \in \omega$, then n is an ordinal.*

Proof We know that $\emptyset$ is an ordinal, so it suffices to show that if n is an ordinal, then n^+ is also an ordinal for every $n \in \omega$.

First, observe that n is the set

$$\{\emptyset, \emptyset^1, \ldots, \emptyset^n\},$$

and n^+ is the set

$$\{\emptyset, \emptyset^1, \ldots, \emptyset^{n+1}\},$$

where the numeral n in $\emptyset^n$ denotes a sequence of n + signs.

It is easy to see that every subset of n^+ has a first element with respect to $\in$. Moreover, if $m \in n^+$, there are two cases: either $m \in n$ or $m = n$. In the first case, since n is an ordinal by hypothesis, we have

$$s(n^+, <, m) = s(n, <, m) = m.$$

In the second case, m is the initial segment

$$\{\emptyset, \emptyset^1, \ldots, \emptyset^n\}$$

of n^+, which is precisely the set

$$n = \{\emptyset, \emptyset^1, \ldots, \emptyset^n\} = \{k \mid k \in n^+ \wedge k < n\} = s(n^+, <, n). \dashv$$

Theorem 1.6.12 *The class Ω of all ordinals is not a set.*

Proof If Ω were a set, Ω itself would be an ordinal. However, by the axiom of regularity, $\Omega \notin \Omega$, which contradicts the assumption that Ω contains all ordinals as elements.$\dashv$

Theorem 1.6.13 *Let A be a class of ordinals. A is a set if and only if A is bounded.*

Proof Assume that A is a set of ordinals and that A is not bounded. Then, $\bigcup_{\alpha \in A} \alpha$ must be equal to Ω. Otherwise, there would be a $\beta \in \Omega$ that does not belong to any $\alpha \in A$, in which case $\beta > \alpha$ for all $\alpha \in A$ (since $\beta \leq \alpha$ implies $\beta \in \alpha$ or $\beta = \alpha$), and β would be an upper bound of A. This contradicts the assumption that A is not bounded.

Now, by the union axiom, $\bigcup_{\alpha \in A} \alpha$ is a set, so it cannot be equal to Ω. It follows that A must be bounded.

If A is bounded and β is an upper bound of A, then $A \subseteq \beta$. Since β is a set, A is also a set. $\dashv$

A *fundamental sequence* of ordinal numbers is a set of ordinal numbers that does not contain an ordinal greater than or equal to all the terms of the set (cf. Kamke, 1950: 89, 90, and 92). For example, ω is a fundamental sequence, whereas $\omega \cup \{\omega\}$ is not. Clearly, A is a fundamental sequence if and only if $\sup A \notin A$.

Theorem 1.6.14 *α is a limit ordinal if and only if α is an ordinal that is a fundamental sequence. Moreover, $\alpha = \sup \alpha$.*

Proof If α is a limit ordinal, α has elements, and $\alpha \neq \beta^+$ for every $\beta \in \alpha$. If α were not a fundamental sequence, there would be an element $\gamma \in \alpha$ such that $\gamma \geq \beta$ for every $\beta \in \alpha$. In such a case, since $\gamma < \alpha$, we would have $\gamma < \gamma^+ < \alpha$, so there cannot be such a γ, and α must be a fundamental sequence.

Since α is a set of ordinals, $\bigcup \alpha$ is an ordinal number and $\sup \alpha = \bigcup \alpha = \alpha$ (by Theorem 1.6.5). $\dashv$

1.7 Transfinite Induction

It is a remarkable fact that the method of mathematical induction, which operates on ω, can be generalized to operate on any ordinal. This is the meaning of the principle of transfinite induction, which is stated below. The idea is simple. Suppose there is a well-ordering (like ω); that is, a set A ordered by a total order R. If we want to prove that all the elements of A have a certain property (represented by set B), we must show that the first element of A is in B, and that the successor of every element of A (with respect to the well-order) is also in B. This implies that all elements of A have the property B. This method is grounded in the next theorem.

Theorem 1.7.1 *If*

(i) $\langle A, R\rangle$ *is a well-ordering,*
(ii) *the first element of* A *is in* B,
(iii) $\forall x\,(x \in A \wedge \forall y\,((y \in A \wedge R(y, x)) \rightarrow y \in B)) \rightarrow x \in B)$,

then $A \subseteq B$.

Proof Assume that $\langle A, R\rangle$ is a well-ordering and let B be a set satisfying conditions (ii) and (iii). Suppose there is an element $x \in A$ such that $x \notin B$, and let $\hat{x}$ be the first element of the set

$$\{x \mid x \in A \wedge x \notin B\}.$$

Then, for every y such that $R(y, \hat{x})$, we have $y \in B$, which, by condition (iii), implies $\hat{x} \in B$, contradicting the assumption. It follows that $A \subseteq B$.$\dashv$

The following theorem is crucial for the proof of Zorn's Lemma, upon which many important constructions in logic and model theory depend. The theorem states that if we have a well-ordering $\langle A, R\rangle$, and a function τ from the set of functions whose domains are the initial segments $s(A, R, x)$ to a set Z, then there is a unique function F from A to Z that can be defined recursively by means of τ. We originally know the action of τ on the set

$$\{f \mid f : s(A, R, x) \rightarrow Z \text{ for some } x \in A\},$$

whose existence proof is left to the reader. The theorem asserts that there exists a unique function F from A to Z such that F can be "defined" in terms of τ. The functions f are used to construct τ by its action on the first element of A and its action on functions constructed from others defined on subsegments of the former. Its application will become clear in the proof of Zorn's Lemma.

Theorem 1.7.2 (Transfinite Recursion) *Let* $\langle A, R\rangle$ *be a well-ordered set, and let* τ *be a function from the set*

$$\{f \mid f : s(A, R, x) \rightarrow Z \textit{ for some } x \in A\}$$

into Z*. Then there exists a unique function* $F : A \rightarrow Z$ *such that*

$$F(x) = \tau(F \restriction s(A, R, x)).$$

Proof The proof strategy is as follows. Let ϕ be the formula defined by the condition:

$$\phi(x, f) \leftrightarrow f \text{ is a function over } s(A, R, x)$$
$$\text{and } \forall y\,(Ryx \rightarrow f(y) = \tau(f \restriction s(A, R, y))).$$

We will show first, using transfinite induction, that $\phi(x, f) = \phi(x, f')$ implies $f = f'$. In the second place, using the substitution axiom, we show that there exists a set B such that

$$f \in B \leftrightarrow \exists x\,(x \in A \wedge \phi(x, f)).$$

Finally, we demonstrate that $F = \bigcup B$ is the desired function.

Step 1: Proving uniqueness of f. Assume $\phi(x, f) = \phi(x, f')$ and that there exists some $y \in A$ such that $f(y) \neq f'(y)$. Since R is a well-ordering, the set

$$\{y \mid y \in A \wedge f(y) \neq f'(y)\}$$

has a least element $\hat{y}$. Then, we have:

$$f \restriction s(A, R, \hat{y}) = f' \restriction s(A, R, \hat{y}),$$

and therefore,

$$f(\hat{y}) = \tau(f \restriction s(A, R, \hat{y})) = \tau(f' \restriction s(A, R, \hat{y})) = f'(\hat{y}),$$

which contradicts the assumption that $f \neq f'$. Hence, we have established that

$$(x \in A \wedge \phi(x, f) \wedge \phi(x, f')) \rightarrow f = f'$$

and, by the replacement theorem, we conclude that

$$\exists B\,\forall f\,(f \in B \leftrightarrow \exists x\,(x \in A \wedge \phi(x, f))).$$

Step 2: Construction of F. Let $F = \bigcup B$, where

$$\mathbf{dom}\, F = \bigcup_{f \in B} \mathbf{dom}\, f, \quad \mathbf{im}\, F = \bigcup_{f \in B} \mathbf{im}\, f,$$

and

$$\langle x, v\rangle \in F \leftrightarrow \exists f \in B\,(v = f(x) = \tau(f \restriction s(A, R, x))).$$

We will show that F satisfies the following conditions:

(1) F is a function.
(2) $F(x) = \tau(F\restriction s(A, R, x))$.
(3) $A = \mathbf{dom}\, F$.
(4) F is unique.

Property 1: F is a function.

If $f, g \in B$, then either $f \subseteq g$ or $g \subseteq f$. Let $s(A, R, y) = \mathbf{dom}\, f$ and $s(A, R, z) = \mathbf{dom}\, g$. Without loss of generality, assume Ryz, so $\mathbf{dom}\, f \subseteq \mathbf{dom}\, g$. If $f \nsubseteq g$, there exists some $v \in \mathbf{dom}\, f$ such that $f(v) \neq g(v)$. Let $\hat{v}$ be the least element of:

$$\{v \mid v \in A \wedge f(v) \neq g(v)\}.$$

Then, we have:

$$f\restriction s(A, R, \hat{v}) = g\restriction s(A, R, \hat{v}),$$

which implies:

$$f(\hat{v}) = \tau(f\restriction s(A, R, \hat{v})) = \tau(g\restriction s(A, R, \hat{v})) = g(\hat{v}),$$

contradicting $f(v) \neq g(v)$. Thus, $f \subseteq g$.

If $f(x), g(x) \in \mathbf{im}\, F$, then (say) $f \subseteq g$, so $\langle x, f(x)\rangle \in g$, and $f(x) = g(x)$. Hence, F is a function.

Property 2: Recursive condition.

If $x \in \mathbf{dom}\, F$, there exists $f \in B$ such that:

$$f(x) = \tau(f\restriction s(A, R, x)).$$

By construction,

$$F(x) = \tau(f\restriction s(A, R, x)) = \tau(F\restriction s(A, R, x)).$$

Property 3: $\mathbf{dom}\, F = A$.

Clearly, $\mathbf{dom}\, F \subseteq A$. Suppose $x \in A$ but $x \notin \mathbf{dom}\, F$. Let $\hat{x}$ be the least element of:

$$\{x \mid x \in A \wedge x \notin \mathbf{dom}\, F\}.$$

There exists $f \in B$ such that $\mathbf{dom}\, f = s(A, R, \hat{x})$. Then,

$$f \cup \{\langle \hat{x}, \tau(f\restriction s(A, R, \hat{x}))\rangle\} \in B,$$

and so $\hat{x} \in \mathbf{dom}\, F$, contradicting $x \notin \mathbf{dom}\, F$. Thus, $A \subseteq \mathbf{dom}\, F$, and $\mathbf{dom}\, F = A$.

Property 4: Uniqueness of F.

Let F' be another function such that $F'(x) = \tau(F' \restriction s(A, R, x))$ for all $x \in A$. By transfinite induction, we show that $F'(x) = F(x)$ for all $x \in A$.

If x_0 is the least element of (A, R), then:

$$F'(x_0) = \tau(F' \restriction s(A, R, x_0)) = \tau(\emptyset) = F(x_0).$$

Suppose $F'(y) = F(y)$ for $y \in s(A, R, x)$. Then:

$$F' \restriction s(A, R, x) = F \restriction s(A, R, x),$$

implying:

$$F'(x) = \tau(F' \restriction s(A, R, x)) = \tau(F \restriction s(A, R, x)) = F(x).$$

By induction, $F'(x) = F(x)$ for all $x \in A$.

Thus, F is unique.⊣

1.8 Zorn's Lemma

The following two theorems serve as preparatory steps for the proof of Zorn's Lemma.[5]

Theorem 1.8.1 (Theorem of Enumeration) *For every set A, there exists an ordinal α such that $A \sim \alpha$.*

Demostración By virtue of the Axiom of Choice, there exists a function E such that for every nonempty subset B of A, $E(B) \in B$. For each ordinal α, there exists a function $F_\alpha : \alpha \to A$ such that, for every $\gamma < \alpha$,

$$F_\alpha(\gamma) = \tau(F_\alpha \restriction s(\alpha, <, \gamma)) = E(A \setminus \mathbf{im}\,(F_\alpha \restriction s(\alpha, <, \gamma))).$$

It is easy to see that if $\beta < \alpha$, then $F_\alpha \restriction \beta = F_\beta$.

Let α be any ordinal, and β an ordinal smaller than α. Define

$$U_\beta = A \setminus \mathbf{im}\,(F_\alpha \restriction s(\alpha, <, \beta)).$$

If $U_\beta \neq \emptyset$, then $F_\alpha(\beta) = E(U_\beta) \in U_\beta \subseteq A$. Therefore, if $\gamma < \beta$, then $F_\alpha(\gamma) \in \mathbf{im}\,(F_\alpha \restriction s(\alpha, <, \beta))$, but $F_\alpha(\beta) \notin \mathbf{im}\,(F_\alpha \restriction s(\alpha, <, \gamma))$ since $\beta \notin s(\alpha, <, \gamma)$. This implies that $F_\alpha(\gamma) \neq F_\alpha(\beta)$, establishing that F_α is injective.

[5] In fact, they are equivalent to the Axiom of Choice. The reader is invited to prove this equivalence as an exercise.

Now suppose that for each ordinal β, there exists an ordinal $\alpha > \beta$ such that $U_\beta \neq \emptyset$. Let $\phi(x)$ be the formula

$$\exists\alpha\exists\beta(x = F_\alpha(\beta) \wedge \mathbf{ord}\,\alpha \wedge \mathbf{ord}\,\beta \wedge \beta < \alpha).$$

Then the set

$$X = \{x \mid x \in A \wedge \phi(x)\}$$

contains all ordinals as elements. This would imply the existence of the set of all ordinals:

$$\Omega = \{x \mid x \in X \wedge \mathbf{ord}\,x\},$$

which is impossible. Therefore, there must exist a β such that $U_\beta = \emptyset$ for some $\alpha > \beta$. Let $\hat{\beta}$ be the smallest such ordinal, and for $\gamma < \hat{\beta}$, we have $U_\gamma \neq \emptyset$. Thus, $F_\alpha \restriction \hat{\beta}$ is injective. Moreover, since $U_{\hat{\beta}} = \emptyset$,

$$A = \mathbf{im}\,(F_\alpha \restriction s(\alpha, <, \hat{\beta})) = \{F_\alpha(\gamma) \mid \gamma < \hat{\beta}\}.$$

This establishes that $F_{\hat{\beta}} = F_\alpha \restriction \hat{\beta}$ is an injective function from $\hat{\beta}$ onto A.⊣

Theorem 1.8.2 (Well-Ordering Theorem) *Every set can be well-ordered.*

Demostración Let A be any set, α an ordinal equinumerous to A, and f a bijection between A and α. Define the relation R on A by the condition

$$Rxy \leftrightarrow f(x) < f(y).$$

It is clear that R is a total ordering on A. If $B \subseteq A$, let $\hat{\beta}$ be the smallest element of $f(B) \subseteq \alpha$. Then $\hat{y} = f^{-1}(\hat{\beta})$ is the least element of B. Indeed,

$$\begin{aligned} y \in B &\Rightarrow f(y) \in f(B) \\ &\Rightarrow \hat{\beta} \leq f(y) \\ &\Rightarrow \hat{y} = f^{-1}(\hat{\beta}) \leq f^{-1}[f(y)] = y. \dashv \end{aligned}$$

The following theorem is one of the cornerstones of logical theory.

Theorem 1.8.3 (Zorn's Lemma) *If $\mathcal{A}$ is a nonempty family of sets such that the union of every nonempty chain of elements of $\mathcal{A}$ belongs to $\mathcal{A}$, then the partially ordered set $\langle \mathcal{A}, \subseteq \rangle$ has a maximal element.*

Proof Let $\mathcal{A}$ be a nonempty family of sets satisfying the given condition. By the Well-Ordering Theorem, there exists a relation R such that $\langle \mathcal{A}, R \rangle$ is well-ordered. We will construct a chain $\mathcal{C}$ (under $\subseteq$) of elements of $\mathcal{A}$ with the following properties:

(1) The first element (under R) of $\mathcal{A}$ is in $\mathcal{C}$.
(2) If $A \in \mathcal{C}$ and $B \subseteq A$ or $A \subseteq B$, then $B \in \mathcal{C}$.

We will show that $\bigcup \mathcal{C}$ is a maximal element of $\langle \mathcal{A}, \subseteq \rangle$.

For each $A \in \mathcal{A}$, let $h_A : \mathrm{s}(\mathcal{A}, R, A) \to \{0, 1\}$ be the canonical function of $\mathcal{C}$ restricted to $\mathrm{s}(\mathcal{A}, R, A)$, defined as follows:

$$h_A(B) = \begin{cases} 1 & \text{if } B \in \mathcal{C}, \\ 0 & \text{otherwise.} \end{cases}$$

Define a function τ over

$$\{f \mid f : \mathrm{s}(\mathcal{A}, R, A) \to \{0, 1\} \text{ for some } A \in \mathcal{A}\}$$

with the property that if h_A is the canonical function of $\mathcal{C}$ restricted to $\mathrm{s}(\mathcal{A}, R, A)$, then

$$\tau(h_A) = \begin{cases} 1 & \text{if } B \subseteq C \text{ or } C \subseteq B \text{ whenever } C \\ & \text{is the first element (under } R\text{) of} \\ & \mathcal{A} \setminus \mathrm{s}(\mathcal{A}, R, A) \text{ and } h_A(B) = 1, \\ 0 & \text{otherwise.} \end{cases}$$

By the Transfinite Recursion Theorem, there exists a unique function F over $\mathcal{A}$ such that

$$F(A) = \tau(F \restriction \mathrm{s}(\mathcal{A}, R, A)).$$

Using this function, we construct the chain $\mathcal{C}$ as

$$\mathcal{C} = \{A \mid A \in \mathcal{A} \text{ and } F(A) = 1\}.$$

Chain Verification.

To verify that $\mathcal{C}$ is a chain, let $A_1, A_2 \in \mathcal{C}$. Without loss of generality, assume RA_1A_2. By hypothesis, $F(A_1) = F(A_2) = 1$, so

$$\tau(F \restriction \mathrm{s}(\mathcal{A}, R, A_2)) = F(A_2) = 1.$$

Thus, A_2 is the first element of $\mathcal{A} \setminus \mathrm{dom}(F \restriction \mathrm{s}(\mathcal{A}, R, A_2))$ (under R). Since $A_1 \in \mathrm{s}(\mathcal{A}, R, A_2)$ and $F(A_1) = 1$, by the definition of τ,

$$A_1 \subseteq A_2 \quad \text{or} \quad A_2 \subseteq A_1.$$

Thus, $\mathcal{C}$ is a chain.

Maximality of $\bigcup \mathcal{C}$.

To show that $\bigcup \mathcal{C}$ is maximal in $\mathcal{A}$, note that by the hypothesis, $\bigcup \mathcal{C} \in \mathcal{A}$. Assume $\bigcup \mathcal{C}$ is not maximal, and let

$$M = \{A \mid A \in \mathcal{A} \text{ and } \bigcup \mathcal{C} \subset A\}.$$

Let $\hat{A}$ be the first element (under R) of M. For any $B \in \mathcal{C}$, we have $B \subseteq \bigcup \mathcal{C} \subset \hat{A}$, so

$$F(\hat{A}) = \tau(F \restriction \mathrm{s}(\mathcal{A}, R, \hat{A})) = 1.$$

This implies $\hat{A} \in \mathcal{C}$, contradicting the assumption that $\bigcup \mathcal{C}$ is not maximal.⊣

We have already seen that for any $n \in \mathbb{N}$, the *finite* Cartesian product

$$A_1 \times \cdots \times A_n$$

is well-defined. Another consequence of the Axiom of Choice is the existence of the Cartesian product of any collection of sets. I will define this notion by leveraging the results on ordinals established earlier.

Definition 1.8.4 If α is an ordinal, the *generalized Cartesian product* of the family (set) of sets $\{X_\beta \mid \beta < \alpha\}$, denoted by $\prod_{\beta<\alpha} X_\beta$, is the family of functions

$$\left\{f \mid f : \alpha \to \bigcup_{\beta<\alpha} X_\beta,\ f(\beta) \in X_\beta\right\}.$$

In other words, $\prod_{\beta<\alpha} X_\beta$ is the set of all functions from α to the union of all X_β that map each index β to its corresponding set X_β.

Theorem 1.8.5 *The Cartesian product $\prod_{\beta<\alpha} X_\beta$ exists for any family of sets $\{X_\beta \mid \beta < \alpha\}$.*

Proof By the Axiom of Choice, there exists a function E that assigns to every nonempty subset $A \subseteq \bigcup_{\beta<\alpha} X_\beta$ an element $E(A)$ in A. In particular, for each $x \in \bigcup_{\beta<\alpha} X_\beta$, $E(\{x\})$ is defined and equals x.

Let $\phi(f)$ be the formula defined by the condition

$$\phi(f) \leftrightarrow f \text{ is a function from } \alpha \text{ to } \bigcup_{\beta<\alpha} X_\beta$$
$$\wedge\ f(i) = E(\{x\}) \text{ for some } x \in X_\beta.$$

By the Replacement Schema Axiom (Theorem 1.4.17), there exists a set F such that

$$f \in F \leftrightarrow \left(f \in \alpha \times \bigcup_{\beta<\alpha} X_\beta \wedge \phi(f) \right).$$

The set F is precisely the Cartesian product $\prod_{\beta<\alpha} X_\beta$.⊣

1.9 Equipotence

The relation of equinumerosity or equipotence between sets (Definition 1.4.20, (7)) has several interesting properties that also facilitate the study of cardinal numbers. The first property is quite obvious: every set is equipotent to itself.

Theorem 1.9.1 *For every set A, $A \sim A$.*

Proof If A is empty, it is equipotent to $\emptyset = 0 = A$. If A is not empty, the identity function $\iota : A \to A$ establishes a bijection between A and itself. ⊣

It is an obvious corollary of Theorem 1.8.1 that $A \precsim A$. One might wonder whether the relation $\precsim$ is connected over the class of all sets; that is, whether we can always compare given sets A and B with each other. The answer is affirmative and is established in the following theorem.

Theorem 1.9.2 *For any sets A and B: $A \precsim B$ or $B \precsim A$.*

Proof By the Theorem of Numeration (Theorem 1.8.1), there exist ordinals α and β such that $A \sim \alpha$ and $B \sim \beta$. Thus, there are respective bijections $f : A \to \alpha$ and $g : B \to \beta$. By Theorem 1.6.6 (4), it holds that $\alpha < \beta$, $\beta < \alpha$, or $\alpha = \beta$. It is the case that:

If $\alpha < \beta$, then $\alpha \subset \beta$, and there exists an embedding $i : \alpha \to \beta$. In this case, the composition $g^{-1} \circ i \circ f$ is an embedding of A into B.
If $\beta < \alpha$, then $\beta \subset \alpha$, and there exists an embedding $j : \beta \to \alpha$. In this case, the composition $f^{-1} \circ j \circ g$ is an embedding of B into A.
Finally, if $\alpha = \beta$, and $\iota : \alpha \to \beta$ is the identity function on α, the composition $g^{-1} \circ \iota \circ f$ is a bijection from A to B. ⊣

It is intuitive that if $A \precsim B$ and $B \precsim A$, then $A \sim B$. However, the proof of this result—known as the Schröder-Bernstein theorem—is not so simple.

Theorem 1.9.3 (Schröder-Bernstein) *If $A \precsim B$ and $B \precsim A$, then $A \sim B$.*

Proof Suppose that $A \precsim B$ and $B \precsim A$. This means that there exist injections $f : A \to B$ and $g : B \to A$. If either of the injections is surjective, it is a bijection, and we have $A \sim B$. However, it may happen that the sets $A \setminus g(B)$ and $B \setminus f(A)$ are non-empty. In this case, we will construct a bijection h between A and B as follows.

We define the sequence of functions φ^k recursively: $\varphi^1 = g \circ f$; $\varphi^{k+1} = \varphi^1 \circ \varphi^k$. Similarly, we define the sequence ψ^k: $\psi^1 = f \circ g$; $\psi^{k+1} = \psi^1 \circ \psi^k$.

If $A_0 = A \setminus g(B)$ and φ^0 is the identity function, we define the set A_A as the union $\bigcup_{k<\omega} \varphi^k(A_0)$. Similarly, if $B_0 = B \setminus f(A)$ and ψ^0 is the identity function, we define the set B_B as the union $\bigcup_{k<\omega} \psi^k(B_0)$. We also define A_B as $\bigcup_{k<\omega} \varphi^k[g(B_0)]$ and B_A as $\bigcup_{k<\omega} \psi^k[f(A_0)]$. Let $h_A = f \restriction A_A$, $h_B = g^{-1} \restriction A_B$, and $h_\infty = f \restriction A_\infty$. We will show that $h = h_A \cup h_B \cup h_\infty$ is a bijection between A and B, for which it is sufficient to establish that $B_A = f(A_A)$, $A_B = g(B_B)$, and $B_\infty = f(A_\infty)$.

First, we have

$$\begin{aligned}
y \in B_A &\Leftrightarrow y \in \psi^k[f(A_0)] \text{ for some } k \\
&\Leftrightarrow y \in \overbrace{(f \circ g) \circ \cdots \circ (f \circ g)}^{k \text{ times}} \circ f(A_0) \\
&\Leftrightarrow y \in f \circ \overbrace{(g \circ f) \circ \cdots \circ (g \circ f)}^{k \text{ times}}(A_0) \\
&\Leftrightarrow y \in f \circ \varphi^k(A_0) = f[\varphi^k(A_0)] \\
&\Leftrightarrow \exists x \in \varphi^k(A_0) \text{ such that } y = f(x) \\
&\Leftrightarrow \exists x \in A_A \text{ such that } y = f(x) \\
&\Leftrightarrow y \in f(A_A).
\end{aligned}$$

Also,

$$\begin{aligned}
x \in A_B &\Leftrightarrow x \in \varphi^k[g(B_0)] \text{ for some } k \\
&\Leftrightarrow x \in \overbrace{(g \circ f) \circ \cdots \circ (g \circ f)}^{k \text{ times}} \circ g(B_0) \\
&\Leftrightarrow x \in g \circ \overbrace{(f \circ g) \circ \cdots \circ (f \circ g)}^{k \text{ times}}(B_0) \\
&\Leftrightarrow x \in g \circ \psi^k(B_0) = g[\psi^k(B_0)] \text{ for some } k \\
&\Leftrightarrow x \in g(B_B).
\end{aligned}$$

Finally, let y be an arbitrary element of B_∞. Then

$$\begin{aligned}
y \in B_\infty &\Leftrightarrow y \in B \setminus (B_A \cup B_B) \\
&\Leftrightarrow y \in (B \setminus B_A) \cap (B \setminus B_B) \\
&\Leftrightarrow y \in B \text{ and } y \notin B_A \text{ and } y \notin B_B.
\end{aligned}$$

Now, first, y must be equal to $f(x)$ for some $x \in A$, because otherwise y would be in $B_0 = B \smallsetminus f(A) \subseteq B_B$. Secondly, this x cannot be in A_A, because in that case, $y = f(x)$ would be in $f(A_A) = B_A$. Thirdly, this x cannot be in $A_B = g(B_B)$, because if it were, we would have $y = f(x) \in f[g(B_B)] \subseteq B_B$. It follows that x must be in A_∞, and hence $f(A_\infty) = B_\infty$.⊣

Other properties of the relations $\precsim$, $\sim$, and $\prec$ are captured in the following theorem, whose proof is left to the reader.

Theorem 1.9.4 *For any sets A, B, and C:*

(1) *If $A \precsim B$ and $B \precsim C$, then $A \precsim C$.*
(2) *If $A \sim B$ and $B \sim C$, then $A \sim C$.*
(3) *If $A \prec B$ and $B \prec C$, then $A \prec C$.*
(4) *If $A \sim B$ and $B \precsim C$, then $A \precsim C$.*
(5) *If $A \sim B$ and $B \prec C$, then $A \prec C$.*
(6) *$A \prec A$ does not hold.*
(7) *If $A \prec B$, then $B \prec A$ does not hold.*

We know that at least the set ω of finite cardinalities—whose existence has already been established—is an infinite set. A curious property of this set, and indeed of any infinite set, is that it can be put into a one-to-one correspondence with some of its proper subsets. This fact is so striking that both Galileo and Bolzano thought it paradoxical (see Bolzano, 1889). However, it is such a natural property of infinite sets that Richard Dedekind used it as the defining characteristic of infinity. It can be demonstrated, of course, that this property is both necessary and sufficient for the infinitude of a set, according to our definition.

Theorem 1.9.5 *A set A is infinite if and only if it satisfies the Dedekind condition; that is, if and only if there exists a bijection between A and some proper subset of itself.*

Proof We will first show that if A is infinite, then it satisfies the Dedekind condition. Indeed, if A is infinite, $A \neq \emptyset$ and there is no bijection between A and any natural number. Therefore, if α is an ordinal such that $A \sim \alpha$, then $\alpha \geq \omega$. Let $\iota : \alpha \to \alpha$ be the identity function and $f : \omega \to \omega$ be the function defined by $f(n) = 2n$. The function $h = f \cup (\iota \restriction \alpha \smallsetminus \omega)$ is a bijection between α and $\beta = \{m \mid m = 2n \text{ for some } n \in \omega\} \cup (\alpha \smallsetminus \omega)$, which is a proper subset of α. If g is a bijection from A to α and i is an injection from β into A, with $B = i(A)$, the function $i \circ h \circ g$ is a bijection from A to B, as illustrated in the following commutative diagram.

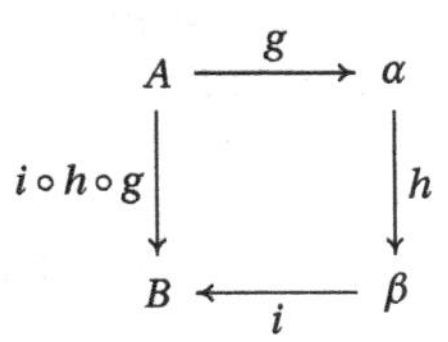

To prove the converse, it suffices to observe that no natural number is equipotent to a proper subset, as this would imply that it is equal to a smaller number. Now suppose that A satisfies the Dedekind condition and that B is a proper subset of A that is equipotent to A. If A were finite, B would also be finite. If $\alpha \sim A$ and $\beta \sim B$ are ordinals that are equipotent to A and B, respectively, α and β would be distinct elements of ω, yet they would also be identical. This contradiction establishes that A cannot be finite.$\dashv$

The question arises whether there exist infinite sets that are not equipotent to ω. The answer was provided by Georg Cantor, who proved the following result.

Theorem 1.9.6 (Cantor) *For every set A, $A \prec \mathcal{P}(A)$.*

Proof Let f be the function $f : A \to \mathcal{P}(A)$ that assigns to each $x \in A$ the singleton set $\{x\}$ in $\mathcal{P}(A)$; this establishes that $A \precsim \mathcal{P}(A)$. Suppose that $A \sim \mathcal{P}(A)$ and let g be a bijection from A to $\mathcal{P}(A)$. Let B be the set $B = \{x \in A \mid x \notin g(x)\}$. Since $B \subseteq A$, we have $B \in \mathcal{P}(A)$, so there exists an $x \in A$ such that $B = g(x)$. The curious thing is that if $x \in B$, then x satisfies the membership condition for B, and thus $x \notin g(x) = B$; moreover, if $x \notin B$, then $x \notin g(x)$ and satisfies the membership condition for B, so $x \in B$. Thus, the assumption that such a bijection g exists leads to the paradox "$x \in B$ if and only if $x \notin B$", which is a contradiction. Therefore, the assumption that there exists such a bijection g cannot be true.$\dashv$

1.10 Cardinal Numbers

A *finite cardinal number* is an element of ω; a *infinite cardinal number* is a limit ordinal. The project of defining the cardinality of an infinite set as the unique limit ordinal that is equipotent to it requires the following proposition.

Theorem 1.10.1 *Every infinite set A is equipotent to a unique limit ordinal.*

Proof Let A be any infinite set. By the numbering theorem (Theorem 1.8.1), there exists an ordinal α such that $A \sim \alpha$. Let B be the set $B = \{\beta \in \alpha \mid \beta \sim A\}$. If B is empty, we can show that α is a limit ordinal. Indeed, if it were not, there would be a β such that $\alpha = \beta^+ = \beta \cup \{\beta\}$. Now, the Dedekind condition implies that α is equipotent to one of its proper subsets $\gamma \subset \alpha$. If $\beta \notin \gamma$, then $\gamma \subseteq \beta$ and α is equipotent to a subset of β, so α is equipotent to β, and therefore $\beta \sim A$, which implies $\beta \in B = \varnothing$. If $\beta \in \gamma$, replace β with any element of $\alpha \setminus \gamma$ to obtain γ', which is still equipotent to α. Again, α is equipotent to β, contradicting the hypothesis. Thus, α must be a limit ordinal.

If B is non-empty, since α is a well-ordered set, B has a first element κ. Let K be the set $K = \{\lambda \in \kappa \mid \lambda \sim A\}$. Since K is empty (because $\lambda \in \kappa$ implies $\lambda \not\sim A$), the argument from the previous paragraph shows that κ is a limit ordinal equipotent to A.$\dashv$

Theorem 1.10.2 *For every limit ordinal α, there exists a next limit ordinal $\beta > \alpha$; that is, a limit ordinal such that if $\alpha \in \gamma \in \beta$, then γ is not a limit ordinal. In fact, the set $G = \{\gamma \in \beta \mid \alpha \neq \gamma\}$ is a fundamental sequence with* $\inf G = \alpha$ *and* $\sup G = \beta$.

Proof Let α be any limit ordinal. By Cantor's Theorem, $\mathcal{P}(\alpha) \succ \alpha$. By the Numbering Theorem 1.8.1, there exists an ordinal λ that is equipotent to $\mathcal{P}(\alpha)$. Let G be the set $G = \{\gamma \in \lambda \mid \gamma \sim \alpha\}$. Since $\gamma \sim \gamma^+$, G is a fundamental sequence, as $\gamma \in G$ implies $\gamma^+ \in G$. The set $\beta = \bigcup G$ is the next limit ordinal we are looking for.$\dashv$

Definition 1.10.3 If A is an infinite set, the *cardinal of* A, $\#A$, is the unique limit ordinal that is equipotent to A. If A is finite, $\#A$ is the natural number that is equipotent to A.

The symbol $\aleph_0$ denotes the first limit ordinal, i.e., ω. $\aleph_{n+1}$ denotes the next limit ordinal, which is well-defined by the previous theorem. The alephs are consecutive, meaning that each cardinal has an immediate successor. The famous continuum hypothesis asserts that the cardinal of the continuum, $\mathfrak{c}$, is $\aleph_1$; that is, $\mathfrak{c}$ is equipotent to $\mathcal{P}(\omega)$. Neither this hypothesis nor its negation are provable within the usual set-theoretic axiomatic systems.

Corollary 1.10.4 *For each infinite cardinal κ, there exists an $n \in \omega$ such that $\kappa = \aleph_n$.*

In other words, the infinite cardinals are precisely the alephs.

Next, we define operations for all cardinals that are analogous to those in the arithmetic of natural numbers, so that, in a sense, cardinal arithmetic can be viewed as a generalization of the latter.

Definition 1.10.5 Let κ and λ be cardinals, and let A and B be sets such that $\kappa \sim A$ and $\lambda \sim B$. We define the following operations:

(1) $\kappa + \lambda$ is $\#(A \cup B)$, provided that A and B are disjoint;
(2) $\kappa \cdot \lambda$ is $\#(A \times B)$;
(3) κ^λ is $\#A^B$, the family of all functions from B to A.

It is left to the reader to show that the operations $+$, $\cdot$, and exponentiation are well-defined, meaning that if $C \sim A$ and $D \sim B$, then $\kappa + \lambda = \#(C \cup D)$, $\kappa \cdot \lambda = \#(C \times D)$, and $\kappa^\lambda = \#C^D$. The proof that they satisfy the associative and commutative laws is straightforward.

Among the cardinal numbers, the relation $\leq$ is defined, as it is defined between ordinals. We have the following laws satisfied by this relation for cardinals.

Theorem 1.10.6 *If $\kappa \leq \mu$ and $\lambda \leq \nu$, then*

(1) $\kappa + \lambda \leq \mu + \nu$;
(2) $\kappa \cdot \lambda \leq \mu \cdot \nu$.

Proof Let A, B, C, and D be sets such that $A \cap B = \varnothing = C \cap D$, $A \sim \kappa$, $B \sim \lambda$, $C \sim \mu$, and $D \sim \nu$. Since $\kappa \leq \mu$ and $\lambda \leq \nu$, there exist injective functions $f : A \to C$ and $g : B \to D$. The function $f \cup g$, defined by the rule:

$$(f \cup g)(x) = \begin{cases} f(x) & \text{if } x \in A \\ g(x) & \text{if } x \in B, \end{cases}$$

is a bijection between $A \cup B$ and a subset of $C \cup D$. In other words, $A \cup B \precsim C \cup D$, which implies that $\kappa + \lambda = \#(A \cup B) \leq \#(C \cup D) = \mu + \nu$.

Similarly, the function $f \times g : A \times B \to C \times D$ defined by

$$(f \times g)(x, y) = (f(x), g(y))$$

is a bijection between $A \times B$ and a subset of $C \times D$. Indeed,

$$\begin{aligned} (f \times g)(x, y) = (f \times g)(x', y') \Rightarrow& \\ &(f(x), g(y)) = (f(x'), g(y')) \\ &\Rightarrow f(x) = f(x') \text{ and } g(y) = g(y') \\ &\Rightarrow x = x' \text{ and } y = y' \\ &\Rightarrow (x, y) = (x', y'). \end{aligned}$$

This shows that $\kappa \cdot \lambda = \#(A \times B) \leq \#(C \times D) = \mu \cdot \nu$.⊣

Theorem 1.10.7 *Let* κ *be any finite cardinal. For all* n*:*

(1) $\kappa + \aleph_n = \aleph_n$;
(2) $\kappa \cdot \aleph_n = \aleph_n$.

Proof To prove (1), let K be any set equipotent to κ such that $K \cap \aleph_n = \varnothing$, $C = \{2m + 1 \mid m \in \kappa\}$, and let $h : K \to C$ be the bijection defined by $h(m) = 2m + 1$ for each $m \in \kappa$. Let $A = \{2m \mid m \in \omega\}$ and let $f : \omega \to A$ be the function defined by $f(m) = 2m$. Clearly, f is a bijection between ω and A. If $B = \aleph_n \setminus \omega$ and $g : B \to B$ is the identity function on B, then the function $f \cup g \cup h$ is a bijection between $A \cup B \cup K$ and $A \cup B \cup C \subset \aleph_n$, so

$$\aleph_n \succ A \cup B \cup C \sim A \cup B \cup K = K \cup \aleph_n \succ \aleph_n.$$

Thus, we conclude that $\kappa + \aleph_n = \aleph_n$.

For (2), let A and B be sets such that $\#(A) = \#(B) = \aleph_n$, with $A \cap B = \varnothing$. If $f : A \simeq B$ is a bijection, we observe that there are as many elements in the family of all sets of the form $\{x, f(x)\}$ as there are in A or B. But $\aleph_n + \aleph_n = \#(A \cup B) = \#\left(\bigcup_{x \in A}\{x, f(x)\}\right) = \aleph_n$. Since $\kappa \cdot \aleph_n = \overbrace{\aleph_n + \cdots + \aleph_n}^{\kappa \text{ times}}$, the conclusion follows by a finite number of substitutions of $\aleph_n + \aleph_n$ by $\aleph_n$.

Another important concept relevant for the development of logical theory is the type of order. We have compared sets with respect to their cardinality, but the type of order of sets deals with a different kind of similarity between ordered sets. Given two ordered sets, that is, two structures $\mathfrak{A} = \langle A, \prec \rangle$ and $\mathfrak{B} = \langle B, \prec \rangle$, a similarity map is an isomorphism between the structures $\mathfrak{A}$ and $\mathfrak{B}$, a function $f : A \to B$ that preserves the order.

Definition 1.10.8 Let $\mathfrak{A} = \langle A, R \rangle$ and $\mathfrak{B} = \langle B, S \rangle$ be strict total orders; that is, both R and S are connected, asymmetric, and transitive. A similarity map is an isomorphism between the structures $\mathfrak{A}$ and $\mathfrak{B}$, a function $f : A \to B$ that preserves the order in the sense that

$$\langle x, y \rangle \in R \quad \text{implies} \quad \langle f(x), f(y) \rangle \in S.$$

If there exists a similarity map between $\mathfrak{A}$ and $\mathfrak{B}$, then we say that $\mathfrak{A}$ and $\mathfrak{B}$ are *similar*. An *order type* μ is the class of all similar strict total orders to a given one. The order type of $\mathfrak{A}$ is denoted as $\ulcorner \mathfrak{A} \lrcorner$.

It is easy to see that $\ulcorner \mathfrak{A} \lrcorner = \ulcorner \mathfrak{B} \lrcorner$ implies that $\#A = \#B$ (but not vice versa). The sum of order types is defined as follows.

Definition 1.10.9 Let $\mathfrak{A} = \langle A, R \rangle$ and $\mathfrak{B} = \langle B, S \rangle$ be strict total orders with $A \cap B = \varnothing$. The *ordered sum* of the structures $\mathfrak{A}$ and $\mathfrak{B}$, denoted as $\mathfrak{A} + \mathfrak{B}$, is the structure $\mathfrak{A} = \langle A \cup B, T \rangle$ with the relation T defined as follows. If $x, y \in A$ with $\langle x, y \rangle \in R$, then $\langle x, y \rangle \in T$; if $x, y \in B$ with $\langle x, y \rangle \in S$, then $\langle x, y \rangle \in T$; if $x \in A$ and $y \in B$, then $\langle x, y \rangle \in T$. If the order type of $\mathfrak{A}$ is μ and the order type of $\mathfrak{B}$ is ν, then the order type $\mu + \nu$ is defined as $\ulcorner \mathfrak{A} + \mathfrak{B} \lrcorner$.

Thus, we conclude the exposition of the cardinal arithmetic results that are required for the development of first-order logic and the subsequent results we aim to establish.

Exercises

1.1 Prove Theorem 1.4.18

1.2 Prove Lemma 1.4.21

1.3 Prove the existence of the set

$$\{f \mid f : s(A, R, x) \to Z \text{ for some } x \in A\},$$

1.4 Prove that the numeration Theorem 1.8.1 and the well-ordering Theorem 1.8.2 are equivalent to the axiom of choice. Hint: derive the axiom of choice from Zorn's lemma.

1.5 Prove Theorem 1.9.4.

1.6 Show that operations $+$, $\cdot$, and exponentiation of Definition 1.10.5 are well-defined.

References

Ackermann, W. (1956). Zur Axiomatik der Mengenlehre. *Mathematische Annalen, 131*(4), 336–345.

Bolzano, B. (1889). *Paradoxien des Unendlichen*. Mayer & Müller.

Gödel, K. (1938), The consistency of the axiom of choice and of the generalized continuum-hypothesis. *Proceedings of the National Academy of Sciences, 24*(12), 556–557.

Hintikka, J. (1988). On the development of the model-theoretic viewpoint in logical theory. *Synthese, 77*, 1–36.

Hintikka, J. (2014). The 'Axiom of Choice' is not an Axiom of Choice. In C. E. Crangle, A. García de la Sienra y H. Longino (2014), pp. 21–24.

Kamke, E. (1950). *Theory of sets*. Dover.

Lévy, A. (1959). On Ackermann's set theory. *The Journal of Symbolic Logic, 24*(2), 154–166.

Lévy, A. (1960). Axiom schemata of strong infinity in axiomatic set theory. *Pacific Journal of Mathematics, 10*, 223–238.

Lévy, A. (1965). A hierarchy of formulas in set theory. *American Mathematical Society memoir* (No. 57). American Mathematical Society.

Lévy, A., & Vaught, R. L. (1961). Principles of partial reflection in the set theories of Zermelo and Ackermann. *Pacific Journal of Mathematics, 11*, 1045–1062.

Reinhardt, W. N. (1970). Ackermann's set theory Equals ZF. *Annals of Mathematical Logic, 2*(2), 189–249.

van Heijenoort, J. (1967). Logic as language and logic as calculus. *Synthese, 17*, 324–330.

Chapter 2
Boolean Algebras

Abstract Chapter 2 provides a fairly comprehensive presentation of the theory of Boolean algebras, including proofs of Tarski's Fixed Point Theorem for lattices and a detailed proof of Stone's Representation Theorem. It begins with an in-depth explanation of the concept of a lattice and develops this concept in order to introduce the notion of a Boolean algebra as a type of lattice, thus emphasizing order as the essence of Boolean algebras. It shows how a lattice can be transformed into a Boolean algebra, how a Boolean algebra can be viewed as a Boolean ring, and thoroughly discusses the concepts of ideals, filters, and ultrafilters—necessary for the development of nonstandard analysis. The chapter concludes with the proof of Stone's Representation Theorem.

Keywords Boolean algebras · Ultrafilters

2.1 Lattices

The starting point of this chapter is the concept of a *partially ordered set (poset)* introduced in Chap. 1, Definition 1.4.20. There, we also discussed the notion of a *total or linear order*.

Building on these concepts, we say that an element x of a poset is *less than* an element y, or that y *dominates* x, if Rxy. For the purposes of this chapter, we will write $x \leq y$ instead of Rxy. The notation $x \geq y$ means the same as $y \leq x$, and $x < y$ means $x \leq y$ but not $y \leq x$. If $X \subseteq A$, an *upper bound* of X in A is an element of A that dominates all elements of X. Similarly, a *lower bound* of X in A is an element of A dominated by all elements of X. An element $x \in A$ is the *least upper bound* (or *supremum*) of X in A if x is an upper bound of X in A and x is a lower bound for the set of all upper bounds of X in A. Similarly, an element $x \in A$ is the *greatest lower bound* (or *infimum*) of X in A if x is a lower bound of X in A and x is an upper bound for the set of all lower bounds of X in A.

An *index set* is a set I of objects used as names for elements of another set: if $i \in I$, we write x_i to denote the i-th element of the set. We adopt the following notational conventions, letting $X = \{x_i \mid i \in I\}$:

A. García de la Sienra, *The Logical Foundations of Nonstandard Analysis*, Synthese Library 530, https://doi.org/10.1007/978-3-032-05677-1_2

(1) $\sup(X)$: the supremum of X,
(2) $\inf(X)$: the infimum of X,
(3) $\bigvee_{i \in I} x_i = \bigvee X = \sup(X)$,
(4) $\bigwedge_{i \in I} x_i = \bigwedge X = \inf(X)$,
(5) For $X = \{x, y\}$, $x \vee y = \sup(X)$ and $x \wedge y = \inf(X)$.

Definition 2.1.1 A *lattice* is a poset $\mathfrak{A} = \langle A, \leq \rangle$ in which every two-element subset has both a supremum and an infimum. The *dual* of the lattice $\mathfrak{A}$ is the lattice $\mathfrak{A}' = \langle A, \geq \rangle$.

An element of a lattice is called *maximal* if it is not dominated by any other element of the lattice. Similarly, an element of a lattice is called *minimal* if it does not dominate any other element of the lattice.

Theorem 2.1.2 *Every lattice has at most one maximal element and at most one minimal element.*

Proof Suppose x and y are maximal elements in a lattice A. This implies, in particular, $x \geq x \vee y$ and $y \geq x \vee y$. Since $x \vee y \geq x$ and $x \vee y \geq y$, it follows that $x = x \vee y = y$. The proof for minimal elements is analogous.⊣

The maximal element of a lattice, if it exists, is called the *greatest element* of the lattice. The minimal element, if it exists, is called the *least element*.

Definition 2.1.3 A *complemented lattice* is a structure $\langle A, \leq, 0, 1 \rangle$ such that $\langle A, \leq \rangle$ is a lattice, 0 and 1 are distinct elements of A, and the following axioms are satisfied for every $x \in A$:

(1) $x \leq 1$,

(2) $x \geq 0$,

(3) There exists $y \in A$ such that $x \vee y = 1$ and $x \wedge y = 0$.

An element y satisfying axiom (3) is called a *complement* of x. Note that an element may have more than one complement in a lattice. However, this does not occur in the special class of lattices that are distributive, as defined below.

Definition 2.1.4 A lattice $\langle A, \leq \rangle$ is *distributive* if the following axioms hold for all x, y, and z in A:

(1) $(x \vee y) \wedge z = (x \wedge z) \vee (y \wedge z)$,

(2) $(x \wedge y) \vee z = (x \vee z) \wedge (y \vee z)$.

Theorem 2.1.5 *In every lattice $\langle A, \leq \rangle$, the following identities hold for all x, y, and z in A:*

(1) *$x \vee y = y \vee x$ and $x \wedge y = y \wedge x$ (commutative laws),*
(2) *$x \vee (y \vee z) = (x \vee y) \vee z$ and $x \wedge (y \wedge z) = (x \wedge y) \wedge z$ (associative laws),*
(3) *$(x \vee y) \wedge y = y$ and $(x \wedge y) \vee y = y$ (absorption laws).*

Proof To prove (1), note that

$$x \vee y = \sup(\{x, y\}) = \sup(\{y, x\}) = y \vee x$$

and

$$x \wedge y = \inf(\{x, y\}) = \inf(\{y, x\}) = y \wedge x.$$

To establish (2), it suffices to show that

$$\sup(\{x, y \vee z\}) = \sup(\{x \vee y, z\})$$

and that

$$\inf(\{x, y \wedge z\}) = \inf(\{x \wedge y, z\}).$$

Let $u = \sup(\{x, y \vee z\})$ and $u' = \sup(\{x \vee y, z\})$. Then u is an upper bound for $\{x, y, z\}$ because

$$x \le u, \qquad y \le y \vee z \le u, \qquad z \le y \vee z \le u.$$

Since $x \vee y$ is the least upper bound of $\{x, y\}$, it follows that $x \vee y \le u$, and hence u is an upper bound for $\{x \vee y, z\}$. This implies $u' \le u$. A similar argument shows that $u \le u'$, so $u = u'$. The other identity is proven analogously.

To prove (3), observe that y is a lower bound for the set $\{x \vee y, y\}$. Moreover, for any lower bound u of the same set, we have $u \le y$, and therefore

$$y = \inf(\{x \vee y, y\}) = (x \vee y) \wedge y.$$

The other identity is proven similarly.⊣

Theorem 2.1.6 *In every complemented lattice $\langle A, \le, 0, 1\rangle$, the following identities hold for all $x \in A$:*

(1) $x \vee 1 = 1$
(2) $x \vee 0 = x$ (0 is the identity element for $\vee$).
(3) $x \wedge 0 = 0$
(4) $x \wedge 1 = x$ (1 is the identity element for $\wedge$).

The proof of this theorem is left as an exercise for the reader.

Theorem 2.1.7 *In a distributive complemented lattice, every element has a unique complement.*

Proof Suppose x has complements y and z. Then:

$$y = y \vee 0 = y \vee (x \wedge z) = (y \vee x) \wedge (y \vee z) = 1 \wedge (y \vee z) = y \vee z.$$

Similarly:

$$z = z \vee 0 = z \vee (x \wedge y) = (z \vee x) \wedge (z \vee y) = 1 \wedge (z \vee y) = z \vee y.$$

Thus, $y = y \vee z = z$.⊣

The unique complement of x in a distributive complemented lattice is denoted by $\overline{x}$.

An important class of lattices is that of *complete lattices*. A lattice A is called a *complete lattice* if and only if every subset of A has both a supremum (sup) and an infimum (inf). Complete lattices are significant in economic theory due to certain results established by Alfred Tarski. These results pertain to the existence of fixed points in lattices and find applications in game theory and utility theory.[1] Below, we proceed to explain and prove these results.

Definition 2.1.8 Let $\mathfrak{A} = \langle A, \leq \rangle$ and $\mathfrak{A}' = \langle A', \leq' \rangle$ be partially ordered sets (posets), and let f be a function from A to A'.

(1) *Isotonicity*: The function f is an *isotony* (or *monotonic*) if and only if, for all $x, y \in A$,

$$f(x) \leq' f(y) \leftrightarrow x \leq y.$$

(2) *Antitonicity*: The function f is an *antitony* (or *antimonotonic*) if and only if, for all $x, y \in A$,

$$f(y) \leq' f(x) \leftrightarrow x \geq y.$$

(3) If $\mathfrak{A} = \mathfrak{A}'$, the function $f : A \to A$ is called *increasing* if and only if f is isotonic.
(4) If $\mathfrak{A} = \mathfrak{A}'$, the function $f : A \to A$ is called *decreasing* if and only if f is antitonic.
(5) If $\mathfrak{A} = \mathfrak{A}'$, the function $f : A \to A$ is called *quasi-increasing* if and only if, for every subset $X \subseteq A$,

$$f(\sup X) \geq \inf f(X) \quad \text{and} \quad f(\inf X) \leq \sup f(X).$$

Condition (5) implies that all discontinuities of the function f are upward jumps.

Definition 2.1.9 Let $\mathfrak{A} = \langle A, \leq \rangle$ be a lattice.

(1) $\mathfrak{A}$ is a *complete lattice* if and only if every subset of A has both a supremum (sup) and an infimum (inf). In particular, if X is empty, then $\sup X = 0$ and $\inf X = 1$, since $\{x \in A : \sup A \leq x \leq \inf A\}$ is clearly empty.

[1] For some applications, see Vives (1990) and Milgrom and Shannon (1994).

(2) $\mathfrak{A}$ is a *totally ordered lattice* if and only if $\geq$ is a total order.
(3) $\mathfrak{A}$ is a *densely ordered lattice* if and only if, for every $x, y \in A$ such that $x < y$, there exists $z \in A$ such that $x < z < y$.
(4) A *fixed point* of the function $f : A \to A$ is an element $x \in A$ such that $x = f(x)$.
(5) An *interval* $[a, b]$ of $\mathfrak{A}$ is the set $\{x \in A : a \leq x \leq b\}$.

In a complete lattice $\mathfrak{A} = \langle A, \leq \rangle$, $\sup A$ is the maximum and $\inf A$ is the minimum of $\mathfrak{A}$. The maximum and minimum elements of a complete lattice are denoted, respectively, by 1 and 0.

Alfred Tarski (1955) proved two central theorems about the existence of fixed points in lattices. The subsequent proofs follow Tarski's original demonstrations, with modern notation and some simplifications.

Theorem 2.1.10 (Tarski's Fixed Point Theorem I) *Let $\langle A, \leq \rangle$ be a complete lattice, and let $f : A \to A$ be an increasing function. Let E be the set of all fixed points of f. Then E is nonempty, and $\langle E, \leq \rangle$ is a complete lattice. In particular, $\sup E$ and $\inf E$ belong to E.*

Proof Let $\langle A, \leq \rangle$ be a complete lattice, and let f be an increasing function from A to A. Let E denote the set of all fixed points of f, and define

$$U = \{x \in A \mid f(x) \geq x\}.$$

The proof consists of showing that

$$u = \sup U \tag{1}$$

is a fixed point (in fact, the supremum of all fixed points) and that both the supremum and infimum of any subset of E exist.

First, it is clear that $x \leq u$ for all $x \in U$. Since f is increasing,

$$x \leq f(x) \leq f(u),$$

implying that $f(u)$ is an upper bound of U. Since $u = \sup U$,

$$u \leq f(u). \tag{2}$$

Furthermore,

$$f(u) \leq f[f(u)],$$

so $f(u) \in U$ and

$$f(u) \leq u. \tag{3}$$

Clearly, (2) and (3) together imply that u is a fixed point of f.

Since any fixed point satisfies $f(x) \geq x$, it follows that $E \subseteq U$, and thus $u \geq x$ for all fixed points x. This implies that $u = \sup E$, and

$$\sup E \in E. \tag{4}$$

The dual lattice $\mathfrak{A}' = \langle A, \geq \rangle$ is also complete, and f is increasing in $\mathfrak{A}'$. The supremum of any set of elements in $\mathfrak{A}'$ coincides with the infimum of the same set in $\mathfrak{A}$. By applying the results established in (4) to $\mathfrak{A}'$, we obtain

$$\inf E \in E. \tag{5}$$

To show that $\mathfrak{E} = \langle E, \leq \rangle$ is a complete lattice, let X be any subset of E. Since $B = [\sup X, 1]$ is an interval in $\mathfrak{E}$, it is immediate that the structure

$$\mathfrak{B} = \langle B, \leq \rangle$$

is a complete lattice. Since B is a set of fixed points of f, it follows that the restriction $g = f{\upharpoonright}B$ is an increasing function from B to B. This implies that $\mathfrak{B}$ along with the function g satisfies the same hypotheses as $\mathfrak{A}$ with the function f. Since the set of fixed points of B is B itself, (5) implies that $v = \inf B \in B$, meaning that the infimum of B is a fixed point of g and, therefore, an upper bound of X.

Since B is the set of upper bounds of X in E, it follows that v is the supremum of X in E. An analogous consideration of the dual lattices $\mathfrak{A}'$ and $\mathfrak{B}'$ shows that X also has an infimum in $\mathfrak{E}$. Since X was an arbitrary subset of E, we conclude that $\mathfrak{E}$ is a complete lattice.$\dashv$

Tarski's second theorem applies to functions with more general characteristics, albeit at the cost of more restrictive assumptions (such as density and total order) regarding the structure of the underlying lattice. The following formulation is a simplification of the original theorem, tailored for its relevance to economic theory. Accordingly, the proof I provide is a simplified variant of Tarski's original proof.

Theorem 2.1.11 (Tarski's Fixed Point Theorem II) *Let $\langle A, \leq \rangle$ be a complete, totally ordered, and densely ordered lattice, and let $f : A \to A$ be a quasi-increasing function. Let E be the set of all fixed points of f. Then E is nonempty, and $\langle E, \leq \rangle$ is a totally ordered complete lattice.*

Proof Let $\langle A, \leq \rangle$ be a complete, totally ordered, and densely ordered lattice, and let f be a quasi-increasing function from A to A. Since $\mathfrak{A}$ is complete, there exist elements $0 = \inf A$ and $1 = \sup A$ in A such that

$$0 \leq x \leq 1$$

for all $x \in A$.

Since $f(x) \in A$, it follows that

$$0 \leq f(x) \leq 1,$$

and in particular,

$$f(1) \leq 1 \quad \text{and} \quad f(0) \geq 0.$$

The proof involves showing that, for any subsets U and V of E (which may be empty),

$$u = \sup\{x \in A \mid f(x) \geq x \text{ and } x \leq \inf U\}$$

and

$$v = \inf\{x \in A \mid f(x) \leq x \text{ and } x \geq \sup V\},$$

the points u and v are fixed points. To do this, we first need to prove that $f(\sup P) \geq \sup P$ and $f(\inf P') \leq \inf P'$ for any nonempty subsets $P \subseteq \{x \in A \mid f(x) \geq x\}$ and $P' \subseteq \{x \in A \mid f(x) \leq x\}$, respectively.

In order to show that $f(\sup P) \geq \sup P$, let P be any subset of A such that

$$f(x) \geq x \quad \text{for all } x \in P. \tag{1}$$

To prove $f(\sup P) \geq \sup P$, assume, for the sake of contradiction, that

$$f(\sup P) < \sup P. \tag{2}$$

If $\sup P = 0$, this would imply $0 > f(\sup P) \geq 0$, which is impossible. Thus,

$$\sup P \neq 0. \tag{3}$$

Since $\mathfrak{A}$ is densely ordered, there exists an element $a \in A$ such that

$$f(\sup P) < a < \sup P. \tag{4}$$

Define now

$$O = \{x \in A \mid x \leq a\}, \tag{5}$$

so that

$$\sup O \leq a < \sup P. \tag{6}$$

Let

$$R = P \wedge \{x \in A \mid \sup O < x\}. \tag{7}$$

If R were empty, we would have $x \leq \sup O$ for all $x \in P$, implying $\sup P \leq \sup O$, contradicting (6). Thus, R is nonempty, and furthermore, $\sup R = \sup P$, because $\sup P$ is an upper bound of R (so $\sup R \leq \sup P$), and $\sup R$ is an upper bound of P since elements of $\{x \in A \mid \sup O < x\}$ dominate elements of $\{x \in A \mid \sup O \geq x\}$, which are complementary subsets in A due to the total ordering of $\mathfrak{A}$.

In order to derive a contradiction, observe that since f is quasi-increasing,

$$f(\sup P) = f(\sup R) \geq \inf f(R),$$

and so, by (4),

$$a > \inf f(R).$$

Thus, there must exist some $x \in R$ such that $f(x) < a$, because if $f(x) \geq a$ for all $x \in R$, then $\inf f(R) \geq a$, contradicting the inequality above.

From (1), we have $x \leq f(x) < a$, and from (7), $\sup O < x$, implying $x \in P$. On the other hand, from (5), $x \in O$, so $x \leq \sup O$, which contradicts $\sup O < x$. This contradiction proves that (2) is false for any nonempty subset P. That is,

$$f(\sup P) \geq \sup P. \tag{8}$$

In the dual lattice $\mathfrak{A}' = \langle A, \geq \rangle$, the analogous result is

$$f(\inf P') \leq \inf P', \tag{9}$$

for any nonempty subset $P' \subseteq \{x \in A \mid f(x) \leq x\}$.

Hence, using (8) and (9), it follows that u and v are fixed points of f. This establishes the existence of fixed points u and v under the given assumptions.

Let U be any subset (empty or not) of:

$$E = \{x \in A \mid f(x) = x\},$$

and let:

$$u = \sup\{x \in A \mid f(x) \geq x \text{ and } x \leq \inf U\}. \tag{10}$$

From (8) and (9), we obtain:

$$f(u) \geq u \quad \text{and} \quad f(\inf U) \leq \inf U. \tag{11}$$

If $u = \inf U$, it follows that:

$$f(u) = u, \tag{12}$$

or $u \in E$.

If $u \neq \inf U$, (1) implies $u \leq \inf U$, because $\inf U$ is an upper bound for the set:

$$\{x \in A \mid f(x) \geq x \text{ and } x \leq \inf U\},$$

and u is the supremum of this set. Since $\mathfrak{A}$ is densely ordered, the set:

$$W = \{x \in A \mid u < x \leq \inf U\}$$

is nonempty, and:

$$u = \inf W. \tag{13}$$

It is also true that $f(x) < x$ for all $x \in W$, because:

$$x \in W \Rightarrow x > u \quad \text{and} \quad x \leq \inf U,$$

which implies:

$$x \notin \{x \in A \mid f(x) \geq x \text{ and } x \leq \inf U\}.$$

Therefore:

$$f(x) < x. \tag{14}$$

This means that W is a nonempty subset of:

$$\{x \in A \mid f(x) \leq x\}.$$

Thus, by (9) and (13):

$$f(u) = f(\inf W) \leq \inf W = u, \tag{15}$$

which, combined with $f(u) \geq u$, implies:

$$f(u) = u. \tag{16}$$

We have thus shown that:

$$u \in E. \tag{17}$$

By duality:

$$v = \inf\{x \in A \mid f(x) \leq x \text{ and } x \geq \sup V\}$$

for any subset $V \subseteq E$, and:

$$v \in E. \tag{18}$$

It is straightforward to see that $u = \sup U$ and $v = \inf V$. Hence, u is the supremum of U in $\langle E, \leq \rangle$, and v is the infimum of V in $\langle E, \leq \rangle$. This implies that $\langle E, \leq \rangle$ is a completely ordered lattice.⊣

2.2 Boolean Algebras

Sometimes a Boolean algebra is defined as a complemented distributive lattice. As we will see, such lattices can be easily converted into Boolean algebras, but a clean treatment of the relationships between the two requires distinguishing their structures clearly. The following definition is quite common:

Definition 2.2.1 A *Boolean algebra* is a sextuple

$$\langle A, \vee, \wedge, ^{-}, 0, 1 \rangle$$

such that A is a nonempty set, $\vee$ and $\wedge$ are binary operations on A, $^{-}$ is a unary operation on A, and 0 and 1 are distinct elements of A, that satisfies the following axioms for all $x, y, z \in A$:

(1) *Commutativity*: $x \vee y = y \vee x$ and $x \wedge y = y \wedge x$.
(2) *Associativity*: $x \vee (y \vee z) = (x \vee y) \vee z$ and $x \wedge (y \wedge z) = (x \wedge y) \wedge z$.
(3) *Absorption*: $(x \vee y) \wedge y = y$ and $(x \wedge y) \vee y = y$.
(4) *Complementarity*: $x \vee \overline{x} = 1$ and $x \wedge \overline{x} = 0$.
(5) *Distributivity*:

$$(x \vee y) \wedge z = (x \wedge z) \vee (y \wedge z) \quad \text{and} \quad (x \wedge y) \vee z = (x \vee z) \wedge (y \vee z).$$

Definition 2.2.2 A *homomorphism of Boolean algebras* is a mapping

$$\varphi : A \to A'$$

from a Boolean algebra $\langle A, \vee, \wedge, ^{-}, 0, 1 \rangle$ to another Boolean algebra $\langle A', \vee', \wedge', ^{-'}, 0', 1' \rangle$ such that, for all $x, y \in A$:

(1) $\varphi(x \vee y) = \varphi(x) \vee' \varphi(y)$.
(2) $\varphi(x \wedge y) = \varphi(x) \wedge' \varphi(y)$.
(3) $\varphi(\overline{x}) = \overline{\varphi}(x)$.
(4) $\varphi(0) = 0'$ and $\varphi(1) = 1'$.

The proof of the following theorem, which establishes an important equivalence between complemented distributive lattices and Boolean algebras, is left as an exercise for the reader.

Theorem 2.2.3 *Let $\langle A, \leq, 0, 1\rangle$ be a complemented distributive lattice, where $\vee$, $\wedge$, and $\neg$ are the operations of supremum, infimum, and complement in the lattice. Then the structure $\langle A, \vee, \wedge, \neg, 0, 1\rangle$ is a Boolean algebra. Conversely, if $\langle A, \vee, \wedge, \neg, 0, 1\rangle$ is a Boolean algebra and $\leq$ is defined by the condition $x \leq y \leftrightarrow x \vee y = y$, then $\langle A, \leq, 0, 1\rangle$ is a complemented distributive lattice.*

From now on, the order relation $\leq$ will be understood to be defined in every Boolean algebra by the condition $x \leq y \leftrightarrow x \vee y = y$. The lattice constituted by this relation on the universe of the algebra will be called the *lattice associated with the Boolean algebra.*

Similarly, every Boolean algebra is associated with a certain ring, as established by the following theorem.

Theorem 2.2.4 *Let $\mathfrak{A} = \langle A, \vee, \wedge, \neg, 0, 1\rangle$ be a Boolean algebra. Define the structure*

$$\mathfrak{R} = \langle A, +, \cdot, 0, 1\rangle$$

as follows: Let $x + y = (x \wedge \overline{y}) \vee (\overline{x} \wedge y)$ for all $x, y \in A$; $\cdot$ is the same operation as $\wedge$; 0 and 1 are the same as in $\mathfrak{A}$. Then $\mathfrak{R}$ is a Boolean ring, i.e., a ring with identity in which every element is idempotent.

Proof Since $\wedge$ is associative, $x \wedge 1 = 1 \wedge x = x$ and $x^2 = x \wedge x = x$ for all $x \in A$ (the elements of A are idempotent). We only need to show that $\langle A, +, 0\rangle$ is an abelian group and that $\cdot$ is distributive over $+$. The most cumbersome part is proving the associativity of $+$, which we proceed to do as follows:

$$\begin{aligned}
x + (y + z) &= \big(x \wedge \overline{[(y \wedge \overline{z}) \vee (\overline{y} \wedge z)]}\big) \vee (\overline{x} \wedge [(y \wedge \overline{z}) \vee (\overline{y} \wedge z)]) \\
&= \big(x \wedge \big[\overline{(y \wedge \overline{z})} \wedge \overline{(\overline{y} \wedge z)}\big]\big) \vee (\overline{x} \wedge [(y \wedge \overline{z}) \vee (\overline{y} \wedge z)]) \\
&= (x \wedge [(\overline{y} \vee z) \wedge (y \vee \overline{z})]) \vee (\overline{x} \wedge [(y \wedge \overline{z}) \vee (\overline{y} \wedge z)]) \\
&= (x \wedge [((\overline{y} \vee z) \wedge y) \vee ((\overline{y} \vee z) \wedge \overline{z})]) \\
&\qquad \vee (\overline{x} \wedge [(y \wedge \overline{z}) \vee (\overline{y} \wedge z)]) \\
&= (x \wedge [(0 \vee (y \wedge z)) \vee ((\overline{y} \wedge \overline{z}) \vee 0)]) \\
&\qquad \vee (\overline{x} \wedge [(y \wedge \overline{z}) \vee (\overline{y} \wedge z)]) \\
&= (x \wedge [(y \wedge z) \vee (\overline{y} \wedge \overline{z})]) \\
&\qquad \vee (\overline{x} \wedge [(y \wedge \overline{z}) \vee (\overline{y} \wedge z)]) \\
&= [x \wedge (y \wedge z)] \vee [x \wedge (\overline{y} \wedge \overline{z})] \\
&\qquad \vee [\overline{x} \wedge (y \wedge \overline{z})] \vee [\overline{x} \wedge (\overline{y} \wedge z)] \\
&= [(x \wedge y) \wedge z] \vee [(x \wedge \overline{y}) \wedge \overline{z}] \\
&\qquad \vee [(\overline{x} \wedge y) \wedge \overline{z}] \vee [(\overline{x} \wedge \overline{y}) \wedge z]
\end{aligned}$$

$$\begin{aligned}
&= [(x \wedge \overline{y}) \wedge \overline{z}] \vee [(\overline{x} \wedge y) \wedge \overline{z}] \\
&\quad \vee [(x \wedge y) \wedge z] \vee [(\overline{x} \wedge \overline{y}) \wedge z] \\
&= ([(x \wedge \overline{y}) \vee (\overline{x} \wedge y)] \wedge \overline{z}) \vee ([(x \wedge y) \vee (\overline{x} \wedge \overline{y})] \wedge z) \\
&= ([(x \wedge \overline{y}) \vee (\overline{x} \wedge y)] \wedge \overline{z}) \\
&\quad \vee ([(0 \vee (x \wedge y)) \vee ((\overline{x} \wedge \overline{y}) \vee 0)] \wedge z) \\
&= ([(x \wedge \overline{y}) \vee (\overline{x} \wedge y)] \wedge \overline{z}) \\
&\quad \vee ([((\overline{x} \wedge x) \vee (y \wedge x)) \vee ((\overline{x} \wedge \overline{y}) \vee (y \wedge \overline{y}))] \wedge z) \\
&= ([(x \wedge \overline{y}) \vee (\overline{x} \wedge y)] \wedge \overline{z}) \\
&\quad \vee ([((\overline{x} \vee y) \wedge x) \vee ((\overline{x} \vee y) \wedge \overline{y})] \wedge z) \\
&= ([(x \wedge \overline{y}) \vee (\overline{x} \wedge y)] \wedge \overline{z}) \vee ([(\overline{x} \vee y) \wedge (x \vee \overline{y})] \wedge z) \\
&= ([(x \wedge \overline{y}) \vee (\overline{x} \wedge y)] \wedge \overline{z}) \vee \left(\left[\overline{(x \wedge \overline{y})} \wedge \overline{(\overline{x} \wedge y)}\right] \wedge z\right) \\
&= ([(x \wedge \overline{y}) \vee (\overline{x} \wedge y)] \wedge \overline{z}) \vee \left(\overline{[(x \wedge \overline{y}) \vee (\overline{x} \wedge y)]} \wedge z\right) \\
&= (x + y) + z.
\end{aligned}$$

Clearly,

$$\begin{aligned}
x + 0 &= \left(x \wedge \overline{0}\right) \vee (\overline{x} \wedge 0) \\
&= x \wedge 1 \\
&= x.
\end{aligned}$$

Interestingly, each element of A is its own additive inverse, since we have

$$\begin{aligned}
x + x &= (x \wedge \overline{x}) \vee (\overline{x} \wedge x) \\
&= 0 \vee 0 \\
&= 0
\end{aligned}$$

This is expressed by saying that the algebra is *of characteristic 2*.

It is obvious that the group is abelian, since

$$\begin{aligned}
x + y &= (x \wedge \overline{y}) \vee (\overline{x} \wedge y) \\
&= (y \wedge \overline{x}) \vee (\overline{y} \wedge x) \\
&= y + x
\end{aligned}$$

Finally,

$$
\begin{aligned}
x\,(y+z) &= x \wedge [(y \wedge \overline{z}) \vee (\overline{y} \wedge z)] \\
&= [x \wedge (y \wedge \overline{z})] \vee [x \wedge (\overline{y} \wedge z)] \\
&= [(x \wedge y) \wedge \overline{z}] \vee [\overline{y} \wedge (x \wedge z)] \\
&= 0 \vee [(x \wedge y) \wedge \overline{z}] \vee 0 \vee [\overline{y} \wedge (x \wedge z)] \\
&= [(x \wedge y) \wedge \overline{x}] \vee [(x \wedge y) \wedge \overline{z}] \\
&\qquad \vee [\overline{x} \wedge (x \wedge z)] \vee [\overline{y} \wedge (x \wedge z)] \\
&= [(x \wedge y) \wedge (\overline{x} \vee \overline{z})] \vee [(\overline{x} \vee \overline{y}) \wedge (x \wedge z)] \\
&= \left[(x \wedge y) \wedge \overline{(x \wedge z)}\right] \vee \left[\overline{(x \wedge y)} \wedge (x \wedge z)\right] \\
&= xy + xz
\end{aligned}
$$

An analogous argument shows that $(y+z)\,x = yx + zx$.⊣

$\mathfrak{R}$ is called the *Boolean ring associated with* $\mathfrak{A}$.

It is interesting to note that the ideals of a Boolean algebra are the ideals of its associated ring, as established below.

Definition 2.2.5 Let $\mathfrak{A} = \langle A, \vee, \wedge, ^-, 0, 1\rangle$ be a Boolean algebra. An *ideal* of $\mathfrak{A}$ is a nonempty set $\mathcal{I} \subseteq A$ that satisfies the following three conditions:

(1) $0 \in \mathcal{I}$;
(2) if $x \in \mathcal{I}$ and $y \leq x$, then $y \in \mathcal{I}$ for every $y \in A$;
(3) if $x, y \in \mathcal{I}$, then $x \vee y \in \mathcal{I}$.

It is interesting to note that the ideals of a Boolean algebra are the ideals of its associated ring, as established below.

Theorem 2.2.6 *An ideal* $\mathcal{I}$ *of a Boolean algebra is an ideal of its associated Boolean ring.*

Proof Let $\mathcal{I}$ be an ideal of the Boolean algebra

$$\langle A, \vee, \wedge, ^-, 0, 1\rangle.$$

We must prove that $\langle \mathcal{I}, +, 0\rangle$ is a subgroup of $\langle A, +, 0\rangle$ and that $x\mathcal{I} \subseteq \mathcal{I}$ for all $x \in A$.

Clearly, $0 = 0x \in \mathcal{I}$ for all $x \in \mathcal{I}$. If $x, y \in \mathcal{I}$, then

$$x + y = (x \wedge \overline{y}) \vee (\overline{x} \wedge y)\,.$$

But $x \wedge \overline{y} \in \mathcal{I}$ and $\overline{x} \wedge y \in \mathcal{I}$, so $(x \wedge \overline{y}) \vee (\overline{x} \wedge y)$ is an element of $\mathcal{I}$. If y is any element of A and $x \in \mathcal{I}$, then $yx \in \mathcal{I}$. This establishes that $x\mathcal{I} \subseteq \mathcal{I}$.⊣

If it is possible to associate a ring with a Boolean algebra, it is also possible to associate a Boolean algebra with a Boolean ring. More precisely,

Theorem 2.2.7 *Let* $\mathfrak{R} = \langle A, +, \cdot, 0, 1\rangle$ *be a Boolean ring. We define the structure*

$$\mathfrak{A} = \langle A, \vee, \wedge, ^{-}, 0, 1\rangle$$

as follows: $x \vee y \equiv x + y + xy$, $x \wedge y \equiv xy$, *and* $\overline{x} \equiv 1 + x$. *Then* $\mathfrak{A}$ *is a Boolean algebra.*

The proof is left to the reader. We call $\mathfrak{A}$ the *Boolean algebra* associated with the Boolean ring $\mathfrak{R}$. In analogy with Theorem 2.2.6, we have the following.

Theorem 2.2.8 *An ideal* $\mathcal{I}$ *of a Boolean ring is an ideal of the Boolean algebra associated with it.*

The proof is left to the reader.

Theorem 2.2.9 *In a Boolean algebra* $\langle A, \vee, \wedge, ^{-}, 0, 1\rangle$, *the following assertions hold for all* $x, y, z \in A$*:*

(1) *If* $x \vee y = x$ *for all* $x \in A$, *then* $y = 0$; *if* $x \wedge y = x$ *for all* $x \in A$, *then* $y = 1$.
(2) *If* $x \wedge y = 0$ *and* $x \vee y = 1$, *then* $y = \overline{x}$.
(3) $x \wedge y = 0$ *if and only if* $y \leq \overline{x}$.
(4) $\overline{x \vee y} = \overline{x} \wedge \overline{y}$; $\overline{x \wedge y} = \overline{x} \vee \overline{y}$.
(5) *If* $x \leq y$, *then* $\overline{y} \leq \overline{x}$.
(6) *Let* $\{y_i\}_{i \in I}$ *be a family of elements of a Boolean algebra such that* $y = \bigvee_{i \in I} y_i$ *exists. If* x *and* z *are elements of the algebra such that* $x \wedge y_i \leq z$ *for all* $i \in I$, *then* $x \wedge y \leq z$.

Proof For (1), suppose that $x \vee y = x$ for all $x \in A$. Then, in particular, $y = 0 \vee y = 0$. If $x \wedge y = 1$ for all x, then $y = 1 \wedge y = 1$.

For (2), suppose that $x \wedge y = 0$ and $x \vee y = 1$. Then it follows that $y = \overline{x}$, since the order relation induced by a Boolean algebra forms a complemented distributive lattice (Theorem 2.2.3) and in such a lattice, each element has a unique complement (Theorem 2.1.7).

For (3), in one direction, we have:

$$\begin{aligned} x \wedge y = 0 &\Rightarrow (x \wedge y) \vee \overline{x} = 0 \vee \overline{x} \\ &\Rightarrow (x \vee \overline{x}) \wedge (y \vee \overline{x}) = \overline{x} \\ &\Rightarrow 1 \wedge (y \vee \overline{x}) = \overline{x} \\ &\Rightarrow y \leq \overline{x}. \end{aligned}$$

In the other direction we get:

$$y \leq \overline{x} \Rightarrow x \wedge y \leq x \wedge \overline{x}$$

$$\Rightarrow x \wedge y = 0.$$

For (4), we will show that

$$(x \vee y) \wedge (\overline{x} \wedge \overline{y}) = 0$$

and also that

$$(x \vee y) \vee (\overline{x} \wedge \overline{y}) = 1$$

because this implies, by (2), that

$$\overline{x \vee y} = (\overline{x} \wedge \overline{y}).$$

We have, indeed,

$$\begin{aligned}
(x \vee y) \wedge (\overline{x} \wedge \overline{y}) &= (\overline{x} \wedge \overline{y}) \wedge (x \vee y) \\
&= [(\overline{x} \wedge \overline{y}) \wedge x] \vee [(\overline{x} \wedge \overline{y}) \wedge y] \\
&= [(x \wedge \overline{x}) \wedge \overline{y}] \vee [\overline{x} \wedge (y \wedge \overline{y})] \\
&= (0 \wedge \overline{y}) \vee (\overline{x} \wedge 0) \\
&= 0 \vee 0 \\
&= 0.
\end{aligned}$$

Moreover,,

$$\begin{aligned}
(x \vee y) \vee (\overline{x} \wedge \overline{y}) &= [(x \vee y) \vee \overline{x}] \wedge [(x \vee y) \vee \overline{y}] \\
&= [(x \vee \overline{x}) \vee y] \wedge [x \vee (y \vee \overline{y})] \\
&= (1 \vee y) \wedge (x \vee 1) \\
&= 1 \wedge 1 \\
&= 1.
\end{aligned}$$

The proof of the dual identity is analogous and is left to the reader as an exercise.

For (5), the following chain of equivalences holds:

$$\begin{aligned}
x \leq y &\Leftrightarrow x \vee y = y \\
&\Leftrightarrow \overline{x \vee y} = \overline{y} \\
&\Leftrightarrow \overline{x} \wedge \overline{y} = \overline{y} \\
&\Leftrightarrow (\overline{x} \wedge \overline{y}) \vee \overline{x} = \overline{y} \vee \overline{x}
\end{aligned}$$

$$\Leftrightarrow (\overline{y} \wedge \overline{x}) \vee \overline{x} = \overline{y} \vee \overline{x}$$
$$\Leftrightarrow \overline{x} = \overline{y} \vee \overline{x}$$
$$\Leftrightarrow \overline{y} \leq \overline{x}$$

For (6), let $\{y_i\}_{i \in I}$ be a family of elements of a Boolean algebra such that $y = \bigvee_{i \in I} y_i$ exists, and let x and z be elements of the algebra such that $x \wedge y_i \leq z$ for all $i \in I$. We have, for any $i \in I$:

$$x \wedge y_i \leq z \leq z \vee \overline{x}$$

y

$$\overline{x} \wedge y_i \leq \overline{x} \leq z \vee \overline{x}.$$

Por lo tanto,

$$y_i = (x \wedge y_i) \vee (\overline{x} \wedge y_i) \leq z \vee \overline{x}.$$

This implies that

$$y = \bigvee_{i \in I} y_i \leq z \vee \overline{x}$$

and, hence,

$$\begin{aligned} x \wedge y &\leq x \wedge (\overline{x} \vee z) \\ &= (x \wedge \overline{x}) \vee (x \wedge z) \\ &= x \wedge z \\ &\leq z. \dashv \end{aligned}$$

We will now prove a series of important Boolean identities. Boolean identities generally come in pairs, with one member of the pair being the "dual" of the other. Since the operations of join and meet are perfectly symmetrical, the proofs of the duals are entirely analogous, and it is common practice in Boolean algebra to take for granted that proving one establishes the other. The reader can convince themselves of this fact by attempting to prove some duals.

Not every family $\{x_i\}_{i \in I}$ of elements in a Boolean algebra has a supremum or an infimum. In accordance with the definition of a complete lattice, a Boolean algebra is said to be *complete* if and only if every subset of it has both a supremum and an infimum. However, if a certain supremum (or infimum) exists and it is proven, based on its definition, to be identical to another term, then the existence of one implies the existence of the other. In what follows, we will demonstrate that the definitions of certain terms imply their identity with others.

Theorem 2.2.10 (Generalized De Morgan's Laws) *Let $\{x_i\}_{i\in I}$ be a set of elements of a Boolean algebra A. Then:*

(1) *The term $\overline{\bigvee_{i\in I} x_i}$ exists if and only if the term $\bigwedge_{i\in I}\overline{x_i}$ exists, and moreover,*

$$\overline{\bigvee_{i\in I} x_i} = \bigwedge_{i\in I}\overline{x_i};$$

(2) *The term $\overline{\bigwedge_{i\in I} x_i}$ exists if and only if the term $\bigvee_{i\in I}\overline{x_i}$ exists, and moreover,*

$$\overline{\bigwedge_{i\in I} x_i} = \bigvee_{i\in I}\overline{x_i}.$$

Proof For (1), suppose that $\overline{\bigvee_{i\in I} x_i}$ exists. Then its complement $x = \bigvee_{i\in I} x_i$also exists in A. It follows that $x_i \leq x$ for all $i \in I$, and therefore, $\overline{x} \leq \overline{x_i}$; that is, $\overline{x}$ is a lower bound for all $\overline{x_i}$. If $\overline{x}$ were not the infimum of these elements, there would exist some y such that $\overline{x} < y \leq \overline{x_i}$, or equivalently, $x_i \leq \overline{y} < x$. However, this would imply that $x$is not the supremum of x_i, which is a contradiction. Thus, if $\overline{\bigvee_{i\in I} x_i}$ exists, then $\bigwedge_{i\in I}\overline{x_i}$ also exists, and moreover,

$$\overline{\bigvee_{i\in I} x_i} = \overline{x} = \bigwedge_{i\in I}\overline{x_i}.$$

Now suppose that $\bigwedge_{i\in I}\overline{x_i}$ exists. We have $\bigwedge_{i\in I}\overline{x_i} \leq \overline{x_i}$ for all $i \in I$, from which it follows that $x_i \leq \overline{\bigwedge_{i\in I}\overline{x_i}}$ for all $i \in I$. Let y be any upper bound for the set x_i, so $x_i \leq y$for all $i \in I$. It then follows that $\overline{y} \leq \overline{x_i}$ for all $i \in I$, and thus $\overline{y} \leq \bigwedge_{i\in I}\overline{x_i}$, since $\bigwedge_{i\in I}\overline{x_i}$ is the infimum of the $\overline{x_i}$. Consequently, $\overline{\bigwedge_{i\in I}\overline{x_i}} \leq y$, showing that $\overline{\bigwedge_{i\in I}\overline{x_i}}$ is the least upper bound of the set x_i.

The proof of the dual is analogous and is left as an exercise for the reader.⊣

Theorem 2.2.11 (Generalized Associative Laws) *Let $\{I_j\}_{j\in J}$ be a family of pairwise disjoint index sets whose union is I, and let $\{x_i\}_{i\in I}$ be a family of elements of a Boolean algebra. Then:*

(1) *If the term $\bigvee_{j\in J}\left(\bigvee_{i\in I_j} x_i\right)$ exists, then $\bigvee_{i\in I} x_i$ also exists, and moreover,*

$$\bigvee_{j\in J}\left(\bigvee_{i\in I_j} x_i\right) = \bigvee_{i\in I} x_i;$$

(2) *If the term* $\bigwedge_{j\in J}\left(\bigwedge_{i\in I_j} x_i\right)$ *exists, then* $\bigwedge_{i\in I} x_i$ *also exists, and moreover,*

$$\bigwedge_{j\in J}\left(\bigwedge_{i\in I_j} x_i\right)=\bigwedge_{i\in I} x_i .$$

Proof For (1), suppose that $\bigvee_{j\in J}\left(\bigvee_{i\in I_j} x_i\right)$ exists; that is, $x_j = \left(\bigvee_{i\in I_j} x_i\right)$ exists for each $j \in J$, and moreover, $\bigvee_{j\in J} x_j$ exists.

Then, if $i \in I_j$, we have

$$x_i \le x_j \le \bigvee_{j\in J} x_j = \bigvee_{j\in J}\left(\bigvee_{i\in I_j} x_i\right),$$

so $x = \bigvee_{j\in J}\left(\bigvee_{i\in I_j} x_i\right)$ is an upper bound of the x_i with $i \in I_j$ for any $j \in J$.

If x were not the supremum of all x_i $(i \in I)$, there would exist some y such that

$$x_i \le y < x = \bigvee_{j\in J} x_j \qquad \text{for each } i \in I,$$

and y would be an upper bound of the x_j because y would be an upper bound, in particular, of the x_i with $i \in I_j$ for each $j \in J$. This implies that

$$x_j = \bigvee_{i\in I_j} x_i \le y \qquad \text{for each } j \in J.$$

But x is the supremum of the x_j, so

$$x \le y < x.$$

This contradiction establishes that x is the supremum of all x_i, so this supremum exists and, moreover,

$$\bigvee_{j\in J}\left(\bigvee_{i\in I_j} x_i\right) = x = \bigvee_{i\in I} x_i .$$

The proof of the dual is left as an exercise for the reader.⊣

Theorem 2.2.12 (Generalized Distributive Laws) *Let* $\{y_i\}_{i\in I}$ *be a set of elements of a Boolean algebra A. Then:*

(1) *If the term* $x \wedge \bigvee_{i \in I} y_i$ *exists, then the term* $\bigvee_{i \in I}(x \wedge y_i)$ *also exists, and moreover,*

$$x \wedge \bigvee_{i \in I} y_i = \bigvee_{i \in I}(x \wedge y_i);$$

(2) *The term* $x \vee \bigwedge_{i \in I} y_i$ *exists if and only if the term* $\bigwedge_{i \in I}(x \vee y_i)$ *exists, and moreover,*

$$x \vee \bigwedge_{i \in I} y_i = \bigwedge_{i \in I}(x \vee y_i).$$

Proof For (1), suppose that $x \wedge \bigvee_{i \in I} y_i$ exists. Then $x \wedge \bigvee_{i \in I} y_i$ is an upper bound of the elements $x \wedge y_i$.

From the order properties of Boolean algebras, we have: $y_i \leq \bigvee_{i \in I} y_i$, $x \wedge y_i \leq y_i$, $x \wedge y_i \leq x$, and $x \wedge y_i \leq \bigvee_{i \in I} y_i$.

Thus, for all i, $x \wedge y_i$ is a lower bound of $\{\bigvee_{i \in I} y_i, x\}$. Consequently,

$$x \wedge y_i \leq \inf\{\bigvee_{i \in I} y_i, x\} = x \wedge \bigvee_{i \in I} y_i.$$

If this were not the case, there would exist some z such that

$$x \wedge y_i \leq z < x \wedge \bigvee_{i \in I} y_i \qquad \text{for each } i \in I.$$

However, this would imply, by Theorem 2.2.12, that

$$x \wedge \bigvee_{i \in I} y_i \leq z < x \wedge \bigvee_{i \in I} y_i,$$

which is a contradiction.

This contradiction establishes that $\bigvee_{i \in I}(x \wedge y_i)$ exists, and moreover,

$$x \wedge \bigvee_{i \in I} y_i = \bigvee_{i \in I}(x \wedge y_i).$$

The proof of the dual is left as an exercise for the reader.⊣

Theorem 2.2.13 *Let* $\{x_i\}_{i \in I}$ *and* $\{y_j\}_{j \in J}$ *be sets of elements in a Boolean algebra A. If the terms* $\bigvee_{i \in I} x_i$ *and* $\bigvee_{j \in J} y_j$ *exist, then the term*

$$\bigvee_{(i,j) \in I \times J} (x_i \wedge y_j)$$

exists, and moreover,

$$\bigvee_{i\in I} x_i \wedge \bigvee_{j\in J} y_j = \bigvee_{(i,j)\in I\times J} (x_i \wedge y_j).$$

Proof Suppose that both $\bigvee_{i\in I} x_i$ and $\bigvee_{j\in J} y_j$ exist. For every $(i, j) \in I \times J$, we have

$$x_i \wedge y_j \leq x_i \leq \bigvee_{i\in I} x_i \qquad \text{and} \qquad x_i \wedge y_j \leq y_j \leq \bigvee_{j\in J} y_j.$$

Thus, $\bigvee_{i\in I} x_i \wedge \bigvee_{j\in J} y_j$ is an upper bound of the set $X = \{x_i \wedge y_j\}_{(i,j)\in I\times J}$.

Let z be any upper bound of X, so $x_i \wedge y_j \leq z$ for all $(i, j) \in I \times J$ and we have:

$$\begin{aligned} x_i \wedge y_j \leq z &\Rightarrow x_i \wedge \bigvee_{j\in J} y_j \leq z \\ &\Rightarrow \bigvee_{j\in J} y_j \wedge x_i \leq z \\ &\Rightarrow \bigvee_{j\in J} y_j \wedge \bigvee_{i\in I} x_i \leq z \\ &\Rightarrow \bigvee_{i\in I} x_i \wedge \bigvee_{j\in J} y_j \leq z. \end{aligned}$$

This establishes that $\bigvee_{i\in I} x_i \wedge \bigvee_{j\in J} y_j$ is the supremum of X.⊣

2.3 Ultrafilters

A filter in a Boolean algebra is a subset of the algebra that is closed under finite intersections and contains all elements that dominate any of its members. In other words:

Definition 2.3.1 Let $\mathfrak{A} = \langle A, \vee, \wedge, ^-, 0, 1\rangle$ be a Boolean algebra. A *filter* on A is a subset $\mathscr{F}$ of A such that, for all $x, y \in A$:

(1) If $x, y \in \mathscr{F}$, then $x \wedge y \in \mathscr{F}$;
(2) If $x \in \mathscr{F}$ and $x \leq y$, then $y \in \mathscr{F}$.

Examples of filters include:

- The *trivial filter* $\mathscr{F} = \{1\}$, whose only element is the maximum of the algebra.
- The *improper filter* $\mathscr{F} = A$.

- The *Fréchet filter* $\mathscr{F} = \{x \in A \mid \overline{x} \text{ is finite}\}$, where A is the power set of a given set, and $\leq$ is interpreted as set inclusion. A set that is the complement of a finite set is called *cofinite*.

If $\mathscr{F} \neq A$, the filter $\mathscr{F}$ is called a *proper filter*. A *generator* of a filter $\mathscr{F}$ is an element $x \in A$ such that $x \leq y$ for all $y \in \mathscr{F}$. More generally, a family F of elements of A is said to *generate* the filter $\mathscr{F}$ if $\mathscr{F}$ is the intersection of all filters containing F:

$$\mathscr{F} = \bigcap \{\mathscr{G} \mid F \subseteq \mathscr{G} \text{ and } \mathscr{G} \text{ is a filter on } A\}.$$

A filter generated by some element of A is called a *principal filter*. A subset $B \subseteq A$ is said to have the *finite intersection property* (FIP) if every finite subset of B has a non-zero meet; that is, $\bigwedge C \neq 0$ for every finite $C \subseteq B$.

Previously, we defined an ideal in a Boolean algebra (Definition 2.2.5) as a nonempty subset $\mathscr{I}$ such that the join of any pair of elements in $\mathscr{I}$ belongs to $\mathscr{I}$, and that contains all elements dominated by any of its members.

If $\mathscr{I} \neq A$, the ideal $\mathscr{I}$ is called a *proper ideal*. A *generator* of an ideal $\mathscr{I}$ is an element $x \in A$ such that $y \leq x$ for all $y \in \mathscr{I}$. An ideal generated by some element of A is called a *principal ideal*. A proper ideal $\mathscr{I}$ is *maximal* if and only if there is no proper ideal on A that contains $\mathscr{I}$. Clearly, the concepts of filter and ideal are dual to each other.

Theorem 2.3.2 *If a subset B of a Boolean algebra has the finite intersection property (*FIP*), then for any element x of the algebra, either $B \vee \{x\}$ or $B \vee \{\overline{x}\}$ has the* FIP.

Proof Let B be a non-empty subset of the Boolean algebra such that B has the FIP, and let $x \notin B$ be any element of the algebra. Suppose that neither $B \vee \{x\}$ nor $B \vee \{\overline{x}\}$ has the FIP. This means there exist elements $x_1, \ldots, x_m, x_{m+1}, \ldots, x_{m+n}$ in B (not necessarily distinct) such that

$$x_1 \wedge \cdots \wedge x_m \wedge x = 0 \quad \text{and} \quad x_{m+1} \wedge \cdots \wedge x_{m+n} \wedge \overline{x} = 0.$$

This implies

$$\left(\bigwedge_{i=1}^{m} x_i\right) \wedge x = 0 \quad \text{and} \quad \left(\bigwedge_{j=m+1}^{m+n} x_j\right) \wedge \overline{x} = 0.$$

Thus, for the elements

$$y = \bigwedge_{i=1}^{m} x_i \quad \text{and} \quad z = \bigwedge_{j=m+1}^{m+n} x_j,$$

we have

$$y \wedge x = 0 \quad \text{and} \quad z \wedge \overline{x} = 0.$$

From these, it follows that

$$y \leq \overline{x} \quad \text{and} \quad z \leq x.$$

Now consider $y \wedge z$:

$$y \wedge z \leq \overline{x} \wedge x = 0,$$

which contradicts the assumption that B has the FIP, as $y, z \in B$ and their meet must be non-zero.

Therefore, the initial supposition is false, and either $B \vee \{x\}$ or $B \vee \{\overline{x}\}$ must have the FIP.⊣

Theorem 2.3.3 *Every subset of a Boolean algebra that has the finite intersection property (*FIP*) can be extended to a proper filter on the algebra.*

Proof Let B be a subset of a Boolean algebra A that has the FIP, and define

$$\mathscr{F} = \{x \in A \mid x_1 \wedge \cdots \wedge x_n \leq x \text{ for some } x_1, \ldots, x_n \in B\}.$$

We will show that $\mathscr{F}$ is a filter on A.

First, since B has the FIP, no finite intersection of elements of B is 0, so $0 \notin \mathscr{F}$.

If $x, x' \in \mathscr{F}$, then there exist $x_1, \ldots, x_m, x'_1, \ldots, x'_n \in B$ such that

$$x_1 \wedge \cdots \wedge x_m \leq x \quad \text{and} \quad x'_1 \wedge \cdots \wedge x'_n \leq x'.$$

It follows that

$$x_1 \wedge \cdots \wedge x_m \wedge x'_1 \wedge \cdots \wedge x'_n \leq x \wedge x',$$

and hence $x \wedge x' \in \mathscr{F}$.

If $x \in \mathscr{F}$ and $x \leq x'$, then there exist $x_1, \ldots, x_n \in B$ such that

$$x_1 \wedge \cdots \wedge x_n \leq x \leq x',$$

so $x' \in \mathscr{F}$.

Thus, $\mathscr{F}$ is a proper filter on A that contains B.⊣

Moreover, any non-empty subset F of a Boolean algebra A can be extended to a filter on A. In other words:

Theorem 2.3.4 *The filter $\mathscr{F}$ generated by a subset F of a Boolean algebra A is indeed a filter. The filter is proper if and only if F has the* FIP*. Specifically,*

$$\mathcal{F} = \{x \in A \mid x_1 \wedge \cdots \wedge x_n \leq x \textit{ for some } x_1, \ldots, x_n \in F\}.$$

Proof The family of filters $\mathcal{G}$ that contain F is non-empty, as it includes at least the improper filter A. Clearly, 1 is in every filter.

If $x, y \in F$, then $x, y \in \mathcal{G}$ for any $\mathcal{G}$, and $x \wedge y \in \mathcal{G}$, so $x \wedge y \in \mathcal{F}$.

If $x \in \mathcal{F}$ and $x \leq y$, then x is in every $\mathcal{G}$, so y must also be in every $\mathcal{G}$, implying $y \in \mathcal{F}$.

Any filter containing F must include all finite meets of elements of F and all elements of A that dominate those meets. Thus, $\mathcal{F}$ is the filter generated by F, and it is proper if and only if F has the FIP.$\dashv$

Definition 2.3.5 Let A be a Boolean algebra, and let $\mathcal{F}$ be the collection of all proper filters on A. Then, $\langle \mathcal{F}, \subseteq \rangle$ is a poset, and an *ultrafilter* on A is a proper filter $\mathcal{F}$ on A that is maximal in $\langle \mathcal{F}, \subseteq \rangle$; that is, there is no filter on A that properly contains $\mathcal{F}$.

It is immediate that an ultrafilter is a filter that has no proper extensions that are proper filters. Compare this definition with that of a maximal ideal.

Theorem 2.3.6 *Every Boolean algebra has at least one ultrafilter. In fact, every filter of the algebra can be extended to an ultrafilter.*

Proof There are many filters on any Boolean algebra A. For example, if x is an element of A, the family $\{y \in A \mid x \leq y\}$ is one of them.

We will first show that the union of any non-empty chain in the poset $\langle \mathcal{F}, \subseteq \rangle$ of all proper filters on A belongs to $\mathcal{F}$. To start, it is clear that if $\mathcal{C}$ is such a chain, $\mathcal{F} = \bigcup \mathcal{C}$ is a family of subsets of A. I will show that $\mathcal{F}$ is a filter on A.

$0 \notin \mathcal{F}$, because if 0 were in $\mathcal{F}$, it would belong to some member of the chain $\mathcal{C}$, contradicting the hypothesis that all elements of $\mathcal{C}$ are proper filters. If $x, y \in \mathcal{F}$, then there is a filter $\mathcal{F}'$ in $\mathcal{C}$ such that $x, y \in \mathcal{F}'$, which implies $x \wedge y \in \mathcal{F}' \subseteq \mathcal{F}$. If $x \in \mathcal{F}$, $y \in A$, and $x \leq y$, there exists an $\mathcal{F}'' \in \mathcal{C}$ such that $x \in \mathcal{F}''$, which implies $y \in \mathcal{F}'' \subseteq \mathcal{F}$. This establishes that $\mathcal{F}$ is a filter.

We have shown that the union of any chain in the poset $\langle \mathcal{F}, \subseteq \rangle$ is in the poset. This implies, by Zorn's Lemma (Theorem 1.8.3), that $\mathcal{F}$ contains a maximal filter. This is an ultrafilter on A.

Now let $\mathcal{F}$ be any filter, and let $\mathcal{G}$ be the family of all filters in A that contain $\mathcal{F}$. By an argument analogous to the previous one, $\mathcal{G}$ has a maximal element $\mathcal{M}$ (since the union of any chain of elements of $\mathcal{G}$ is a filter that contains $\mathcal{F}$). I will show that $\mathcal{M}$ is also maximal in $\mathcal{G}$. For if it were not, there would exist $\mathcal{F}' \in \mathcal{F}$ such that $\mathcal{M} \subset \mathcal{F}'$. Since $\mathcal{F} \subseteq \mathcal{M}$, because $\mathcal{M} \in \mathcal{G}$, we would have $\mathcal{F} \subset \mathcal{F}'$, from which it follows that $\mathcal{F}' \in \mathcal{G}$ and $\mathcal{M}$ would not be maximal in $\mathcal{G}$. This contradiction establishes that $\mathcal{M}$ is maximal in $\mathcal{G}$, and therefore that there exists an ultrafilter on A that contains $\mathcal{F}$.$\dashv$

The fundamental property of ultrafilters, which makes them important to distinguish from mere filters, is that for any element x of the algebra, if x does not belong to an ultrafilter $\mathcal{U}$, then its complement does belong to $\mathcal{U}$.

Theorem 2.3.7 (Fundamental Theorem of Ultrafilters) *If $\mathcal{U}$ is an ultrafilter on the algebra A and $x \in A$, then $x \in \mathcal{U}$ or $\overline{x} \in \mathcal{U}$, but not both.*

Proof We can assume, without loss of generality (as the following argument is valid *mutatis mutandis*, by swapping the roles of x and $\overline{x}$), that $\overline{x} \notin \mathcal{U}$. Let $\mathcal{F}$ be the set

$$\{z \in A \mid x \wedge y \leq z \text{ for some } y \in \mathcal{U}\}.$$

Since $x \wedge y \leq x$ for any $y \in \mathcal{U}$, $x \in \mathcal{F}$, $\mathcal{F} = \mathcal{U}$, and thus $x \in \mathcal{U}$.

Indeed, first note that if $0 \in \mathcal{F}$, there would exist $y \in \mathcal{U}$ such that $x \wedge y \leq 0$, which immediately implies $y \leq \overline{x}$. But this would mean $\overline{x} \in \mathcal{U}$, contradicting the hypothesis. Therefore, $0 \notin \mathcal{F}$. If $u, v \in \mathcal{F}$, then $x \wedge y_1 \leq u$ and $x \wedge y_2 \leq v$ for $y_1, y_2 \in \mathcal{U}$, which gives $x \wedge (y_1 \wedge y_2) \leq u \wedge v$; since $y_1 \wedge y_2 \in \mathcal{U}$, $u \wedge v \in \mathcal{F}$. Similarly, if $u \in \mathcal{F}$ and $u \leq v \in A$, we have $x \wedge y \leq u$ for $y \in \mathcal{U}$, so $x \wedge y \leq v$, and hence $v \in \mathcal{F}$. We have shown that $\mathcal{F}$ is a proper filter.

Finally, if $u \in \mathcal{U}$, since $x \wedge u \leq u$, we have $u \in \mathcal{F}$. Thus, $\mathcal{F}$ is a proper filter containing $\mathcal{U}$. Since $\mathcal{U}$ is maximal, there cannot be any filter that properly contains $\mathcal{U}$, and therefore $\mathcal{F} = \mathcal{U}$. It follows that $x \in \mathcal{U}$.⊣

An analogous property of ultrafilters for ideals is expressed in the following theorem.

Theorem 2.3.8 *An ideal $\mathcal{I}$ on the Boolean algebra A is maximal if and only if, for every $x \in A$, $x \in \mathcal{I}$ or $\overline{x} \in \mathcal{I}$, but not both.*

The proof is analogous to the theorem on the fundamental property of ultrafilters and is left as an exercise for the reader. An immediate consequence of both theorems is the following.

Corollary 2.3.9 *$\mathcal{I}$ is a maximal ideal on the Boolean algebra A if and only if $\mathcal{U} = A \setminus \mathcal{I}$ is an ultrafilter on A.*

Proof It is easy to see that $\mathcal{U}$ is a filter on A. Indeed,

$$\begin{aligned} x, y \in \mathcal{U} &\Rightarrow \overline{x}, \overline{y} \in \mathcal{I}, \\ &\Rightarrow \overline{x} \vee \overline{y} \in \mathcal{I}, \\ &\Rightarrow \overline{(x \wedge y)} \in \mathcal{I}, \\ &\Rightarrow x \wedge y \in \mathcal{U}. \end{aligned}$$

Also,

$$\begin{aligned} x \in \mathcal{U} \text{ and } x \leq y &\Rightarrow \overline{x} \in \mathcal{I} \text{ and } \overline{y} \leq \overline{x}, \\ &\Rightarrow \overline{y} \in \mathcal{I}, \\ &\Rightarrow y \in \mathcal{U}. \end{aligned}$$

If $x \notin \mathcal{U}$, then $x \in \mathcal{J}$ and hence $\overline{x} \in \mathcal{U}$. This proves that $\mathcal{U}$ is an ultrafilter.⊣

Theorem 2.3.10 *If $\mathcal{U}$ is an ultrafilter on a Boolean algebra A, and $x \vee y \in \mathcal{U}$, then $x \in \mathcal{U}$ or $y \in \mathcal{U}$.*

Proof Suppose $x \vee y \in \mathcal{U}$ and that neither x nor y are in $\mathcal{U}$. Since $\mathcal{U}$ is an ultrafilter, $\overline{x}$ and $\overline{y}$ must be in $\mathcal{U}$. Then, $\overline{(x \vee y)} = \overline{x} \wedge \overline{y} \in \mathcal{U}$. But this is impossible, as an element and its complement cannot both belong to an ultrafilter.⊣

A *power algebra* is a Boolean algebra whose underlying set A is the power set $\mathcal{P}(I)$ of some non-empty set I. The following theorems concern such algebras.

Theorem 2.3.11 *If I is a non-empty set, an ultrafilter $\mathcal{U}$ on I is principal if and only if $\mathcal{U}$ is generated by a finite set; in fact, by a singleton set. If I is finite, then all ultrafilters on I are principal.*

Proof If the ultrafilter $\mathcal{U}$ on I is principal, then it is generated by a set $X \subseteq I$. Let x be any element of X, and consider the filter $\mathcal{F}$ generated by $\{x\}$. Since $\{x\} \subseteq X$, we have $\mathcal{U} \subseteq \mathcal{F}$, which implies $\mathcal{F} = \mathcal{U}$, as an ultrafilter cannot be properly contained in another filter. It follows that $X = \{x\}$, and thus X is a singleton.

Now suppose X is a finite element of the filter $\mathcal{U}$, and consider the family

$$\mathcal{F} = \{Y \wedge X \mid Y \in \mathcal{U}\},$$

which consists of all intersections of elements of $\mathcal{U}$ with X. Note that $\mathcal{F}$ has a finite number of elements. Since the intersection of any family of finite sets is finite, it follows that $\bigcap \mathcal{F}$ is finite. Moreover, $\bigcap \mathcal{F}$ is in $\mathcal{U}$, as it is the intersection of a finite family of elements of $\mathcal{U}$. Thus,

$$\bigcap \mathcal{U} = X \wedge \bigcap_{Y \in \mathcal{U}} Y = \bigcap_{Y \in \mathcal{U}} (Y \wedge X) = \bigcap \mathcal{F} \in \mathcal{U},$$

because $X \in \mathcal{U}$. Since $\bigcap \mathcal{U} \subseteq Y$ for all $Y \in \mathcal{U}$, it follows that $\bigcap \mathcal{U}$ is a generator of $\mathcal{U}$, and hence $\mathcal{U}$ is principal.

Clearly, in a finite algebra, every filter contains finite elements (as all its elements are finite).⊣

Note that if $\mathcal{U}$ is a non-principal ultrafilter, then $\bigcap \mathcal{U} = \varnothing$. Indeed, if $\bigcap \mathcal{U} \neq \varnothing$, the set $\mathcal{F} = \{X \subseteq I \mid \bigcap \mathcal{U} \subseteq X\}$ is a proper filter such that $\mathcal{U} \subseteq \mathcal{F}$, which implies $\mathcal{F} = \mathcal{U}$ because $\mathcal{U}$ is maximal. In fact, the generator of a principal filter must always be a set of the form $\{x\}$ where $x \in I$. Indeed, let G be the generator of the principal ultrafilter $\mathcal{U}$, and let $x \in G$. Then $\mathcal{U}$ is contained in the filter $\mathcal{F}$ generated by $\{x\}$. Since $\mathcal{U}$ is an ultrafilter, it is maximal, and therefore $\mathcal{U}$ cannot be properly contained in $\mathcal{F}$. It follows that $\mathcal{U} = \mathcal{F}$, and thus the generator G of $\mathcal{U}$ is $\{x\}$.

Notice that this implies that non-principal ultrafilters cannot contain finite elements. For otherwise the intersection of all finite sets would be a finite member of the ultrafilter contained in all elements of the ultrafilter and so it would be a generator of the ultrafilter.

Theorem 2.3.12 *If I is an infinite set, then there exists an ultrafilter $\mathcal{U}$ and a countable partition $\{I_n \mid n \in \omega\}$ of I such that $I_n \notin \mathcal{U}$ for all $n \in \omega$.*

Proof Since I is infinite, there exists a multitude of surjective functions $f : I \to \omega$. For any such function and for each $n \in \omega$, define I_n as the set $\{x \in I \mid f(x) = n\}$. It is evident that the sequence $\{I_n\}_{n\in\omega}$ is a partition of I. Now consider the sequence $\{F_n\}_{n\in\omega}$ where $F_n = I \backslash I_n$. Since $I_n \neq \varnothing$ for all $n \in \omega$, the finite union $I_{n_1} \vee \cdots \vee I_{n_m}$ is a proper subset of I for any finite subsequence $I_{n_1}, \ldots, I_{n_m}$. Thus,

$$F_{n_1} \wedge \cdots \wedge F_{n_m} = (I \setminus I_{n_1}) \wedge \cdots \wedge (I \setminus I_{n_m}) = I \setminus (I_{n_1} \vee \cdots \vee I_{n_m}) \neq \varnothing.$$

This establishes that the set of F_n has the finite intersection property (FIP). Therefore, there exists an ultrafilter $\mathcal{U}$ such that $F_n \in \mathcal{U}$ but $I_n \notin \mathcal{U}$ for all $n \in \omega$.

Incidentally, it is clear that $\mathcal{U}$ is non-principal because

$$\bigcap_{n\in\omega} F_n = I \setminus \bigcup_{n\in\omega} I_n = \varnothing \notin \mathcal{U}.$$

Indeed, if it were principal, $\bigcap \mathcal{U} \in \mathcal{U}$, but this is impossible because $\bigcap \mathcal{U} \subseteq \bigcap_{n\in\omega} F_n$.⊣

An ultrafilter $\mathcal{U}$ on an infinite set I is said to be *countably incomplete* if there exists a countable partition $\{I_n \mid n \in \omega\}$ of I such that $I_n \notin \mathcal{U}$ for all $n \in \omega$. Theorem 2.3.12 asserts that for every infinite set, there exists a countably incomplete ultrafilter on it. An immediate corollary of the preceding theorem is the following:

Theorem 2.3.13 *There exists a non-principal ultrafilter on every infinite set I.*

Proof Let I be infinite, and let $\mathcal{U}$ be an ultrafilter on I. By Theorem 2.3.12, there exists a partition $\{I_n \mid n \in \omega\}$ of I such that $I_n \notin \mathcal{U}$ for all $n \in \omega$. If $\mathcal{U}$ were principal, it would be generated by a set of the form $\{x\}$ with $x \in I$. Since $\{I_n \mid n \in \omega\}$ is a partition of I, there exists precisely one $n \in \omega$ such that $x \in I_n$. In that case, $\{x\} \subseteq I_n$, and since $\mathcal{U}$ is a filter, $I_n \in \mathcal{U}$. This contradicts the assumption that $I_n \notin \mathcal{U}$, and it follows that $\mathcal{U}$ is not principal.⊣

The following proposition is relevant for the construction of nonstandard analysis. First, an important definition:

Definition 2.3.14 Let α be an infinite cardinal. A proper filter $\mathcal{F}$ on I is said to be *α-regular* if and only if there exists a set $E \subseteq \mathcal{F}$ of cardinality $|E| = \alpha$ such that each $i \in I$ belongs to only finitely many elements $e \in E$.

The existence of α-regular filters—in fact, ultrafilters—is established in the following proposition.

Theorem 2.3.15 *For any set I of infinite cardinality α, there exists an α-regular ultrafilter $\mathcal{U}$ on I. Every such ultrafilter is non-principal.*

Proof Let α be an infinite cardinal, and let $J = \mathcal{P}_\omega(\alpha)$, the family of all finite subsets of α. We will show that J has an α-regular ultrafilter, which also induces an α-regular ultrafilter on I.

For each ordinal $\beta \in \alpha$, define the set $\hat{\beta}$ of all subsets $j \in J$ that contain β, i.e.,

$$\hat{\beta} = \{j \in J \mid \beta \in j\}.$$

Let E be the set of all such sets:

$$E = \{\hat{\beta} \mid \beta \in \alpha\}.$$

The set E has cardinality α, since the function $f : E \to \alpha$ defined by $f(\hat{\beta}) = \beta$ is injective. Indeed, if $\hat{\beta} \neq \hat{\gamma}$, then there exists some $j \in J$ such that $j \in \hat{\beta}$ and $j \notin \hat{\gamma}$, meaning $\beta \in j$ and $\gamma \notin j$, which is impossible if $\beta = \gamma$.

Since j is finite and $j \in \hat{\beta}$ implies $\beta \in j$, only a finite number of β can belong to j, and hence only a finite number of j can belong to $\hat{\beta}$.

Therefore, any proper filter $\mathcal{F}$ that includes E is α-regular. Since any finite subset $B = \{\hat{\beta}_1, \ldots, \hat{\beta}_n\}$ has a non-empty intersection (as $\beta_1 \wedge \cdots \wedge \beta_n \in B$), it follows that E has the finite intersection property (FIP). Consequently, $\mathcal{F}$ can be extended to an ultrafilter $\mathcal{U}$ on J, and it follows that $\mathcal{U}$ is an α-regular ultrafilter.⊣

Theorem 2.3.16 *Every non-principal ultrafilter on I contains all cofinite subsets of I.*

Proof As a consequence of Theorem 2.3.11, a non-principal ultrafilter does not contain finite sets. By the fundamental property of ultrafilters (Theorem 2.3.7), it must then contain the complements of finite sets, which are precisely the cofinite subsets.⊣

Let $\mathcal{F}$ be a filter on a Boolean algebra A and X a non-empty subset of A. We say that $\mathcal{F}$ *preserves the product* $\bigwedge X$ if and only if

$$\bigwedge X \in \mathcal{F} \quad \text{if and only if} \quad \{x\} \in \mathcal{F} \text{ for all } x \in X.$$

An interesting result about filters, with particular utility in model theory, is the following:

Theorem 2.3.17 *Let $\bigwedge X_0, \bigwedge X_1, \ldots$ be a countable sequence of products of elements in the Boolean algebra $\mathfrak{A}$. There exists an ultrafilter on $\mathfrak{A}$ that preserves each product.*

Proof We will first show that there exists a sequence $x_n \in X_n$ such that no finite product of elements of the form $(\bigwedge X_n) \vee \overline{x}_n$ equals 0.

For any $n \in \omega$, if $X_n = \{x_n\}$ is a singleton, then $\bigwedge X_n = x_n$, and thus $(\bigwedge X_n) \vee \overline{x}_n = x_n \vee \overline{x}_n = 1$. If X_0 is not a singleton, we can choose x_0 such that it is not equal to 1. In any case (since $\overline{x}_0 \neq 0$), $(\bigwedge X_0) \vee \overline{x}_0 \neq 0$.

Now suppose there exist $x_0, \ldots, x_n$ ($x_k \in X_k$ for $k = 0, \ldots, n$) such that

$$y = \left[\left(\bigwedge X_0\right) \vee \overline{x}_0\right] \wedge \left[\left(\bigwedge X_1\right) \vee \overline{x}_1\right] \wedge \cdots \wedge \left[\left(\bigwedge X_n\right) \vee \overline{x}_n\right] \neq 0.$$

If $y \leq x$ for all $x \in X_{n+1}$, then $y \leq \bigwedge X_{n+1}$, and hence $y \wedge \bigwedge X_{n+1} = y \neq 0$. Thus,

$$y \wedge \left[\left(\bigwedge X_{n+1}\right) \vee \overline{x}_{n+1}\right] = \left[y \wedge \bigwedge X_{n+1}\right] \vee \left[y \wedge \overline{x}_{n+1}\right] \neq 0$$

for any choice of $x_{n+1} \in X_{n+1}$.

Conversely, if there exists $x_{n+1} \in X_{n+1}$ such that $x_{n+1} < y$, then $\overline{y} < \overline{x}_{n+1}$ and $\overline{y} < \left[(\bigwedge X_{n+1}) \vee \overline{x}_{n+1}\right]$. Since $y \wedge \left[(\bigwedge X_{n+1}) \vee \overline{x}_{n+1}\right] = 0$ if and only if $\left[(\bigwedge X_{n+1}) \vee \overline{x}_{n+1}\right] \leq \overline{y}$, it follows that $y \wedge \left[(\bigwedge X_{n+1}) \vee \overline{x}_{n+1}\right] \neq 0$.

Let $(\bigwedge X_{n_1}) \vee x_{n_1}, \ldots, (\bigwedge X_{n_k}) \vee x_{n_k}$ be any finite family of such elements, and let $n = \max\{n_1, \ldots, n_k\}$. Then,

$$0 < \bigcap_{i=0}^{n} \left[\left(\bigwedge X_i\right) \vee \overline{x}_i\right] \leq \left[\left(\bigwedge X_{n_1}\right) \vee \overline{x}_{n_1}\right] \wedge \cdots \wedge \left[\left(\bigwedge X_{n_k}\right) \vee \overline{x}_{n_k}\right].$$

Let C be the set of all finite products of the form

$$\left[\left(\bigwedge X_{n_1}\right) \vee \overline{x}_{n_1}\right] \wedge \cdots \wedge \left[\left(\bigwedge X_{n_k}\right) \vee \overline{x}_{n_k}\right].$$

Since C has the finite intersection property (FIP), C can be extended to a filter and subsequently to an ultrafilter $\mathcal{U}$.

We will show that $\mathcal{U}$ preserves each product $\bigwedge X_n$.

For any $n \in \omega$, let I be an index set such that $X_n = \{x_i\}_{i \in I}$. If $\bigwedge_{i \in I} x_i \in \mathcal{U}$, then since $\bigwedge_{i \in I} x_i \leq x_i$ for all $i \in I$ and $\mathcal{U}$ is a filter, $x_i \in \mathcal{U}$. Conversely, if $x_i \in \mathcal{U}$ for each $i \in I$, then by the construction of $\mathcal{U}$, $(\bigwedge_{i \in I} x_i) \vee \overline{x}_n \in \mathcal{U}$, and $\overline{x}_n \notin \mathcal{U}$ (since $x_n \in \mathcal{U}$ and $\mathcal{U}$ is an ultrafilter). Thus, $\bigwedge_{i \in I} x_i \in \mathcal{U}$.⊣

2.4 Stone's Representation Theorem

Stone's Representation Theorem states that every Boolean algebra is isomorphic to some field of sets. In other words, up to isomorphism, the only Boolean algebras are set algebras. This is a very convenient way to conceptualize Boolean algebras, which also has pedagogical value. In this section, we aim to provide an explanation and proof of this remarkable result. We begin with an assertion about the relationship between proper ideals and Boolean homomorphisms.

Theorem 2.4.1 (Homomorphism Theorem) *Every proper ideal in a Boolean algebra is the kernel of some epimorphism. In particular, if the ideal is maximal, then the epimorphism is onto the algebra 2.*

Proof Let $\mathscr{I}$ be an ideal on the Boolean algebra

$$\mathfrak{A} = \langle A, \vee, \wedge, ^{-}, 0, 1\rangle,$$

and let

$$\mathfrak{R} = \langle A, +, \cdot, 0, 1\rangle$$

be the associated Boolean ring of $\mathfrak{A}$. From Theorem 2.2.5, $\mathscr{I}$ is also an ideal in $\mathfrak{R}$. Let

$$\mathfrak{R}' = \langle A', +', \cdot', 0', 1'\rangle$$

be the quotient ring modulo $\mathscr{I}$, $A/\mathscr{I}$. The elements of A' are of the form $x + \mathscr{I} = \{x + y \mid y \in \mathscr{I}\}$. The operation $+'$ is defined by:

$$(x + \mathscr{I}) +' (y + \mathscr{I}) = (x + y) + \mathscr{I},$$

and $\cdot'$ is defined by:

$$(x + \mathscr{I}) \cdot' (y + \mathscr{I}) = xy + \mathscr{I}.$$

Thus, $0'$ is $\mathscr{I}$ itself because for any $x \in A+$:

$$(x + \mathscr{I}) +' \mathscr{I} = (x + \mathscr{I}) +' (0 + \mathscr{I}) = (x + 0) + \mathscr{I} = x + \mathscr{I}.$$

Similarly, $1' = 1 + \mathscr{I} = \{1 + x \mid x \in \mathscr{I}\} = \{\overline{x} \mid x \in \mathscr{I}\}$.

If the ideal is not maximal, there exist elements $x \in A$ such that neither $x \in \mathscr{I}$ nor $\overline{x} \in \mathscr{I}$. In this case, x and $\overline{x}$ are not in $1'$. When $\mathscr{I}$ is maximal, $1'$ is an ultrafilter, and the ring has only two elements, $0'$ and $1'$. In this case, the Boolean algebra

$$\mathfrak{A}' = \langle A', \vee', \wedge', ^{-'}, 0', 1'\rangle$$

associated with this ring is the two-element Boolean algebra.

In general, the elements of $A/\mathscr{I}$ are of the form $x + \mathscr{I}$. The map $\varphi : A \to A'$ defined by $x \mapsto x + \mathscr{I}$ is surjective, and hence an epimorphism from the Boolean algebra $\mathfrak{A}$ onto the associated Boolean algebra $\mathfrak{A}'$. Moreover, the image of $\mathscr{I}$ under φ is $0'$, since:

$$\varphi(x \wedge y) = xy + \mathscr{I} = (x + \mathscr{I}) \cdot' (y + \mathscr{I}) = \varphi(x) \cdot' \varphi(y).$$

Also,

$$\varphi(x \vee y) = \varphi(x + y + xy) = (x + y + xy) + \mathcal{I} = \varphi(x) \vee' \varphi(y).$$

Finally, if $x \in \mathcal{I}$, then $x + y \in \mathcal{I}$ for all $y \in \mathcal{I}$, so:

$$\varphi(x) = x + \mathcal{I} = \mathcal{I}. \dashv$$

Recall that a *Hausdorff topological space* is a space X such that for every pair of points $x, y \in X$, there exist disjoint neighborhoods of x and y. A subset A of a topological space X is *disconnected* if there exist non-empty disjoint open sets H, K such that $A = H \cup K$. For $x \in X$, the largest connected subset of X containing x is called the *component* of x. X is *totally disconnected* if the components of X are points, i.e., the only non-empty connected sets are the singleton sets of elements of X. A *cover* of a subset A of X is a family $\{A_\xi\}_{\xi<\alpha}$ of subsets of X such that $A \subseteq \bigcup_{\xi<\alpha} A_\xi$. A subset $A \subseteq X$ is *compact* if every open cover of A has a finite subcover. A *basis* for a topology τ on a set X is a family $\mathcal{B}$ of subsets of X such that for every element $x \in X$ and every $V \in \tau$, there exists a $W \in \mathcal{B}$ such that $x \in W \subseteq V$.

The pair $\langle 2, \sigma \rangle$, where $2 = \{0, 1\}$ and $\sigma = \mathcal{P}(2)$ (the power set of 2), forms a topological space (where σ is the *discrete topology* on 2). Let I be any set. For each $\alpha \in I$, let $2_\alpha = 2$, so that 2^I is precisely the Cartesian product $\prod_{\alpha \in I} 2_\alpha$, i.e., the set of all functions $f : I \to 2$. Let τ be the *product topology* (or *Tychonoff topology*) on 2^I. τ is generated by taking as a basis for open sets those of the form $\prod U_\alpha$, where: U_α is open in 2 for each $\alpha \in I$; and $U_\alpha = 2$ for all but finitely many coordinates.

Since 2^I (with topology τ) is the product of compact spaces, it is immediate that 2^I is also compact. A space of the form 2^I is called a *Cantor space*. A *field of sets* C on a non-empty set H is a family of sets closed under union, intersection, and complementation with respect to H. The field C is *separating* if for every pair of points $x, y \in H$, there exists a set $P \in C$ such that $x \in P$ and $y \in H \setminus P$.

Definition 2.4.2 A *Boolean space* is a compact Hausdorff space that is totally disconnected; that is, a Hausdorff space such that every open set is the union of the clopen (closed and open) sets it contains.

We need to make an assertion about the set of all homomorphisms from an arbitrary Boolean algebra into the two-element algebra 2. The following lemma guarantees that this set is never empty, but rather abundant.

Lemma 2.4.3 *For every non-zero element x of a Boolean algebra A, there exists a homomorphism f with values in 2 such that $f(x) = 1$.*

Proof Let $\mathcal{I}$ be the ideal generated by $\overline{x}$. By Zorn's Lemma, there exists a maximal ideal $\mathcal{J}$ in A containing $\overline{x}$ and, therefore, not containing x. The Homomorphism Theorem (2.4.1) implies the existence of an epimorphism from A onto 2 with kernel $\mathcal{J}$.$\dashv$

Lemma 2.4.4 *The set H of all homomorphisms h valued in 2 on a Boolean algebra A is a compact Hausdorff space.*

Proof First, note that for every $x \in A$, the sets $\{f \mid f(x) = 0\}$ and $\{f \mid f(x) = 1\}$ are clopen in 2^A. Indeed, if $\pi_x : 2^A \to 2$ is the function projecting any map f onto $f(x)$, then $\{f \mid f(x) = 0\} = \pi_x^{-1}(\{0\})$. Since $\{0\}$ is clopen in 2 and π_x is continuous, it follows that $\{f \mid f(x) = 0\}$ is closed. A similar argument shows that $\{f \mid f(x) = 1\}$ is closed.

The set H is the intersection of the following subsets of 2^A:

$\{f \mid f(x \vee y) = f(x) \vee f(y)\}$,
$\{f \mid f(x \wedge y) = f(x) \wedge f(y)\}$,
$\{f \mid f(1) = 1\}$.

Using similar arguments as above, these sets are all closed, so H is a closed subset of 2^A and thus a compact subspace of 2^A.⊣

Lemma 2.4.5 *If H is a compact Hausdorff space and C is a separating field of clopen subsets of H, then H is a Boolean space, and C is the field of all clopen subsets of H.*

Proof It is left as an exercise for the reader to show that H is a Boolean space. We will show that C separates points and closed sets.

Suppose F is a closed set and $x \notin F$. For each $y \in F$, separate y from x using an appropriate set $F_y \in C$. Since the elements of C are clopen, the sets F_y form an open cover of F. By the compactness of H, there exists a finite subcover of F by sets in C, none of which contain x. As C is a field, the union of these sets is an element of C that separates F from x.

The above implies that C is a basis for H. Indeed, for any open set F in H and $x \in F$, there exists $P \in C$ such that $x \in P$ and P separates x from $H \setminus F$. Note that this implies $P \subseteq F$. Since the elements of C are open, $\mathcal{P} = \{P_x\}_{x \in F}$ is an open cover of F, and therefore there exists a finite subfamily $P_1, \ldots, P_n$ of elements of $\mathcal{P}$ that covers F. Since $P_i \subseteq F$ for each $i = 1, \ldots, n$, it follows that $F = P_1 \vee \cdots \vee P_n$. Thus, F is a finite union of sets in C. Since C is closed under finite unions, $F \in C$.⊣

Theorem 2.4.6 (Stone's Representation Theorem) *Every Boolean algebra A is isomorphic to a field of sets. More precisely, A is isomorphic to the algebra of all clopen subsets of the set of homomorphisms from A to 2.*

Proof Let $\varphi : A \to \mathcal{P}(H)$ be the map defined by:

$$\varphi(x) = \{f \in H \mid f(x) = 1\},$$

where H is the set of all homomorphisms from A to 2, and let C be the image of A under φ. We will show that C is a Boolean algebra and that φ is an isomorphism from A onto C. Note that by Lemma 2.4.3, $\varphi(x) \neq \varnothing$ if $x \neq 0$.

We have:

$$\varphi(x \vee y) = \{f \in H \mid f(x \vee y) = 1\} = \{f \in H \mid f(x) \vee f(y) = 1\},$$

which simplifies to:

$$\{f \in H \mid f(x) = 1\} \vee \{f \in H \mid f(y) = 1\} = \varphi(x) \vee \varphi(y).$$

Similarly:

$$\varphi(x \wedge y) = \{f \in H \mid f(x \wedge y) = 1\} = \{f \in H \mid f(x) \wedge f(y) = 1\},$$

which simplifies to:

$$\{f \in H \mid f(x) = 1\} \wedge \{f \in H \mid f(y) = 1\} = \varphi(x) \wedge \varphi(y).$$

Since $f(0) = 0$ for all homomorphisms f, $\varphi(0) = \varnothing$. Finally:

$$\varphi(\overline{x}) = \{f \in H \mid f(\overline{x}) = 1\} = \{f \in H \mid f(x) = 0\} = \overline{\varphi}(x),$$

so in particular, $\varphi(1) = H$. Since the homomorphic image of any Boolean algebra is itself a Boolean algebra, it is immediate that C is a field of sets.

By the Homomorphism Theorem (2.4.1), if $x, y \in A$ are distinct, and one does not dominate the other (or, without loss of generality, y dominates x), the principal ideal generated by x does not contain y and is the kernel of some epimorphism f such that $f(x) = 0$ and $f(y) = 1$. This establishes that φ is injective.

If f and g are distinct elements of H, there exists at least one $x \in A$ such that $f(x) \neq g(x)$. Let $X = \varphi(x)$. Then $g(x) = 0$, so $g \notin X$. This shows that C separates points in A. By Lemma 2.4.5, C is the field of all clopen subsets of H.⊣

This concludes the proof of Stone's Representation Theorem. In the next chapter, we will use results obtained here to develop the semantic theory of first-order logic and provide an unconventional proof of the completeness theorem. As we will see in Chap. 4, the theory of ultrafilters is essential in the development of nonstandard analysis.

Exercises

2.1 Prove Theorem 2.1.6.

2.2 Prove Theorem 2.2.3.

2.3 Prove Theorem 2.2.7

2.4 Prove dual of Theorem 2.2.9.

2.5 Prove dual of Theorem 2.2.10

2.6 Prove dual of Theorem 2.2.11.

2.7 Prove dual of Theorem 2.2.12.

2.8 Prove Theorem 2.3.8.

2.9 Prove that H in the proof of Theorem 2.3.8 is a Boolean space.

References

Milgrom P., & Shannon, C. (1994). Monotone comparative statics. *Econometrica, 62*(1), 157–180.

Tarski, A. (1955). A Lattice-theoretical fixpoint theorem and its applications. *Pacific Journal of Mathematics, 5*, 285–309.

Vives, X. (1990). Nash equilibrium with strategic complementarities. *Journal of Mathematical Economics, 19*, 305–321.

Chapter 3
First-Order Logic

Abstract This chapter begins with a detailed exposition of protothetics—the calculus of sentences—and presents a thorough proof of the Compactness Theorem for this calculus as a purely topological result. It then moves on to first-order logic, starting from the notion of a structure, since first-order languages were introduced for the purpose of studying such entities. The chapter includes a detailed presentation of the concept of a model of a first-order language and the related notions of satisfaction and truth. It provides detailed proofs of the Completeness Theorems of Tarski and Gödel, the Compactness Theorem (also known as Maltsev's Lemma), the Löwenheim–Skolem–Tarski Theorems, and uses the Compactness Theorem to demonstrate the existence of a number field that includes infinitesimal numbers. The chapter concludes with explanations of key concepts in Model Theory such as omitting types, elementary equivalence of models, elementary embeddings, and elementary chains.

Keywords First order logic · Models · Compactness · Completeness

3.1 The Sentential Calculus

The sentential calculus (SC) is the most elementary part of logic. It is also known as 'protothetic' because it contains the basic rules for the formulation of any theory. It is concerned solely with the structure of compound statements, without analyzing the internal structure of simple or categorical statements. For example, a statement like "John is Greek", which follows a subject-predicate form, will not be analyzed by SC as such but will merely be represented "as a block" using a capital letter like 'P'. On the other hand, SC will reflect the structure determined by logical connectives, such that it represents a statement like "If John is a man, then John is mortal" as '$P \to Q$', where 'P' now represents "John is a man" and 'Q' represents "John is mortal". Thus, the non-logical symbols of the language of SC are essentially capital letters representing statements. These need not be simple statements; in fact, certain formulas of predicate calculus with quantifiers—the so-called prime formulas—will be represented by these letters (in the following section).

A. García de la Sienra, *The Logical Foundations of Nonstandard Analysis*,
Synthese Library 530, https://doi.org/10.1007/978-3-032-05677-1_3

3.2 The Language of Sentential Calculus

The language of SC (like all logical languages) consists of two categories of symbols: logical symbols, which are constant across all languages, and non-logical symbols or parameters, which depend on the specific language in question. The *logical symbols* of SC are four: negation ($\neg$), the conditional ($\rightarrow$), and the grouping symbols (and). The non-logical symbols of SC are infinite in number, as they include all uppercase letters from A to Z, with or without subscripts. That is, for every positive integer n, the letters can include or exclude the numeral n as a subscript, as illustrated by A_n. These letters are called *sentential letters*. Summarizing, the classification of the symbols in SC is as follows:

Logical symbols: $\neg$, $\rightarrow$, (,).
Non-logical symbols or parameters
(sentential letters):
$A, B, \ldots, Z, A_1, B_1, \ldots, Z_1, \ldots, A_n, B_n, \ldots, Z_n, \ldots$

3.3 The Formulas of Sentential Calculus

Using the symbols of SC, it is possible to construct complex expressions according to certain rules, which are intuitively motivated by the structure of the statements they aim to represent. First, we define a *formula* as any finite sequence of symbols from SC. Thus, sequences such as:

$$) \rightarrow AB_{132}(\neg$$

are formulas but, clearly, these sequences do not serve to represent the structure of statements. For this reason, it is necessary to distinguish those formulas that are well-formed, as follows:

(1) If ϕ is a sentential letter, then ϕ is a *well-formed formula* (wff).

(2) If ϕ and ψ are wffs, then both $(\neg\phi)$ and $(\phi \rightarrow \psi)$ are wffs.

(3) Only formulas constructed according to (1) and (2) are wffs.

Other connectives are usually introduced in SC, namely conjunction ($\wedge$), disjunction ($\vee$), and biconditional ($\leftrightarrow$). These are defined here, based on the primitives introduced earlier, through the following definitional equivalences:

$(\phi \vee \psi)$ is equivalent to $((\neg\phi) \rightarrow \psi)$.
$(\phi \wedge \psi)$ is equivalent to $(\neg(\phi \rightarrow (\neg\psi)))$.
$(\phi \leftrightarrow \psi)$ is equivalent to $((\phi \rightarrow \psi) \wedge (\psi \rightarrow \phi))$.

3.4 The Semantics of Sentential Calculus

The well-formed formulas (wffs) of SC represent statements, but what concerns us here is only that these statements can be either true or false. In other words, if we represent truth with the numeral '1' and falsehood with '0', we are only interested in knowing that, given any statement P, it is either true (assigned the value 1) or false (assigned the value 0), but not both (it cannot be assigned both numbers simultaneously). The numbers 0 and 1 are called *truth values*.

Since logic is centrally concerned with the concept of logical consequence, we are interested in examining the relationships between the possible assignments of truth values to the wffs. That is, we want to know what truth values are assigned to certain formulas when specific truth values are assigned to others. For this purpose, it is essential to introduce the concept of an assignment.

An *assignment* is a function v that associates each sentential letter with a truth value; in other words, it is a function from the set S of sentential letters to 2, the set {0, 1}. Any assignment is fully determined by specifying which values it assigns to the sentential letters. In other words, if $v : \mathcal{S} \to 2$ is an assignment of veritative values to the sentential letters (the elements of $\mathcal{S}$), then v can be uniquely extended to a function $\hat{v}$ from the set $\mathcal{F}$ of all well-formed formulas to 2, according to the following rules:

$$\hat{v}(\phi) = \begin{cases} v(\phi) & \text{si } \varphi \in \mathcal{S} \\ 1 - \hat{v}(\psi) & \text{si } \varphi = (\neg\psi); \\ 1 - \hat{v}(\psi) + \hat{v}(\psi) \cdot \hat{v}(\theta) & \text{si } \varphi = (\psi \to \theta). \end{cases}$$

It is easy to see, furthermore, that the assignment must satisfy the following equations for the other connectives:

$$\hat{v}(\phi) = \begin{cases} \hat{v}(\psi) + \hat{v}(\theta) - \hat{v}(\psi) \cdot \hat{v}(\theta) & \text{si } \phi = \psi \vee \theta; \\ \hat{v}(\psi) \cdot \hat{v}(\theta) & \text{si } \phi = \psi \wedge \theta; \\ 1 - \hat{v}(\psi) - \hat{v}(\theta) + 2\hat{v}(\psi) \cdot \hat{v}(\theta) & \text{si } \phi = \psi \leftrightarrow \theta. \end{cases}$$

In terms of the concept of an assignment, it is possible to define the concepts of tautology, contingent formula, and contradictory formula. A *tautology* is a wff that takes the value 1 under every assignment. A wff is *contingent* if it takes the value 1 under some assignments and the value 0 under others. A wff is *contradictory* if it takes the value 0 under every assignment. The *truth table* of the formula ϕ is the collection of all possible assignments of veritative values to the sentential letters of ϕ and, as a result, to ϕ itself.

3.5 Derivations in Sentential Calculus

The only inference rule in SC is *modus ponendo ponens* (or simply, *modus ponens*): from $\phi \rightarrow \psi$ and ϕ, deduce ψ. If Σ is a set of statements and ϕ is a statement, a *derivation* or *deduction* of ϕ from Σ in SC is a sequence $\phi_0, \ldots, \phi_n$ of wffs such that $\phi_n = \phi$, and for each $k \leq n$:

(1) ϕ_k is either a tautology,

(2) $\phi_k \in \Sigma$, or
(3) for some $i, j \leq k$, ϕ_i is $\phi_j \rightarrow \phi_k$, and ϕ_k is obtained from ϕ_i and ϕ_j by *modus ponens*.

If there exists a derivation of ϕ from Σ, we say that ϕ is *derivable* from Σ and write $\Sigma \vdash \phi$.

A statement ϕ is said to be a *substitution instance* of a statement ψ if and only if ϕ results from substituting the sentential letters of ψ with statements, under the condition that occurrences of the same letter are replaced by the same statement.

A set of statements Σ is *inconsistent* if and only if there exists a derivation from Σ of a formula of the form $\phi \wedge \neg\phi$; otherwise, we say that Σ is *consistent*.

3.6 Compactness in Sentential Calculus

The *compactness theorem* in SC is usually derived as a corollary of Gödel's completeness theorem. However, within SC, it is possible to reverse this order and prove the compactness theorem without relying on the completeness theorem, revealing its inherently *topological* nature. From this theorem, completeness theorems can be obtained as corollaries.

Consider, in particular, the space $2^{\mathcal{S}}$, defined as follows. Let $\mathcal{S} \equiv \{S_\alpha \mid \alpha < \omega\}$ be the set of all sentential letters. For each $\alpha < \omega$, let 2_α be the set 2 itself (the set of "truth values"), endowed with the topology $\tau_\alpha = \mathcal{P}(2)$, the discrete topology on 2. The space $2^{\mathcal{S}}$ is the Cartesian product $V = \bigtimes_{\alpha<\omega} 2_\alpha$; that is, the set of all functions $v : \mathcal{S} \rightarrow 2$. Each of these functions is a truth value assignment to all sentential letters, so $2^{\mathcal{S}}$ is nothing more than the set of such assignments.

The canonical projection corresponding to the letter $S_\alpha \in \mathcal{S}$ is the function $\pi_\alpha : V \rightarrow 2$ that maps each element $v \in V$ to its component in 2. We denote this component by v_α, where $v_\alpha = v(S_\alpha) \in 2$. A *cylinder set* is the preimage of a canonical projection or a finite intersection of such preimages. That is, a set of the form:

$$\bigcap_{i=1}^{n} \pi_{\alpha_i}^{-1}(A_{\alpha_i}) = \left\{ v \in V \mid v_{\alpha_1} \in A_{\alpha_1} \text{ and } \cdots \text{ and } v_{\alpha_n} \in A_{\alpha_n} \right\}$$

for any n and subsets $A_{\alpha_i} \subseteq 2$ $(1 \leq i \leq n)$.

The sets $\pi_\alpha^{-1}(A_\alpha)$ form a subbasis for the product topology τ, meaning that the cylinder sets constitute a basis for this topology. Since the factor spaces 2_α are compact with respect to the discrete topology, the Tychonoff Theorem (due to A. N Tychonoff) implies that the space $\langle V, \tau \rangle$ is compact.

If the formula ϕ is satisfiable (i.e., there exists at least one assignment v such that $v(\phi) = 1$) and the sentential letter S_α appears in ϕ, there are three possible cases for S_α:

(1) every assignment that satisfies ϕ also satisfies S_α;
(2) (ii) every assignment that satisfies ϕ assigns 0 to S_α;
(3) (iii) some assignments that satisfy ϕ assign 1 to S_α, and others assign 0.

If we denote the case corresponding to S_α as $A_\alpha(\phi)$ (i.e., $A_\alpha(\phi)$ equals 2, $\{1\}$, or $\{0\}$), then $\pi_\alpha^{-1}(A_\alpha(\phi))$ is the set of all assignments that satisfy ϕ whose α-th component, v_α, belongs to $A_\alpha(\phi)$. Thus, we can define $A_\alpha(\phi)$ as the set:

$$A_\alpha(\phi) = \left\{ v_\alpha \in 2 \mid \hat{v}(\phi) = 1 \right\}.$$

We will simply write A_α when the context makes it clear which wff ϕ this expression refers to.

If $S_{\alpha_1}, \ldots, S_{\alpha_n}$ are all the sentential letters that appear in ϕ, we now show that the set $\ulcorner \phi \urcorner$ of all assignments that satisfy ϕ is equal to the intersection $\bigcap_{i=1}^n \pi_{\alpha_i}^{-1}(A_{\alpha_i})$. Indeed, if $\ulcorner \phi \urcorner = \varnothing$, each of the A_{α_i} is empty, and *a fortiori* so is the intersection. Furthermore,

$$\begin{aligned} v \in \ulcorner \phi \urcorner &\Leftrightarrow v_{\alpha_1} \in A_{\alpha_1} \text{ y } \cdots \text{ y } v_{\alpha_n} \in A_{\alpha_n} \\ &\Leftrightarrow v \in \pi_{\alpha_1}^{-1}(A_{\alpha_1}) \text{ y } \cdots \text{ y } v \in \pi_{\alpha_n}^{-1}(A_{\alpha_n}) \\ &\Leftrightarrow v \in \bigcap_{i=1}^{n} \pi_{\alpha_i}^{-1}(A_{\alpha_i}). \end{aligned}$$

We see, then, that the sets $\ulcorner \phi \urcorner$ are precisely the cylinder sets determined by the canonical projections π_α. These sets form the basic sets of the product topology τ, and therefore they are open in V. However, they are also closed because each set of the form $\ulcorner \neg\phi \urcorner$ is also basic and is the complement of the basic set $\ulcorner \phi \urcorner$. In other words, the sets $\ulcorner \phi \urcorner$ are *clopen* (both closed and open) in the product topology τ.

A set of statements is *satisfiable* if and only if there exists a $v \in V$ such that $\hat{v}(\varphi) = 1$ for every $\varphi \in \Sigma$. Let $\ulcorner \Sigma \urcorner$ denote the set:

$$\ulcorner \Sigma \urcorner = \left\{ v \in V \mid \hat{v}(\varphi) = 1 \text{ for all } \varphi \in \Sigma \right\}.$$

Since $\ulcorner \Sigma \urcorner = \bigcap_{\varphi \in \Sigma} \ulcorner \varphi \urcorner$, we obtain the following result:

Lemma 3.6.1 *Σ is satisfiable if and only if $\bigcap_{\varphi \in \Sigma} \ulcorner \varphi \urcorner$ is non-empty.*

We are now in a position to prove the following theorem:

Theorem 3.6.2 (Compactness Theorem for SC) *Let Σ be a set of statements in SC. Σ is satisfiable if and only if every finite subset of Σ is satisfiable.*

Proof Assume that every finite subset Σ_0 of Σ is satisfiable. This means that for each such subset Σ_Γ, there exists an assignment v such that $\hat{v}(\varphi) = 1$ for all $\varphi \in \Sigma_0$. In other words,

$$\ulcorner \Sigma_0 \urcorner = \bigcap_{\varphi \in \Sigma_0} \ulcorner \varphi \urcorner \neq \varnothing.$$

This implies that the intersection of any finite subfamily of the set $\mathcal{H} = \{\ulcorner \varphi \urcorner \mid \varphi \in \Sigma\}$ is non-empty. Therefore, $\mathcal{H}$ possesses the finite intersection property (FIP).

By the Tychonoff Theorem, we know that the topological space V is compact. In a compact space, a family of closed sets satisfying the FIP has a nonempty intersection. Thus,

$$\bigcap \mathcal{H} = \bigcap_{\varphi \in \Sigma} \ulcorner \varphi \urcorner \neq \varnothing.$$

By Lemma 3.6.1, it follows that Σ is satisfiable.$\dashv$

An immediate consequence of Theorem 3.6.2 is the following result:

Theorem 3.6.3 (Completeness for SC) *Let Σ be a set of statements in SC. If Σ is consistent, then Σ is satisfiable.*

Proof Assume that Σ is consistent, and let

$$\Sigma_0 = \{\varphi_\Delta, \ldots, \varphi_m\}$$

be an arbitrary finite subset of Σ. Let $\varphi = \varphi_1 \wedge \cdots \wedge \varphi_m$ be the conjunction of the elements in Σ_0, and suppose that Σ is not satisfiable. This implies that $\hat{v}(\varphi) = 0$ for every $v \in 2^{\mathcal{S}}$, so that

$$\hat{v}(\varphi \rightarrow (S \wedge \neg S)) = 1$$

for every v; that is, $\varphi \rightarrow (S \wedge \neg S)$ is a tautology. Since $\Sigma \vdash \varphi$, it follows that $\mathit{\Sigma} \vdash S \wedge \neg S$. However, this would mean that $\mathit{\Sigma}$ is inconsistent, contradicting the assumption. Therefore, Σ_0 is satisfiable.

We have thus shown that Σ is finitely satisfiable. By the compactness theorem, Σ is satisfiable.$\dashv$

Corollary 3.6.4 *If $\Sigma \models \phi$, then there exists a finite subset Σ_0 of Σ such that $\Sigma_0 \models \phi$.*

Proof If Σ is not satisfiable, there exists a finite subset Σ_0 of Σ that is not satisfiable. This means there is no assignment v such that v satisfies Σ_0 but does not satisfy ϕ. Hence, $\Sigma_0 \models \phi$.

If Σ is satisfiable, $\Sigma \models \phi$ implies that $\Sigma \cup \{\neg\phi\}$ is not satisfiable. By the Compactness Theorem for SC, there exists a finite subset $\Sigma_0 \cup \{\neg\phi\}$ of $\Sigma \cup \{\neg\phi\}$ that is not satisfiable. Let v be any assignment that satisfies Σ_0 (which is satisfiable). Then v does not satisfy $\neg\phi$, and therefore v satisfies ϕ. This means that $\Sigma_\Gamma \models \phi$.⊣

A wff ϕ is said to be a *tautological consequence* of the set Σ of wffs, or that Σ tautologically implies ϕ, written as $\Sigma \models \phi$, if and only if every assignment that satisfies Σ also satisfies ϕ.

A common convention among logicians is to abbreviate a metalanguage formula of the form

$$\phi \to (\psi \to \theta)$$

as

$$\phi \to \psi \to \theta.$$

We adopt this convention here and proceed with the following rule.

Lemma 3.6.5 (Rule T) *If $\Sigma \vdash \phi_\Delta, \ldots, \Sigma \vdash \phi_n$, and $\Gamma = \{\phi_\Delta, \ldots, \phi_n\}$ tautologically implies ψ, then $\Sigma \vdash \psi$.*

Proof Since Γ tautologically implies ψ, the statement

$$\phi_1 \to \cdots \to \phi_n \to \psi$$

is a tautology. From Σ, we can construct a derivation:

$$\phi_{11}, \ldots, \phi_{m_1 1}, \ldots, \phi_{1n}, \ldots, \phi_{m_n n}, \phi_1 \to \cdots \to \phi_n \to \psi, \ldots, \psi.$$

Here, $\phi_{m_1 1} = \phi_1, \ldots, \phi_{m_n n} = \phi_n$, and ψ is obtained through successive applications of *modus ponens*. This establishes that $\Sigma \vdash \psi$.⊣

An immediate consequence of Rule T is the important Deduction Theorem. We abbreviate $\Sigma \cup \{\phi\}$ as $\Sigma; \phi$.

Theorem 3.6.6 (Deduction Theorem) *If $\Sigma; \phi \vdash \psi$, then $\Sigma \vdash \phi \to \psi$.*

Proof If Σ is inconsistent or $\phi \to \psi$ is a tautology, there is nothing to prove. We can therefore assume that $\phi \neq \psi$ and that Σ is consistent. If $\Sigma \vdash \psi$, since ψ tautologically implies $\psi \to (\phi \to \psi)$, we have $\Sigma \vdash \phi \to \psi$ by Rule T.

In the remaining case, ψ is obtained by *modus ponens* from θ and $\theta \to \psi$. By the inductive hypothesis, since θ and $\theta \to \psi$ appear earlier in the derivation concluding with ψ, we have $\Sigma \vdash \phi \to \theta$ and $\Sigma \vdash \phi \to (\theta \to \psi)$. Since $\{\phi \to \theta, \phi \to (\theta \to \psi)\}$ tautologically implies $\phi \to \psi$, it follows by Rule T that $\Sigma \vdash \phi \to \psi$.⊣

It is easy to see that if $\Sigma; \neg\phi$ is inconsistent, then $\Sigma \vdash \phi$. Indeed,

$$\begin{aligned}\Sigma; \neg\phi \text{ inconsistent} &\Rightarrow \Sigma; \neg\phi \vdash \psi \wedge \neg\psi \text{ for every } \psi\\ &\Rightarrow \Sigma \vdash \neg\phi \rightarrow (\psi \wedge \neg\psi)\\ &\Rightarrow \Sigma \vdash \phi.\end{aligned}$$

by Rule T, since $\neg\phi \rightarrow (\psi \wedge \neg\psi)$ tautologically implies ϕ.

Theorem 3.6.7 (Completeness for SC II) *Let Γ be a set of statements in* SC. *If $\Gamma \models \phi$, then there is a deduction of ϕ from Γ.*

Proof Assume that $\Gamma \models \phi$ and, contrary to the claim, that $\Gamma; \neg\phi$ is consistent. Then $\Gamma; \neg\phi$ is satisfiable, and hence there exists an assignment v such that v satisfies both Γ and $\neg\phi$. However, the assumption that $\Gamma \models \phi$ implies that this same v also satisfies ϕ, leading to the conclusion that v both satisfies and does not satisfy ϕ.

This contradiction establishes that $\Gamma; \neg\phi$ is inconsistent, and therefore, $\Gamma \vdash \phi$.$\dashv$

3.7 First Order Logic

The formalized logical languages were introduced to provide foundations for arithmetic and, more generally, to address the properties of certain mathematical structures whose general definition is as follows.

Definition 3.7.1 A *structure* or *relational system* is a sequence

$$\mathfrak{A} = \langle A, R_\xi, G_\zeta, a_\varsigma \rangle_{\xi \in \mathfrak{a}, \zeta \in \mathfrak{b}, \varsigma \in \mathfrak{c}},$$

where $\mathfrak{a}$, $\mathfrak{b}$, and $\mathfrak{c}$ are order types (pairwise disjoint well-ordered sets) whose respective cardinalities are α, β, γ; A is a non-empty set; for each $\xi \in \mathfrak{a}$, R_ξ is a relation on A; for each $\zeta \in \mathfrak{b}$, G_ζ is a function on A; and for each $\varsigma \in \mathfrak{c}$, a_ς is an element of A. Any of the sets $\mathfrak{a}$, $\mathfrak{b}$, $\mathfrak{c}$ may be empty, but $\mathfrak{a}$ and $\mathfrak{b}$ cannot both be empty simultaneously. If $\mathfrak{a} = \varnothing$, $\mathfrak{A}$ must involve at least one function (i.e., $\mathfrak{b}$ must be non-empty); in this case, $\mathfrak{A}$ is called an *algebraic structure*.

The *order* of the structure, $\mathfrak{y}$, is the sum $\mathfrak{a} + \mathfrak{b} + \mathfrak{c}$ of the order types $\mathfrak{a}$, $\mathfrak{b}$, and $\mathfrak{c}$. Clearly, as each of the orders $\mathfrak{a}$, $\mathfrak{b}$, $\mathfrak{c}$ has a first element, we can map the first element of $\mathfrak{b}$ to $\alpha + 1$ and obtain an isomorphism between $\mathfrak{b}$ and $\mathfrak{b}' = \{\zeta \in \alpha + \beta \mid \alpha + 1 \leq \zeta < \alpha + \beta\}$. Similarly, we can obtain an isomorphism between $\mathfrak{c}$ and $\mathfrak{c}' = \{\varsigma \in \alpha + \beta + \gamma \mid \alpha + \beta + 1 \leq \varsigma < \alpha + \beta + \gamma\}$. From now on, we will identify the set $\mathfrak{a}$ with α, $\mathfrak{b}$ with $\mathfrak{b}'$, and $\mathfrak{c}$ with $\mathfrak{c}'$.

The *similarity type* of the structure $\mathfrak{A}$ is the function $\mu \in \omega^{\mathfrak{y}}$ such that

$$\mu(\iota) = \begin{cases} n & \text{if } \iota \in \mathfrak{a} \text{ and } R_\iota \subset A^n, \\ m & \text{if } \iota \in \mathfrak{b} \text{ and } G_\iota \text{ is a function from } A^m \text{ to } A, \\ 0 & \text{if } \iota \in \mathfrak{c}. \end{cases}$$

If $\iota \in \mathfrak{a}$, the number $\mu(\iota)$ is called the *arity* of the relation R_ι. If $\iota \in \mathfrak{b}$, the function G_ι is an application with $\mu(\iota)$ arguments. A is the *domain* or *universe* of the structure $\mathfrak{A}$, and the R_ξ are the *fundamental relations* of the structure. The elements of A are also called the *elements* of $\mathfrak{A}$. Two relational systems are similar if they have the same similarity type. The class of structures with similarity type μ is called the *similarity type* of μ.

Let $\mathfrak{A} = \langle A, R_\xi, G_\zeta, a_\varsigma \rangle$ and $\mathfrak{B} = \langle B, S_\xi, H_\zeta, b_\varsigma \rangle$ be two similar structures. $\mathfrak{A}$ is a *substructure* or *subsystem* of $\mathfrak{B}$ ($\mathfrak{A} \subseteq \mathfrak{B}$) if and only if (1) $A \subseteq B$; (2) for each ξ, $R_\xi = S_\xi \cap A^n$, where $n = \mu(\xi)$; (3) G_ζ is the restriction of H_ζ to A; (4) $a_\varsigma = b_\varsigma$. $\mathfrak{B}$ is also called an *extension* (or *enlargement*) of $\mathfrak{A}$.

A *homomorphism* of $\mathfrak{A}$ into $\mathfrak{B}$ is a function $f : A \to B$ that preserves relations and functions in the following sense:

(1) $(x_1, \ldots, x_{\mu(\xi)}) \in R_\xi$ if and only if $(f(x_1), \ldots, f(x_{\mu(\xi)})) \in S_\xi$.
(2) $f(G_\zeta(x_1, \ldots, x_{\mu(\zeta)})) = H_\zeta(f(x_1), \ldots, f(x_{\mu(\zeta)}))$.
(3) $f(a_\varsigma) = b_\varsigma$.

If f is injective, the homomorphism is called a *monomorphism*; if it is surjective, an *epimorphism*; and if it is both injective and surjective, an *isomorphism*.

We say that $\mathfrak{B}$ is a *homomorphic image* of $\mathfrak{A}$ if there exists an epimorphism from A to B. A structure $\mathfrak{A}$ is said to be *embedded* in the system $\mathfrak{B}$ if $\mathfrak{A}$ is isomorphic to a subsystem of $\mathfrak{B}$; this isomorphic image of $\mathfrak{A}$ in $\mathfrak{B}$ is called the *embedding* of $\mathfrak{A}$ in $\mathfrak{B}$. An isomorphism of a structure with itself is called an *automorphism*. If $\mathfrak{A}$ is isomorphic to $\mathfrak{B}$, we write $\mathfrak{A} \cong \mathfrak{B}$.

3.8 The Language of First Order Logic

To express statements about a structure, a language is required. For each similarity type μ, a language is introduced, called the *language of similarity type* μ, denoted $\mathcal{L}(\mu)$ or simply $\mathcal{L}$ if there is no ambiguity.

The class $\mathcal{M}$ of structures of similarity type μ is associated with the *language of the similarity class*, $\mathcal{L}(\mu)$, or simply $\mathcal{L}$ if no confusion can arise. Any element of $\mathcal{M}$ is said to be a *model* of the language $\mathcal{L}$. These languages, called *first-order languages*, consist of logical constants and parameters.

The *logical constants* include countably many individual variables:

$$x_1, x_2, x_3, \ldots;$$

logical connectives and (optionally) the identity symbol: $\neg, \rightarrow, =$; and grouping symbols:

$$(\quad \text{and} \quad)$$

(sometimes combined with brackets [and]).

Other connectives and the existential quantifier, such as

$$\wedge, \vee, \leftrightarrow, \exists,$$

can be defined in terms of these symbols.

The *parameters* include the following types of symbols:

(1) The universal quantifier $\forall$.
(2) *Predicate symbols*: For each $\xi \in \mathfrak{a}$, a predicate symbol P_ξ corresponding to the relation R_ξ, with arity $\mu(\xi)$.
(3) *Function symbols*: For each $\zeta \in \mathfrak{b}$, a function symbol F_ζ corresponding to the function G_ζ, with $\mu(\zeta)$ arguments.
(4) *Individual constant symbols*: For each $\varsigma \in \mathfrak{c}$, a constant symbol c_ς corresponding to the element $a_\varsigma \in A$.

A *formula* of first-order logic is any finite sequence of symbols. The formation rules stipulate how to construct well-formed formulas from given formulas. There are two classes of expressions: terms and formulas. A *term* is an individual variable, a constant, or an n -ary function symbol followed by n terms.

More precisely, the *formation rules for terms* are as follows:

(1) If τ is an individual variable x, then τ is a term;
(2) If τ is an individual constant symbol c, then τ is a term;
(3) If F is an n -ary function symbol and $\tau_1, \ldots, \tau_n$ are terms, then $F(\tau_1 \cdots \tau_n)$ is a term;
(4) Only expressions constructed using the three preceding rules are terms.

The concept of a *well-formed formula* (wff) is defined analogously to that of a term, starting from basic or atomic formulas, which are of the form $P\tau_1 \ldots \tau_n$, where n is the arity of the predicate P, and of the form $\tau_1 = \tau_2$ if the identity symbol is included. The precise definition of a well-formed formula is as follows:

(1) If ϕ is an expression of the form $P\tau_1 \ldots \tau_n$ or of the form $\tau_1 = \tau_2$, then ϕ is a wff.
(2) If ϕ is a wff, then $(\neg\phi)$ is a wff.
(3) If ϕ and ψ are wffs, then $(\phi \rightarrow \psi)$ is a wff.
(4) If ϕ is a wff and x is an individual variable, then $\forall x\, \phi$ is a wff.
(5) Only a formula constructed using the four preceding rules is a wff.

The other connectives and the existential quantifier can be defined by the following conditions. For arbitrary formulas ϕ, ψ, and variable x:

$(\phi \vee \psi)$ abbreviates $((\neg\phi) \to \psi)$
$(\phi \wedge \psi)$ abbreviates $\neg(\neg\phi \vee \neg\psi)$
$(\phi \leftrightarrow \psi)$ abbreviates $(\phi \to \psi) \wedge (\psi \to \phi)$
$\exists x\, \phi$ abbreviates $\neg\forall x \neg\phi$.

A variable x can occur free in a formula. This concept is defined recursively as follows:

(1) For atomic ϕ, x *occurs free in* ϕ if and only if x occurs in (is a symbol of) ϕ.
(2) x *occurs free in* $\neg\phi$ if and only if x occurs free in ϕ.
(3) x *occurs free in* $(\phi \to \psi)$ if and only if x occurs free in ϕ or in ψ.
(4) x *occurs free in* $\exists x'\phi$ if and only if x occurs free in ϕ and $x \neq x'$.

A *sentence* is a well-formed formula (wff) in which no variable occurs free.

A special class of formulas is that of the prime formulas. A *prime formula* is either an atomic formula or a formula of the form $\forall x\, \phi$. It is evident that the set of all formulas can be obtained from the prime formulas using the formula construction operations. From now on, prime formulas will be interpreted as the atomic formulas of the sentential calculus.

The *interpretation rules* of a language $\mathcal{L}$ for a similarity class $\mathcal{M}$ are precisely those that connect this language with the universe of any structure in $\mathcal{M}$ and its fundamental relations, as previously explained. The *cardinality* of language $\mathcal{L}$, denoted $||\mathcal{L}||$, is $\omega + \text{card}(\mathfrak{y})$.

An *expansion* of the language $\mathcal{L}$ results from adding a set of symbols. If the symbols are constants not appearing in $\mathcal{L}$, the expansion is called *simple*. If $\mathcal{L}'$ is an expansion of $\mathcal{L}$, then $\mathcal{L}$ is called a *reduction* of $\mathcal{L}'$. If the added symbols were not originally in $\mathcal{L}$, the structure $\mathfrak{A}'$ resulting from adding interpretations of the new symbols to a model $\mathfrak{A}$ of $\mathcal{L}$ is a model of $\mathcal{L}'$. The structure $\mathfrak{A}$ is called the *expansion* of $\mathfrak{A}$ to $\mathcal{L}'$, and $\mathfrak{A}$ is called the *reduct* of $\mathfrak{A}'$ to $\mathcal{L}$.

Given a sentence ϕ in the language $\mathcal{L}$ and a structure $\mathfrak{A} \in \mathcal{M}$, it is of primary importance in logical applications to ask whether ϕ is true in $\mathfrak{A}$. To give a precise response to this question, the concept of truth must be properly defined. To handle the occurrence of variables—which may be free—in formulas, we introduce the family of all functions s that map the set of variables into the universe A. Each of these functions assigns an element of the universe to each variable. If s is one such assignment, $s(x_i|a)$ denotes the function that is identical to s except that it assigns the element a of the universe to x_i.

Using these assignments s, we define the notion of satisfaction with an assignment s. The satisfaction of a formula by a model and an assignment s is defined as follows:

Definition 3.8.1 A well-formed formula ϕ is *satisfied by* $\mathfrak{A}$ *with* s (denoted as $\mathfrak{A} \models \phi\,[s]$) if and only if, for arbitrary formulas ϕ, ψ and variable x_i:

(1) $\mathfrak{A} \models \tau_1 = \tau_2\,[s]$ if and only if $s(\tau_1) = s(\tau_2)$, where τ_1 and τ_2 are terms.
(2) If P_ξ is an n-place predicate symbol,

$$\mathfrak{A} \models P_\xi \tau_1 \cdots \tau_n\ [s] \quad \text{if and only if} \quad (s(\tau_1), \ldots, s(\tau_n)) \in R_\xi$$

(3) $\mathfrak{A} \models \neg\phi\ [s]$ if and only if $\mathfrak{A} \not\models \phi\ [s]$.
(4) $\mathfrak{A} \models (\phi \to \psi)\ [s]$ if and only if $\mathfrak{A} \not\models \phi\ [s]$ or $\mathfrak{A} \models \psi\ [s]$ or both.
(5) $\mathfrak{A} \models \forall x_i\ \phi\ [s]$ if and only if $\mathfrak{A} \models \phi\ [s(x_i|a)]$ for every $a \in |\mathfrak{A}|$.

Functions s assign an element of the universe to each variable in the language. However, sometimes we may want to focus specifically on the variables occurring in a given formula. If ϕ is a formula, we may write $\phi(x_1 \cdots x_n)$ to indicate that the free variables of ϕ form a subset of $\{x_1, \ldots, x_n\}$. If τ is a term, $\tau(x_1 \cdots x_n)$ is simply τ itself.

Additionally, we will denote by $[\![a_1 \cdots a_n]\!]$ any s such that $s(x_i) = a_i$ and by $\mathfrak{A} \models \phi[\![a_1 \cdots a_n]\!]$ the fact that $\mathfrak{A}$ satisfies ϕ with any such s (or with that particular assignment of denotations to the variables $x_1, \ldots, x_n$). If τ is the constant c, then $\tau[\![a_1 \cdots a_n]\!]$ denotes the same as c; if τ is the variable x_i, then $\tau[\![a_1 \cdots a_n]\!]$ denotes the same as $s(x_i)$.

A well-formed formula ϕ is *satisfiable* if and only if there exists a class of structures $\mathcal{M}$ such that ϕ is a formula of $\mathcal{L}(\mathcal{M})$ and there is a structure $\mathfrak{A} \in \mathcal{M}$ along with a function s such that $\mathfrak{A} \models \phi\ [s]$. A set of formulas $\Gamma \subset \mathcal{L}(\mathcal{M})$ is *satisfiable* if and only if there exists a structure $\mathfrak{A} \in \mathcal{M}$ and a function s such that $\mathfrak{A}$ satisfies every member of Γ with s.

The crucial concept of *implication* or *logical consequence* can now be defined. Let Γ be a set of well-formed formulas (wffs) and let ϕ be a wff. We say that Γ *logically implies* ϕ, or that ϕ is a *logical consequence* of Γ ($\Gamma \models \phi$), if and only if, for every structure $\mathfrak{A} \in \mathcal{M}$ and every function s mapping the set of terms to the universe of $\mathfrak{A}$: if $\mathfrak{A}$ satisfies every formula in Γ with s, then $\mathfrak{A}$ also satisfies ϕ with s.

We say that ϕ is *valid* in $\mathfrak{A}$ if ϕ is satisfied by $\mathfrak{A}$ with every s. A formula ϕ is *logically valid* if and only if ϕ is satisfied by every $\mathfrak{A}$ with every s. A set Γ of formulas is valid in $\mathfrak{A}$ if every element of Γ is valid in $\mathfrak{A}$; in this case, we say that $\mathfrak{A}$ is a *model* of Γ.

It can be proven that if ϕ is a sentence and $\mathfrak{A}$ is any structure in $\mathcal{M}$, then either ϕ is satisfied by $\mathfrak{A}$ with *every* s, or it is not satisfied with *any* s at all. If the former holds, we say that ϕ is *true* in $\mathfrak{A}$, and we say that $\mathfrak{A}$ is a model of ϕ. Thus, in particular, if ϕ is a sentence and Γ is a set of sentences, then ϕ is a logical consequence of Γ if and only if every model of Γ is also a model of ϕ.

Given a structure $\mathfrak{A} \in \mathcal{M}$, we define the *theory of* $\mathfrak{A}$ (Th $\mathfrak{A}$) as the set of all sentences of $\mathcal{L}(\mathcal{M})$ that are true in $\mathfrak{A}$. Similarly, the *theory of* $\mathcal{K}$, where $\mathcal{K} \subset \mathcal{M}$ (Th $\mathcal{K}$), is the set of all sentences that are true in every element of $\mathcal{K}$. A *theory* is defined as a set of sentences that is closed under implication; that is, T is a *theory* if and only if

$$T \models \phi \quad \text{implies that} \quad \phi \in T.$$

A set of sentences Γ is *consistent* if and only if there is no formula ϕ such that Γ derives both ϕ and $\neg\phi$. Let Σ be a consistent set of sentences, and let Mod Σ denote the class of all models of Σ. The set of *consequences* of Σ (Cn Σ) is defined as Th Mod Σ. It is easy to see that a set of sentences T is a theory if and only if $T = \text{Cn}\, T$.

Broadly speaking, a set Σ of formulas is *decidable* if there is an algorithm that can determine, in a finite number of steps, whether a given sentence is or is not an element of Σ.[1] A theory T is *axiomatizable* if and only if there exists a decidable set of sentences Σ such that $T = \text{Cn}\, \Sigma$. In particular, if Σ is finite, then T is said to be *finitely axiomatizable*.

A *basic valuation* is an assignment v of truth values to the prime formulas of $\mathscr{L}(\mathcal{M})$; that is, a function from prime formulas to $\{0, 1\}$. Each basic valuation can be uniquely extended to a valuation $\hat{v}$ of the set $\mathcal{F}$ of all well-formed formulas in $\{0, 1\}$, according to the following rules:

$$\hat{v}(\phi) = \begin{cases} v(\phi) & \text{if } \phi \text{ is prime;} \\ 1 - \hat{v}(\psi) & \text{if } \phi = (\neg\psi); \\ 1 - \hat{v}(\psi) + \hat{v}(\psi)\cdot\hat{v}(\theta) & \text{if } \phi = (\psi \to \theta). \end{cases}$$

Additionally, it is easy to see that the assignment must satisfy the following equations for the other connectives:

$$\hat{v}(\phi) = \begin{cases} \hat{v}(\psi) + \hat{v}(\theta) - \hat{v}(\psi)\cdot\hat{v}(\theta) & \text{if } \phi = \psi \vee \theta; \\ \hat{v}(\psi)\cdot\hat{v}(\theta) & \text{if } \phi = \psi \wedge \theta; \\ 1 - \hat{v}(\psi) - \hat{v}(\theta) + 2\hat{v}(\psi)\cdot\hat{v}(\theta) & \text{if } \phi = \psi \leftrightarrow \theta. \end{cases}$$

Each model $\mathfrak{A}$ and assignment s determines a valuation $\hat{v}$ for all formulas of $\mathscr{L}(\mathcal{M})$, which aligns perfectly with the definition of satisfaction. That is, the formulas satisfied by $\hat{v}$ are precisely those satisfied by $\mathfrak{A}$ with s, provided that the basic valuation determining $\hat{v}$ assigns 1 precisely to the prime formulas satisfied by $\mathfrak{A}$ with s.

Theorem 3.8.2 *Let $\mathfrak{A} \in \mathcal{M}$ and let s be a function from the set of variables of $\mathscr{L}(\mathcal{M})$ into $|\mathfrak{A}|$. Let v be a basic valuation such that*

$$v(\phi) = 1 \quad \textit{if and only if} \quad \mathfrak{A} \models \phi\,[s]$$

for every prime formula ϕ of $\mathscr{L}(\mathcal{M})$, and let $\hat{v}$ be the extended valuation associated with v. Then, for every well-formed formula ϕ of $\mathscr{L}(\mathcal{M})$,

$$\hat{v}(\phi) = 1 \quad \textit{if and only if} \quad \mathfrak{A} \models \phi\,[s].$$

[1] For a precise definition of decidability, see Boolos et al. (2002).

Proof Since formulas of the form $\forall x\ \psi$ are prime, it suffices to prove the statement for negations and conditionals. The proof proceeds by induction on the complexity of formulas, with the base case of the induction being the set of prime formulas.

If ϕ is of the form $(\neg\psi)$, we have

$$\begin{aligned}\hat{v}(\phi) = 1 \quad &\Leftrightarrow \quad \hat{v}(\neg\psi) = 1\\ &\Leftrightarrow \quad 1 - \hat{v}(\psi) = 1\\ &\Leftrightarrow \quad \hat{v}(\psi) = 0\\ &\Leftrightarrow \quad \mathfrak{A} \not\models \psi\ [s]\\ &\Leftrightarrow \quad \mathfrak{A} \models \neg\psi\ [s]\\ &\Leftrightarrow \quad \mathfrak{A} \models \phi\ [s].\end{aligned}$$

If ϕ is of the form $(\psi \rightarrow \theta)$,

$$\begin{aligned}\hat{v}(\phi) = 1 \quad &\Leftrightarrow \quad 1 - \hat{v}(\psi) + \hat{v}(\psi)\cdot\hat{v}(\theta) = 1\\ &\Leftrightarrow \quad \hat{v}(\psi) - \hat{v}(\psi)\cdot\hat{v}(\theta) = 0\\ &\Leftrightarrow \quad \hat{v}(\psi)[1 - \hat{v}(\theta)] = 0\\ &\Leftrightarrow \quad \hat{v}(\psi) = 0 \quad \text{or} \quad 1 - \hat{v}(\theta) = 0\\ &\Leftrightarrow \quad \hat{v}(\psi) = 0 \quad \text{or} \quad \hat{v}(\theta) = 1\\ &\Leftrightarrow \quad \mathfrak{A} \not\models \psi\ [s] \quad \text{or} \quad \mathfrak{A} \models \theta\ [s]\\ &\Leftrightarrow \quad \mathfrak{A} \models (\psi \rightarrow \theta)\ [s]. \dashv\end{aligned}$$

A *tautology* in predicate calculus is a well-formed formula (wff) that is true under every valuation. It follows that a tautology is logically valid, but not every logically valid formula is a tautology. For example, the formula

$$(\forall x\ \phi \vee \forall x\ \theta) \rightarrow \forall x\ (\phi \vee \theta)$$

is logically valid but certainly not a tautology, as shown by the valuation

$$\hat{v}(\forall x\phi) = 1, \quad \hat{v}(\forall x\theta) = 1, \quad \hat{v}(\forall x\ (\phi \vee \theta)) = 0.$$

The *rules of derivation* stipulate which formulas can be inferred from given formulas. There are many possible sets of derivation rules, but it is possible to select a rather large set of logical axioms (for example, one that includes all tautologies) and only use *modus ponendo ponens* as the sole rule of inference. For this purpose, if ϕ is a formula, we say that a *generalization* of ϕ is a statement of the form

$$\forall x_{i_1} \cdots x_{i_n}\ \phi$$

for $n \in \omega$.

The expression ϕ^x_τ denotes the result of substituting in the formula ϕ all free occurrences of x by τ. However, not all such substitutions are admissible, since it may happen that a free variable in τ, when substituted for x in ϕ, becomes bound by a quantifier appearing in ϕ. The term τ is *admissible* for x in the formula ϕ only in the following cases:

(1) ϕ is atomic.
(2) $\phi = \neg\psi$ and τ is admissible for x in ψ.
(3) $\phi = (\psi \to \theta)$ and τ is admissible for x in both ψ and θ.
(4) $\phi = \forall x'\ \psi$ and either x does not occur free in $\forall x'\ \psi$ or x' does not occur in τ and τ is admissible for x in ψ.

The set Λ of axioms for $\mathcal{L}(\mathcal{M})$ consists of the generalizations of the following formulas, where τ is admissible for x in ϕ:

(A1) All tautologies.
(A2) $\forall x\ \phi \to \phi^x_\tau$.
(A3) $\forall x(\phi \to \psi) \to (\forall x\phi \to \forall x\psi)$.
(A4) $\phi \to \forall x\phi$, where x does not occur free in ϕ.

If ϕ is any wff and Γ is a set of wffs, a *deduction* or *derivation* of ϕ from Γ is a sequence $\phi_0, \ldots, \phi_n$ of wffs such that $\phi_n = \phi$ and for each $i \leq n$, either:

(1) ϕ_i is in $\Gamma \cup \Lambda$, or
(2) for some natural numbers j and k less than i, ϕ_i is obtained by *modus ponens* from ϕ_j and ϕ_k, where $\phi_k = \phi_j \to \phi_i$.

A formula ϕ is called a *theorem* of Γ (denoted as $\Gamma \vdash \phi$) if and only if ϕ belongs to the set generated from $\Gamma \cup \Lambda$ by *modus ponens*. It follows that there is a derivation of ϕ from Γ if and only if ϕ is a theorem of Γ. Thus, it is justified to say that ϕ is *deducible* from Γ if and only if $\Gamma \vdash \phi$.

Some useful theorems implied by these definitions are the following.

Theorem 3.8.3 *$\Sigma \vdash \phi$ if and only if $\Sigma \cup \Lambda$ tautologically implies ϕ.*

Proof It is obvious that $\{\phi, \phi \to \psi\}$ tautologically implies ϕ. Let v be an assignment that satisfies all elements of $\Gamma \cup \Lambda$, and let $\phi_1, \ldots, \phi_n$ (where $\phi_n = \phi$) be a derivation of ϕ from Γ. For each $k \leq n$, if $\phi_k \in \Gamma$, then v satisfies ϕ_k. If ϕ_k is derived by *modus ponens* from formulas $\phi_j \to \phi_k$ and ϕ_j, and if v satisfies both $\phi_j \to \phi_k$ and ϕ_j, then v also satisfies ϕ_k.

If $\Sigma \cup \Lambda$ tautologically implies ϕ, the *Compactness Theorem* 3.6.2 implies that there is a finite subset Σ_0 of Σ such that $\Sigma_0 \models \phi$.$\dashv$

Another useful result in logical theory development is the following.

Theorem 3.8.4 (Generalization Theorem) *If $\Gamma \vdash \phi$ and x does not occur free in any formula of Γ, then $\Gamma \vdash \forall x\ \phi$.*

Proof Suppose $\Gamma \vdash \phi$ and x does not occur free in any formula in Γ. We have three possible cases (note that x may occur free in ϕ):

Case (i): ϕ is a logical axiom. Then $\forall x\ \phi$ is also an axiom, and $\Gamma \vdash \forall x\ \phi$.

Case (ii): $\phi \in \Gamma$, meaning that x does not occur free in ϕ. Then $\phi \to \forall x\ \phi$ is an axiom in group A4, and we have the following derivation:

$$\Gamma \vdash \phi$$
$$\Gamma \vdash \phi \to \forall x\ \phi \quad \text{(A4)}$$
$$\Gamma \vdash \forall x\ \phi.$$

Case (iii): There is a derivation of ϕ from Γ, and ϕ is obtained by *modus ponens* from formulas ψ and $\psi \to \phi$ in that derivation. By the inductive hypothesis,

$$\Gamma \vdash \forall x\ \psi \quad \text{and} \quad \Gamma \vdash \forall x\ (\psi \to \phi).$$

Thus, we obtain:

$$\Gamma \vdash \forall x\ \psi$$
$$\Gamma \vdash \forall x\ (\psi \to \phi)$$
$$\Gamma \vdash \forall x\ (\psi \to \phi) \to (\forall x\ \psi \to \forall x\ \phi) \quad \text{(A3)}$$
$$\Gamma \vdash \forall x\ \psi \to \forall x\ \phi$$
$$\Gamma \vdash \forall x\ \phi. \dashv$$

The derivation rules in our language have been appropriately chosen so that the following theorem holds. The proof is left to the reader.

Theorem 3.8.5 (Soundness Theorem) *For any set Γ of formulas in $\mathcal{L}$ and any formula ϕ in this language:*

$$\Gamma \vdash \phi \quad \Rightarrow \quad \Gamma \models \phi.$$

An interesting connection between satisfiability and consistency, based on the previous theorem, is the following.

Theorem 3.8.6 *For any set Γ of formulas in $\mathcal{L}$ and any formula ϕ in this language:*

$$\Gamma \vdash \phi \Leftrightarrow \Gamma \models \phi \Leftrightarrow \Gamma \textit{ is satisfiable} \Leftrightarrow \Gamma \textit{ is consistent}.$$

Proof Suppose $\Gamma \vdash \phi$ implies $\Gamma \models \phi$, and suppose Γ is satisfiable. Assume, for contradiction, that Γ is inconsistent. Then there exists some formula ψ such that $\Gamma \vdash \psi \wedge \neg\psi$. It follows that

$$\Gamma \models \psi \wedge \neg\psi.$$

But

$$\Gamma \models \psi \wedge \neg\psi \quad \Leftrightarrow \quad \Gamma \models \psi \quad \text{and} \quad \Gamma \models \neg\psi.$$

Thus, there exists a structure $\mathfrak{A}$ and an assignment s such that

$$\mathfrak{A} \models \psi\ [s] \quad \text{and} \quad \mathfrak{A} \models \neg\psi\ [s].$$

This contradiction establishes that Γ is consistent.

For the converse, suppose the satisfiability of Γ implies its consistency, and assume $\Gamma \vdash \phi$ and that some $\langle \mathfrak{A}, s \rangle$ satisfies Γ. It is easy to see that $\Gamma; \neg\phi$ is inconsistent, since $\Gamma; \neg\phi \vdash \phi$ and $\Gamma; \neg\phi \vdash \neg\phi$. It follows that $\Gamma; \neg\phi$ is not satisfiable, and thus $\mathfrak{A} \not\models \neg\phi\ [s]$. Therefore, $\mathfrak{A} \models \phi\ [s]$.$\dashv$

Corollary 3.8.7 *For any set Γ of formulas in $\mathcal{L}$ and any formula ϕ in this language:*

If Γ is satisfiable, then Γ is consistent.

We have now described a *first-order logic*—the language $\mathcal{L}$ of the class of structures $\mathcal{M}$—in a concise but complete manner. The most important result of first-order logic is *Gödel's Completeness Theorem*, which states that every logical consequence of a set Γ of formulas is also derivable from Γ. This result implies the *Compactness Theorem*, which states that if every finite subset of Γ is satisfiable, then Γ itself is satisfiable. The next section is dedicated to these topics.

3.9 Completeness and Compactness

Following our somewhat algebraic approach, in this section we provide a proof of Gödel's Completeness Theorem for predicate calculus based on the concept of a Lindenbaum Algebra. We first demonstrate that every consistent set of statements Γ is satisfiable and, from this, we derive as a corollary that every logical consequence of Γ is also derivable. An immediate consequence of this result is the Compactness Theorem: A set Γ of sentences is satisfiable if and only if every *finite subset* of Γ is satisfiable. These are among the most important results in first-order logic, as they provide a gateway into model theory, and they also form the foundation of infinitesimal analysis.

Theorem 3.9.1 *Let $\mathcal{L}$ be a first-order language, and let T be a consistent set of statements in $\mathcal{L}$ that is closed under $\vdash$ (provability). Define the equivalence relation $\sim$ on formulas of $\mathcal{L}$ by:*

$$\phi \sim \psi \quad \textit{if and only if} \quad T \vdash (\phi \leftrightarrow \psi).$$

Define the following equivalence classes and sets:

$[\phi] = \{\psi \mid \psi \sim \phi\}$ *(the equivalence class of* ϕ*).*
$B_T = \{[\phi] \mid \phi$ *is a formula in* $\mathcal{L}\}$
(the set of all equivalence classes of formulas).
$1_T = \{\phi \mid T \vdash \phi\}$ *(the class of all provable formulas).*
$0_T = \{\phi \mid T \vdash \neg\phi\}$ *(the class of all refutable formulas).*

Now, define the following operations:

$[\phi] \cup [\psi] = [\phi \vee \psi]$ *(join, disjunction).*
$[\phi] \cap [\psi] = [\phi \wedge \psi]$ *(meet, conjunction).*
$\overline{[\phi]} = [\neg\phi]$ *(complement, negation).*

Then, the structure:

$$\mathfrak{B}_T = \langle B_T, \cup, \cap, ^{-}, 0_T, 1_T \rangle$$

is a Boolean algebra.

Proof First, we need to show that the operations are *well-defined*, meaning that their results do not depend on the chosen representatives. That is, if $\phi, \psi, \gamma, \theta$ are formulas such that $\gamma \sim \phi$ and $\theta \sim \psi$, then:

$$[\phi \vee \psi] = [\gamma \vee \theta], \quad [\phi \wedge \psi] = [\gamma \wedge \theta], \quad \text{and} \quad [\neg\phi] = [\neg\gamma].$$

Since the hypotheses imply $T \vdash (\gamma \vee \theta) \leftrightarrow (\phi \vee \psi)$ and $T \vdash (\neg\gamma) \leftrightarrow (\neg\phi)$, we obtain:

$$[\phi] \cup [\psi] = [\phi \vee \psi] = [\gamma \vee \theta] = [\gamma] \cup [\theta],$$

and similarly for conjunction:

$$[\phi] \cap [\psi] = [\phi \wedge \psi] = [\gamma \wedge \theta] = [\gamma] \cap [\theta].$$

For negation:

$$\overline{[\gamma]} = [\neg\gamma] = [\neg\phi] = \overline{[\phi]}.$$

Now, we verify that $\mathfrak{B}_T$ satisfies the *axioms of Boolean algebra*. For example, the *commutativity of disjunction* holds because:

$$T \vdash (\phi \vee \psi) \leftrightarrow (\psi \vee \phi),$$

so we have:

$$[\phi] \cup [\psi] = [\phi \vee \psi] = [\psi \vee \phi] = [\psi] \cup [\phi].$$

The verification of the remaining Boolean algebra axioms (associativity, distributivity, identity elements, and De Morgan's laws) is left as an exercise for the reader. $\dashv$

The algebra $\mathfrak{B}_T$ is called the *Lindenbaum Algebra* of T. It is interesting to examine the meaning of the order relation in this algebra. From Theorem 3.9.1, we have:

$$[\phi] \leq [\psi] \quad \text{if and only if} \quad [\phi] \cup [\psi] = [\psi].$$

That is,

$$T \vdash (\phi \vee \psi) \leftrightarrow \psi.$$

This, in turn, is equivalent to:

$$T \vdash \phi \rightarrow \psi,$$

since:

$$T \vdash ((\phi \vee \psi) \leftrightarrow \psi) \leftrightarrow (\phi \rightarrow \psi).$$

Hence, the equivalence class of ϕ precedes the equivalence class of ψ in the order of $\mathfrak{B}_T$ iff the sentence $\phi \rightarrow \psi$ is derivable from T.

Thus, the greatest lower bound (infimum), $\inf \mathscr{E}$, of a family of equivalence classes of formulas $\mathscr{E} = \{[\phi] \mid \phi \in \Gamma\}$ is the *equivalence class* $[\theta]$ such that $[\theta]$ implies every element in Γ, i.e.,

$$T \vdash \theta \rightarrow \phi \quad \text{for every } \phi \in \Gamma;$$

and $[\theta]$ is implied by any other equivalence class that implies every element of Γ, i.e.,

$$\text{if } T \vdash \psi \rightarrow \phi \text{ for every } \phi \in \Gamma, \text{ then } T \vdash \psi \rightarrow \theta.$$

This means that $[\theta] \leq [\phi]$ for every $\phi \in \Gamma$, and if $[\psi] \leq [\phi]$ for all $\phi \in \Gamma$, then $[\psi] \leq [\theta]$.

Inspired by William of Ockham and the nominalists, we recall that universal quantification over a predicate Px (i.e., "x is a P") was understood as equivalent to the conjunction of all statements of the form Pc, where c is a name of one of the entities satisfying the predicate; that is to say,

$$\forall x\ Px \leftrightarrow \quad (Pc_1 \wedge Pc_2 \wedge \cdots$$

Using this idea, we want to show that if x_i appears free in ϕ, then it is (almost always) the case that:

$$T \vdash \forall x_i\, \phi \to \phi^{x_i}_{x_k}$$

for every $k < \omega$. However, the problem is that the variable x_k might not be admissible for x_i.

To address this, we take an alphabetic variant ϕ_k for each $k < \omega$ and demonstrate that, in the Lindenbaum algebra $\mathfrak{B}_T$:

$$\bigwedge_{k \in \omega} [(\phi_k)^{x_i}_{x_k}] = [\forall x_i\ \phi]$$

and

$$\bigvee_{k \in \omega} [(\phi_k)^{x_i}_{x_k}] = [\exists x_i\ \phi].$$

The alphabetic variant $\forall x_k\ [(\phi_k)^{x_i}_{x_k}]$ is obtained from $\forall x_i\ \phi$ by replacing all bound occurrences of x_k in ϕ with x_j, the first variable in the sequence $x_1, x_2, \ldots$ that does not appear in ϕ, and then replacing all free occurrences of x_i with x_k.

It is easy to verify that the formula:

$$\forall x_k\ [(\phi_k)^{x_i}_{x_k}]$$

is logically equivalent to $\forall x_i\ \phi$.

Lemma 3.9.2 *Let T be a consistent set of statements and let θ be a formula in which the variable x_i does not appear free. If*

$$T \vdash \theta \to [(\phi_k)^{x_i}_{x_k}]$$

for all $k < \omega$, then

$$T \vdash \theta \to \forall x_i\ \phi.$$

Proof Suppose that $T \vdash \theta \to [(\phi_k)^{x_i}_{x_k}]$ for all $k \in \omega$, and choose x_k such that it does not appear in θ. Since no element of T has free variables, we proceed as follows:

$$T; \theta \vdash \theta \to [(\phi_k)^{x_i}_{x_k}]$$

$$T; \theta \vdash \theta$$

$$T; \theta \vdash [(\phi_k)^{x_i}_{x_k}]$$

$$T;\theta \vdash \forall x_k[(\phi_k)^{x_i}_{x_k}] \quad \text{(Generalization Theorem)}$$

$$T;\theta \vdash \forall x_i\ \phi$$

$$T \vdash \theta \to \forall x_i\ \phi.$$

⊣

Lemma 3.9.3 *In the algebra* $\mathfrak{B}_T$*:*

(1) $\bigwedge_{k\in\omega}[(\phi_k)^{x_i}_{x_k}] = [\forall x_i\ \phi]$
(2) $\bigvee_{k\in\omega}[(\phi_k)^{x_i}_{x_k}] = [\exists x_i\ \phi]$.

Proof Condition (1) states that the equivalence class of the formula $\forall x_i\ \phi$ is the infimum of the set

$$\{[(\phi_k)^{x_i}_{x_k}] \mid k < \omega\}.$$

This means that $T \vdash \forall x_i\ \phi \to [(\phi_k)^{x_i}_{x_k}]$ for every $k < \omega$, and moreover, if $T \vdash \theta \to [(\phi_k)^{x_i}_{x_k}]$ for all $k < \omega$, then

$$T \vdash \theta \to \forall x_i\ \phi.$$

This last part follows from Lemma 3.9.2. Since $\forall x_i\ \phi$ is equivalent to $\forall x_k[(\phi_k)^{x_i}_{x_k}]$, we have:

$$\forall x_i\ \phi \vdash \forall x_k[(\phi_k)^{x_i}_{x_k}]$$

$$\forall x_i\ \phi \vdash \forall x_k[(\phi_k)^{x_i}_{x_k}] \to [(\phi_k)^{x_i}_{x_k}] \quad \text{(Axiom A2)}$$

$$\forall x_i\ \phi \vdash [(\phi_k)^{x_i}_{x_k}]$$

$$\vdash \forall x_i\ \phi \to [(\phi_k)^{x_i}_{x_k}].$$

Dually, condition (2) states that the equivalence class of the existential quantifier is the supremum of the set

$$\{[(\phi_k)^{x_i}_{x_k}] \mid k < \omega\}.$$

In other words, this means that $T \vdash [(\phi_k)^{x_i}_{x_k}] \to \exists x_i\ \phi$ for all $k < \omega$, and moreover, if

$$T \vdash [(\phi_k)^{x_i}_{x_k}] \to \theta$$

for all $k < \omega$, then

$$T \vdash \exists x_i\ \phi \to \theta.$$

Since the following formula is a case of Axiom (A2):

$$\forall x_k \neg[(\phi_k)^{x_i}_{x_k}] \rightarrow \neg[(\phi_k)^{x_i}_{x_k}],$$

and $\forall x_i \neg\phi$ is equivalent to $\forall x_k \neg[(\phi_k)^{x_i}_{x_k}]$, we have:

$$T \vdash \forall x_i \neg\phi \rightarrow \neg[(\phi_k)^{x_i}_{x_k}]$$

$$T \vdash [(\phi_k)^{x_i}_{x_k}] \rightarrow \neg\forall x_i \neg\phi$$

$$T \vdash [(\phi_k)^{x_i}_{x_k}] \rightarrow \exists x_i\, \phi.$$

Finally,

$$T \vdash \neg\theta \rightarrow \forall x_i \neg\phi$$

$$T \vdash \neg\forall x_i \neg\phi_i \rightarrow \theta$$

$$T \vdash \exists x_i\, \phi \rightarrow \theta. \dashv$$

Let $\mathcal{F}$ be a filter on $\mathfrak{B}_T$. We say that $\mathcal{F}$ *preserves products* $\bigwedge_{k<\omega}[(\phi_k)^{x_i}_{x_k}]$ if and only if

$$\bigwedge_{k<\omega}[(\phi_k)^{x_i}_{x_k}] \in \mathcal{F} \quad \text{is equivalent to} \quad [(\phi_k)^{x_i}_{x_k}] \in \mathcal{F} \text{ for all } k < \omega.$$

We saw in Chap. 2 that in every Boolean algebra, there exists an ultrafilter that preserves the products of a countable sequence of products (Theorem 2.3.17). If the cardinality of the language of T is ω and T is consistent, this result implies that there exists an ultrafilter in $\mathfrak{B}_T$ that preserves all products of the form $\bigwedge_{k<\omega}[(\phi_k)^{x_i}_{x_k}]$.

In the case where the cardinality of the language of T is uncountable, there are uncountably many formulas ϕ, so the number of such products is also uncountable. The good news is that, although Theorem 2.3.17 cannot be generalized to uncountable sequences of products in arbitrary Boolean algebras, it is a significant fact that this generalization is possible in the particular case of Lindenbaum algebras.

Lemma 3.9.4 *If T is consistent, then there exists an ultrafilter $\mathcal{U}$ in the algebra $\mathfrak{B}_T$ that preserves all products of the form $\bigwedge_{k<\omega}[(\phi_k)^{x_i}_{x_k}]$.*

Proof Let $\mathcal{F} = \{[\phi] \mid T \vdash \phi\}$. We will show that $\mathcal{F}$ is a filter that preserves all products of the form $\bigwedge_{k<\omega}[(\phi_k)^{x_i}_{x_k}]$.

In order to show that $\mathcal{F}$ is a filter on $\mathfrak{B}_T$, suppose first that $[\phi] \leq [\psi]$ and that $[\phi] \in \mathcal{F}$. Then $T \vdash \phi \rightarrow \psi$, $T \vdash \phi$, and by *modus ponens*, $T \vdash \psi$. Hence, $[\psi] \in \mathcal{F}$.

Secondly, if $[\phi]$ and $[\psi]$ are in $\mathcal{F}$, then $T \vdash \phi$ and $T \vdash \psi$, so $T \vdash \phi \wedge \psi$, since the formula "$\phi \rightarrow (\psi \rightarrow (\phi \wedge \psi))$" is a tautology. It follows that $[\phi] \cap [\psi] = [\phi \wedge \psi] \in \mathcal{F}$.

Now assume that $[(\phi_k)^{x_i}_{x_k}] \in \mathcal{F}$ for all $k < \omega$. Then $T \vdash (\phi_k)^{x_i}_{x_k}$ for all $k < \omega$, and therefore $T \vdash \forall x_i\ \phi$. This establishes that

$$\bigwedge_{k<\omega} [(\phi_k)^{x_i}_{x_k}] = [\forall x_i\ \phi] \in \mathcal{F}.$$

Conversely, if $T \vdash \forall x_i\ \phi$, then, as shown in the proof of Theorem 3.9.3, we have $T \vdash \forall x_i \phi \rightarrow ((\phi_k)^{x_i}_{x_k})$ for all $k < \omega$. Thus, $[(\phi_k)^{x_i}_{x_k}] \in \mathcal{F}$ for all $k < \omega$, establishing that $\mathcal{F}$ preserves products.

Moreover, $\mathcal{F}$ is a proper filter because T is consistent. Then, by Theorem 2.3.6, $\mathcal{F}$ can be extended to an ultrafilter $\mathcal{U}$.⊣

Lemma 3.9.5 *If the ultrafilter $\mathcal{U}$ in the algebra $\mathfrak{B}_T$ preserves all products of the form $\bigwedge_{k<\omega}[(\phi_k)^{x_i}_{x_k}]$, then it also preserves all unions of the form $\bigvee_{k<\omega}[(\phi_k)^{x_i}_{x_k}]$.*

Proof Assume that $\mathcal{U}$ preserves $\bigwedge$. Then, we have the following chain of equivalences:

$$\begin{aligned}
\bigvee_{k<\omega} [(\phi_k)^{x_i}_{x_k}] \in \mathcal{U} &\Leftrightarrow \overline{\bigvee_{k<\omega} [(\phi_k)^{x_i}_{x_k}]} \notin \mathcal{U} \\
&\Leftrightarrow \bigwedge_{k<\omega} \overline{[(\phi_k)^{x_i}_{x_k}]} \notin \mathcal{U} \\
&\Leftrightarrow \overline{[(\phi_k)^{x_i}_{x_k}]} \notin \mathcal{U} \quad \text{for some } k < \omega \\
&\Leftrightarrow [(\phi_k)^{x_i}_{x_k}] \in \mathcal{U} \quad \text{for some } k < \omega. \dashv
\end{aligned}$$

The previous lemmas have now put us in a position to present an unusual proof of the Completeness Theorem.

Theorem 3.9.6 (Tarski's Completeness Theorem) *If Γ is a consistent set of sentences in $\mathcal{L}$, then Γ is satisfiable.*

Proof Let $T = \text{Cn}\ \Gamma$ and let $\mathcal{U}$ be an ultrafilter on $\mathfrak{B}_T$ that preserves products. We will construct a model $\mathfrak{A}$ of T as follows.

First, we define the following equivalence relation among variables of $\mathcal{L}$:

$$x_i \sim x_j \quad \text{if and only if} \quad T \vdash x_i = x_j.$$

The elements of the universe A of $\mathfrak{A}$ will be the equivalence classes of variables modulo $\sim$. The equivalence class of the variable x_i will be denoted as $\mathbf{x}_i$.

If c is a constant symbol of $\mathcal{L}$, since $\vdash (\exists x_1)(c = x_1)$ and $\mathcal{U}$ preserves unions, it follows that $[c = x_i] \in \mathcal{U}$ for some i. Let the class $\mathbf{x}_i$ be the interpretation of c.

If τ is any term in $\mathcal{L}$, let x_i be a variable not appearing in τ. Then $\vdash \exists x_i(\tau = x_i)$. Since $\mathcal{U}$ preserves unions, $[\tau = x_j] \in \mathcal{U}$ for some j. Let $\mathbf{x}_j$ be the interpretation of the term τ.

If P is an n-ary predicate symbol in $\mathcal{L}$, we define the relation R by the condition:

$$(\mathbf{x}_{i_1}, \ldots, \mathbf{x}_{i_n}) \in R \quad \text{if and only if} \quad [Px_{i_1} \cdots x_{i_n}] \in \mathcal{U}.$$

To establish that $\mathfrak{A}$ is a model of T, we will show by induction on formulas in $\mathcal{L}$ that

$$\mathfrak{A} \models \phi[\![\mathbf{x}_1 \cdots \mathbf{x}_n]\!] \quad \text{if and only if} \quad [\phi(x_1 \cdots x_n)] \in \mathcal{U}.$$

Case (i): ϕ is of the form $\tau_1 = \tau_2$, where $\tau_1(x_1 \cdots x_n)$ and $\tau_2(x_1 \cdots x_n)$ are terms in $\mathcal{L}$, we have

$$\mathfrak{A} \models \phi[\![\mathbf{x}_1 \cdots \mathbf{x}_n]\!] \Leftrightarrow \tau_1[\![\mathbf{x}_1 \cdots \mathbf{x}_n]\!] = \tau_2[\![\mathbf{x}_1 \cdots \mathbf{x}_n]\!].$$

That is, if $\mathfrak{A} \models \phi[\![\mathbf{x}_1 \cdots \mathbf{x}_n]\!]$, then the terms have a common interpretation. Let $\mathbf{x}_i$ be the common interpretation of τ_1 and τ_2. We know that $[\tau_1 = x_i]$ and $[\tau_2 = x_i]$ are in $\mathcal{U}$. Hence,

$$[(\tau_1 = x_i) \wedge (\tau_2 = x_i)] = [\tau_1 = x_i] \cap [\tau_2 = x_i] \in \mathcal{U},$$

since $\mathcal{U}$ is a filter. Furthermore, since $\vdash ((\tau_1 = x_i) \wedge (\tau_2 = x_i) \rightarrow (\tau_1 = \tau_2))$, it follows that $[\tau_1 = \tau_2] \in \mathcal{U}$.

Conversely, if $[\tau_1 = \tau_2] \in \mathcal{U}$, let $\mathbf{x}_i = \tau_1[\mathbf{x}_1 \cdots \mathbf{x}_n]$ and $\mathbf{x}_j = \tau_2[\mathbf{x}_1 \cdots \mathbf{x}_n]$. Then,

$$[x_i = \tau_1] \in \mathcal{U} \quad \text{and} \quad [x_j = \tau_2] \in \mathcal{U}.$$

Thus,

$$[(\tau_1 = \tau_2) \wedge (x_i = \tau_1) \wedge (x_j = \tau_2)] \in \mathcal{U}.$$

But since $\vdash ((\tau_1 = \tau_2) \wedge (x_i = \tau_1) \wedge (x_j = \tau_2)) \rightarrow (x_i = x_j))$, we get $[x_i = x_j] \in \mathcal{U}$. This implies $\tau_1 = x_i = x_j = \tau_2$, meaning $\mathfrak{A} \models \tau_1 = \tau_2$.

Case (ii): ϕ is of the form $P\tau_1 \cdots \tau_n$ where R is the interpretation of P, we have

$$\mathfrak{A} \models \phi[\![\mathbf{x}_1 \cdots \mathbf{x}_q]\!] \Leftrightarrow (\tau_1[\mathbf{x}_1 \cdots \mathbf{x}_q], \ldots, \tau_n[\mathbf{x}_1 \cdots \mathbf{x}_q]) \in R.$$

Suppose $\mathfrak{A} \models \phi[\![\mathbf{x}_1 \cdots \mathbf{x}_q]\!]$ and let

$$\mathbf{x}_{i_1} = \tau_1[\mathbf{x}_1 \cdots \mathbf{x}_q], \ldots, \mathbf{x}_{i_n} = \tau_n[\mathbf{x}_1 \cdots \mathbf{x}_q].$$

Then $[x_{i_j} = \tau_j] \in \mathcal{U}$ for $j = 1, \ldots, n$, and

$$(\mathbf{x}_{i_1}, \ldots, \mathbf{x}_{i_n}) \in R.$$

Thus,

$$[Px_{i_1} \cdots x_{i_n}] \in \mathcal{U}.$$

Since $\mathcal{U}$ is a filter,

$$[Px_{i_1} \cdots x_{i_n} \wedge (x_{i_1} = \tau_1) \wedge \cdots \wedge (x_{i_n} = \tau_n)] \in \mathcal{U}.$$

Since

$$\vdash (Px_{i_1} \cdots x_{i_n} \wedge (x_{i_1} = \tau_1) \wedge \cdots \wedge (x_{i_n} = \tau_n) \to P\tau_1 \cdots \tau_n),$$

it follows that $[P\tau_1 \cdots \tau_n] \in \mathcal{U}$.

Conversely, if $[P\tau_1 \cdots \tau_n] \in \mathcal{U}$, let $\mathbf{x}_{i_j} = \tau_j[\mathbf{x}_1 \cdots \mathbf{x}_q]$ for $j = 1, \ldots, n$. Then, by (i),

$$[x_{i_j} = \tau_j] \in \mathcal{U}.$$

Since

$$[Px_{i_1} \cdots x_{i_n}] \in \mathcal{U} \Leftrightarrow (\mathbf{x}_{i_1}, \ldots, \mathbf{x}_{i_n}) \in R \Leftrightarrow (\tau_1[\![\mathbf{x}_1 \cdots \mathbf{x}_q]\!], \ldots, \tau_n[\![\mathbf{x}_1 \cdots \mathbf{x}_q]\!]) \in R,$$

we conclude that $\mathfrak{A} \models P\tau_1 \cdots \tau_n[\![\mathbf{x}_1 \cdots \mathbf{x}_q]\!]$.

Inductive Hypothesis: Suppose, as the inductive hypothesis, that ψ and γ are statements such that

$$\mathfrak{A} \models \psi[\![\mathbf{x}_1 \cdots \mathbf{x}_n]\!] \quad \text{if and only if} \quad [\psi(x_1 \cdots x_n)] \in \mathcal{U}$$

and

$$\mathfrak{A} \models \gamma[\![\mathbf{x}_1 \cdots \mathbf{x}_n]\!] \quad \text{if and only if} \quad [\gamma(x_1 \cdots x_n)] \in \mathcal{U}.$$

Case (iii): ϕ is $\neg\psi$

$$\begin{aligned} \mathfrak{A} \models \phi[\![\mathbf{x}_1 \cdots \mathbf{x}_n]\!] &\Leftrightarrow \mathfrak{A} \not\models \psi[\![\mathbf{x}_1 \cdots \mathbf{x}_n]\!] \\ &\Leftrightarrow [\psi(x_1 \cdots x_n)] \notin \mathcal{U} \\ &\Leftrightarrow \neg[\psi(x_1 \cdots x_n)] \in \mathcal{U} \\ &\Leftrightarrow [\neg\psi(x_1 \cdots x_n)] \in \mathcal{U} \\ &\Leftrightarrow [\phi(x_1 \cdots x_n)] \in \mathcal{U}. \end{aligned}$$

Case (iv): ϕ is $\psi \wedge \gamma$

$$\begin{aligned}\mathfrak{A} \models \phi[\![\mathbf{x}_1 \cdots \mathbf{x}_n]\!] &\Leftrightarrow \mathfrak{A} \models \psi[\![\mathbf{x}_1 \cdots \mathbf{x}_n]\!] \text{ and } \mathfrak{A} \models \gamma[\![\mathbf{x}_1 \cdots \mathbf{x}_n]\!] \\ &\Leftrightarrow [\psi(x_1 \cdots x_n)] \in \mathcal{U} \text{ and } [\gamma(x_1 \cdots x_n)] \in \mathcal{U} \\ &\Leftrightarrow [\psi(x_1 \cdots x_n)] \cap [\gamma(x_1 \cdots x_n)] \in \mathcal{U} \\ &\Leftrightarrow [\psi(x_1 \cdots x_n) \wedge \gamma(x_1 \cdots x_n)] \in \mathcal{U} \\ &\Leftrightarrow [\phi(x_1 \cdots x_n)] \in \mathcal{U}.\end{aligned}$$

Case (v): ϕ is $\forall x_i \psi$ for some $i \leq n$. Note that $\psi(x_1 \cdots x_{i-1} x_k x_{i+1} \cdots x_n)$ is $(\psi_k)^{x_i}_{x_k}$, so we have:

$$\begin{aligned}\mathfrak{A} \models \phi[\![\mathbf{x}_1 \cdots \mathbf{x}_n]\!] &\Leftrightarrow \mathfrak{A} \models \psi[\![\mathbf{x}_1 \cdots \mathbf{x}_{i-1}\mathbf{x}_k\mathbf{x}_{i+1} \cdots \mathbf{x}_n]\!] \quad \forall k < \omega \\ &\Leftrightarrow [\psi(x_1 \cdots x_{i-1} x_k x_{i+1} \cdots x_n)] \in \mathcal{U} \quad \forall k < \omega \\ &\Leftrightarrow [(\psi_k)^{x_i}_{x_k}] \in \mathcal{U} \quad \forall k < \omega \\ &\Leftrightarrow \bigwedge_{k<\omega} [(\psi_k)^{x_i}_{x_k}] \in \mathcal{U} \\ &\Leftrightarrow [\forall x_i\ \psi] \in \mathcal{U} \\ &\Leftrightarrow [\phi(x_1 \cdots x_{i-1} x_i x_{i+1} \cdots x_n)] \in \mathcal{U}. \dashv\end{aligned}$$

Theorem 3.9.7 *A consistent set of formulas* Γ *is satisfiable if and only if* $\Gamma \models \phi$ *implies* $\Gamma \vdash \phi$ *for every set of statements* Γ *and every well-formed formula (wff)* ϕ.

Proof Assume that $\Gamma \models \phi$ implies $\Gamma \vdash \phi$ for every set of statements Γ and every wff ϕ, and assume that Γ is consistent. Then, either $\Gamma \nvdash \phi$ or $\Gamma \nvdash \neg\phi$. If the former holds, then it follows that $\Gamma \not\models \phi$, which means there exists a structure $\langle \mathfrak{A}, s \rangle$ that satisfies Γ but not ϕ. If the latter holds, an analogous argument shows that there exists a structure $\langle \mathfrak{B}, s \rangle$ that satisfies Γ but not ϕ. In either case, Γ is satisfiable.

Now, suppose that the consistency of Γ implies its satisfiability, and that $\Gamma \models \phi$. Assume, for contradiction, that $\Gamma; \neg\phi$ is consistent. Then $\Gamma; \neg\phi$ is satisfiable, meaning there exists a structure $\langle \mathfrak{A}, s \rangle$ that satisfies both Γ and $\neg\phi$. However, the assumption that $\Gamma \models \phi$ implies that this same structure $\langle \mathfrak{A}, s \rangle$ satisfies ϕ, leading to the contradiction that $\mathfrak{A} \models \phi\ [s]$ and $\mathfrak{A} \not\models \phi\ [s]$. This contradiction establishes that $\Gamma; \neg\phi$ is inconsistent and, therefore, $\Gamma \vdash \phi$.$\dashv$

Theorem 3.9.8 (Gödel's Completeness Theorem, 1930) *Let* Γ *be a set of formulas in* $\mathcal{L}$ *and* ϕ *a formula in this language. If* $\Gamma \models \phi$, *then* $\Gamma \vdash \phi$.

Proof This follows immediately from Theorems 3.9.6 and 3.9.7.$\dashv$

Another corollary of Theorem 3.9.6 is the first (downward) Löwenheim-Skolem-Tarski (LST) theorem.

Theorem 3.9.9 (LST Downward Theorem) *Every consistent theory* T *in the language* $\mathcal{L}$ *has a model of at most cardinality* $||\mathcal{L}||$.

Proof In the proof of Theorem 3.9.6, we constructed a model using the variables of $\mathscr{L}$, which form a set of cardinality ω. Thus, the cardinality of this model cannot exceed $||\mathscr{L}|| = \omega \cup \alpha$, where α is the order type of the structures of $\mathscr{L}$.⊣

A result that follows directly from the completeness theorem is the compactness theorem. Recall that a set of statements Γ is finitely satisfiable if and only if every finite subset of Γ is satisfiable.

Theorem 3.9.10 (Compactness Theorem (Maltsev, 1936)) *Let Γ be a set of formulas in L. Γ is satisfiable if and only if Γ is finitely satisfiable. In particular, a set of statements Γ has a model if and only if every finite subset of Γ has a model.*

Proof It is obvious that if Γ is satisfiable, then it is finitely satisfiable. Suppose that Γ is finitely satisfiable but, for contradiction, that Γ is not satisfiable. Then, by Theorem 3.9.8, Γ is inconsistent, meaning there exists a formula ϕ such that $\Gamma \vdash \phi \wedge \neg\phi$. This implies that there is a derivation of $\phi \wedge \neg\phi$ from Γ, i.e., a finite sequence $\langle \psi_0, \ldots, \psi_n \rangle$ of wffs such that $\psi_n = \phi \wedge \neg\phi$ and, for each $i \leq n$, either ψ_i is a logical axiom, belongs to Γ, or is obtained from previous formulas in the sequence by *modus ponens*. Consequently, there exists a finite subset Γ_0 of Γ such that $\Gamma_0 \vdash \phi \wedge \neg\phi$, meaning that Γ_0 is inconsistent. By the Soundness Theorem, $\Gamma_0 \vDash \phi \wedge \neg\phi$, which means that Γ_0 is not satisfiable (as it would otherwise satisfy the contradictory formula $\phi \wedge \neg\phi$). This contradiction shows that Γ is satisfiable. ⊣

The Compactness Theorem implies that a theory that has arbitrarily large finite models must also have an infinite model. This leads to the following theorem.

Theorem 3.9.11 *If a theory T has arbitrarily large finite models, then it has an infinite model.*

Proof Let T be a theory in the language $\mathscr{L}$ that has arbitrarily large finite models, and let $\tilde{\mathscr{L}}$ be the simple expansion $\mathscr{L} \cup \{c_n \mid n \in \omega\}$. Consider the set Δ of formulas in $\tilde{\mathscr{L}}$ defined by

$$\Delta = T \cup \{\neg(c_n = c_m) \mid n < m < \omega\}.$$

Any finite subset Δ' of Δ involves at most the constants $c_1, \ldots, c_m$ for some m. Let $\mathfrak{A}$ be a model of T with at least $m + 1$ elements, and let $a_1, \ldots, a_m$ be a list of $m + 1$ distinct elements of $\mathfrak{A}$. The model $(\mathfrak{A}, a_1, \ldots, a_m)$ for the finite expansion $\mathscr{L}' = \mathscr{L} \cup \{c_1, \ldots, c_m\}$ of $\mathscr{L}$ is a model of Δ' because the elements $a_1, \ldots, a_m$ are distinct and thus satisfy $\{\neg(c_k = c_l) \mid k, l \leq m\}$. Since Δ' is an arbitrary finite subset of Δ, the Compactness Theorem 3.9.10 implies that Δ is satisfiable by some model $\tilde{\mathfrak{B}} = (\mathfrak{B}, \{a_n\}_{n<\omega})$. The reduction of $\tilde{\mathfrak{B}}$ to $\mathscr{L}$ gives the model $\mathfrak{B}$, which is infinite.⊣

Another important result, which clearly explains why there exist non-standard models of familiar theories such as arithmetic, follows from this.

Theorem 3.9.12 (LST Upward Theorem) *If a theory T has an infinite model of cardinality λ, then it has models of any cardinality $\alpha \geq \lambda$.*

Proof The proof proceeds by constructing a theory T' obtained by expanding the language $\mathcal{L}$ of T with a set of distinct constants $\{c_\xi\}_{\xi<\alpha}$ and adding to T the family Γ of all statements of the form "$c_\xi \neq c_\zeta$" for $\xi < \zeta < \alpha$. If $\Delta = T \cup \Gamma$, any finite subset Δ_Γ of Δ contains only finitely many instances of constants from $\{c_\xi\}_{\xi<\alpha}$, say $c_1, \ldots, c_m$ for some $m < \omega$. If $\mathfrak{A}$ is a model of T of cardinality λ, the expanded model $\langle \mathfrak{A}, a_1, \ldots, a_m \rangle$ satisfies Δ_0. This means that Δ is finitely satisfiable, so it has a model $\mathfrak{B}$. The cardinality of $\mathfrak{B}$ is at least α.⊣

Gottfried Wilhelm Leibniz (1646–1716), the famous German philosopher who developed an interesting philosophical and theological system, was also a prominent mathematician who introduced the concept of the infinitesimal number, a number greater than zero in absolute value but smaller than any positive real number. He developed differential and integral calculus based on this concept.

Leibniz argued that infinitesimal numbers (distinct from zero) were "fictitious" numbers that should be treated as ordinary numbers because they followed "the same laws" as the latter. However, we now know this is not entirely accurate: for example, infinitesimal numbers do not satisfy Archimedes' axiom, which states that for any two positive numbers x and y with $x < y$, no matter how small x is, there exists a positive integer n such that $nx > y$. Since nx remains infinitesimal for all n when x is infinitesimal, it follows that $nx < y$ for any positive real number y.

The question of which laws infinitesimals do and do not obey remained unanswered until the early 1960s, when Abraham Robinson (1918–1974), using methods from model theory, founded the field now known as 'nonstandard analysis'. The first result that demonstrated that Leibniz's intuitions could be rigorously defined is the following theorem.

Theorem 3.9.13 *There exists a nonstandard model of the theory of the field of real numbers. In this model, there exist infinitesimal numbers distinct from zero.*

Proof Let $\mathfrak{R} = \langle \mathbb{R}, +, \cdot, 0, 1, \leq \rangle$ be the ordered field of real numbers, and let $\mathcal{L} = \{+, \cdot, 0, 1, \leq\}$ be the language of this structure. Consider the expansion $\mathcal{L}' = \mathcal{L} \cup \{c_r\}_{r\in\mathbb{R}}$, where c_r denotes the real number r. Let Γ be the set of all formulas of the form "$0 < x < |c_r|$," and let Δ be the union of the theory of $\mathfrak{R}$ with Γ. Clearly, any finite subset of Δ is satisfiable, so by the Compactness Theorem 3.9.10, Δ itself is satisfiable by some model $\mathfrak{S}'$. The reduct $\mathfrak{S}$ of $\mathfrak{S}'$ to $\mathcal{L}$ is an ordered field in which there exists a positive number smaller than any positive real number–that is, a nonzero infinitesimal.⊣

It is possible to construct a logical and set-theoretic framework in which it is feasible to practice nonstandard analysis, i.e., a version of mathematical analysis using infinitesimals. The next chapter will be dedicated to establishing the foundations of this framework using the concept of ultrafilter.

In the proof of Theorem 3.9.13, we considered an expansion $\mathcal{L}_\mathbb{R}$ of the language $\mathcal{L}$ of the structure $\mathfrak{R}$ by introducing a constant c_r for each real number r. This also gave us an expanded model $\mathfrak{R}_\mathbb{R}$. We then introduced the set Γ of all formulas of the form "$0 < x < |c_r|$". If we had instead introduced the set Δ_R of all atomic statements and negations of atomic statements in $\mathcal{L}_{\mathbb{R}}$ that are true in the model $\mathfrak{R}_{\mathbb{R}}$, we would have obtained the diagram of $\mathfrak{R}$.

In general, the *diagram* of $\mathfrak{A}$, denoted Δ_A, is the set of all atomic statements and negations of atomic statements in $\mathcal{L}_A$ that are true in the model $\mathfrak{A}_A$. The *elementary diagram* of $\mathfrak{A}$ is the theory $\text{Th}(\mathfrak{A}_A)$, consisting of all statements in $\mathcal{L}_A$ that are true in $\mathfrak{A}_A = (\mathfrak{A}, a)_{a\in A}$.

A *type* in the variables $x_1, \ldots, x_n$ is a maximally consistent set of formulas in these variables. Given any model $\mathfrak{A}$ and an n-tuple $a_1, \ldots, a_n \in A$, the set $\Gamma(x_\Delta, \ldots, x_n)$ of all formulas $\gamma(x_1, \ldots, x_n)$ satisfied by $a_1, \ldots, a_n$ is a type, and in fact, it is the only type realized by $a_1, \ldots, a_n$. It is called the *type of* $a_1, \ldots, a_n$ *in* $\mathfrak{A}$.

Let Σ be a set of formulas in the variables $x_1, \ldots, x_n$, and let $\mathfrak{A}$ be a model for $\mathcal{L}$. We say that $\mathfrak{A}$ *realizes* Σ if and only if some n-tuple of elements from A satisfies Σ in $\mathfrak{A}$. $\mathfrak{A}$ *omits* Σ if and only if $\mathfrak{A}$ does not realize Σ. The phrase 'Σ is satisfiable in $\mathfrak{A}$' has exactly the same meaning as "$\mathfrak{A}$ realizes Σ". Σ is *consistent* if and only if Σ is satisfiable in some model.

In particular, if Σ is a type $\Gamma(x_1, \ldots, x_n)$, we say that $\mathfrak{A}$ realizes the type $\Gamma(x_1, \ldots, x_n)$.

If $\mathfrak{A}$ and $\mathfrak{B}$ are models of the language $\mathcal{L}$, we say that $\mathfrak{A}$ and $\mathfrak{B}$ are *elementarily equivalent*, denoted $\mathfrak{A} \equiv \mathfrak{B}$, if and only if they satisfy the same statements. Clearly, $\mathfrak{A} \cong \mathfrak{B}$ (isomorphism) implies $\mathfrak{A} \equiv \mathfrak{B}$, but the converse is false unless the models are finite.

If $\mathfrak{A}$ and $\mathfrak{B}$ are models of $\mathcal{L}$, we say that $\mathfrak{A}$ is *elementarily embedded* in $\mathfrak{B}$, or that $\mathfrak{B}$ is an *elementary extension* of $\mathfrak{A}$, denoted $\mathfrak{A} \precsim \mathfrak{B}$, if $\mathfrak{A} \subseteq \mathfrak{B}$, $\mathfrak{A}$ is embedded in $\mathfrak{B}$, and $\mathfrak{A} \equiv \mathfrak{B}$. If $\mathfrak{A}$ is not isomorphic to $\mathfrak{B}$, we write $\mathfrak{A} \prec \mathfrak{B}$.

Let $\mathfrak{A}$ and $\mathfrak{B}$ be models of the language $\mathcal{L}$. We say that $\mathfrak{A}$ is *elementarily embedded* in $\mathfrak{B}$, or that $\mathfrak{B}$ is an *elementary extension* of $\mathfrak{A}$, denoted $\mathfrak{A} \preceq \mathfrak{B}$, if $\mathfrak{A} \subseteq \mathfrak{B}$ (i.e., $\mathfrak{A}$ is a substructure of $\mathfrak{B}$), $\mathfrak{A}$ is embedded in $\mathfrak{B}$, and $\mathfrak{A}$ and $\mathfrak{B}$ are *elementarily equivalent*, meaning they satisfy the same first-order sentences. If $\mathfrak{A}$ is not isomorphic to $\mathfrak{B}$, we write $\mathfrak{A} \prec \mathfrak{B}$.

Theorem 3.9.14 *Let $\mathfrak{A}$ and $\mathfrak{B}$ be models of language $\mathcal{L}$ and let $f : A \to B$. Then the following three statements are equivalent:*

(1) *$\mathfrak{A}$ can be isomorphically embedded into $\mathfrak{B}$.*
(2) *$\mathfrak{B}$ is isomorphic to an extension of $\mathfrak{A}$.*
(3) *$\mathfrak{B}$ can be expanded to a model of the diagram of $\mathfrak{A}$.*

The proof is left as an exercise for the reader.

A *chain of structures* is a sequence $\langle \mathfrak{A}_\lambda : \lambda < \alpha \rangle$ such that $\mathfrak{A}_\lambda \subseteq \mathfrak{A}_\mu$ whenever $\lambda \leqslant \mu < \alpha$. If, for each $\lambda < \alpha$,

$$\mathfrak{A}_\lambda = \langle A_\lambda, R_\xi^\lambda, G_\zeta^\lambda \rangle,$$

then the *union* of the chain $\langle \mathfrak{A}_\lambda : \lambda < \alpha \rangle$ is the structure:

$$\langle \bigcup_{\lambda<\alpha} A_\lambda, \bigcup_{\lambda<\alpha} R_\xi^\lambda, G_\zeta^\lambda \rangle.$$

This is an extension of each of the structures $\mathfrak{A}_\lambda$, but it does not necessarily retain all their elementary properties. For example, consider the chain $\mathfrak{A}_n = \langle A_n, < \rangle$, where A_n is the initial segment $\{0, 1, 2, \ldots, n\}$ of the nonnegative integers. The statement asserting that $<$ has a maximum element is true in each $\mathfrak{A}_n$ but false in the union.

However, if the chain is *elementary*—that is, if $\mathfrak{A}_\lambda \prec \mathfrak{A}_\mu$ whenever $\lambda < \mu < \alpha$—then we have the following result.

Theorem 3.9.15 *The union of an elementary chain is an elementary extension of each structure in the chain.*

Proof Let $\mathfrak{A}$ be the union of the chain $\langle \mathfrak{A}_\lambda : \lambda < \alpha \rangle$. We must show that, for any formula $\phi(x_1, \ldots, x_n)$, any $\xi < \alpha$, and any elements $a_1, \ldots, a_n$ in A_ξ,

$$\mathfrak{A} \models \phi[\![a_1 \cdots a_n]\!] \quad \text{if and only if} \quad \mathfrak{A}_\xi \models \phi[\![a_1 \cdots a_n]\!]. \tag{*}$$

We proceed by induction on the complexity of formulas.

The base case is atomic formulas, where (*) is trivial.

If (*) holds for ϕ and ψ, then it also holds for $\neg\phi$ and $\phi \wedge \psi$ (see Exercise 3.5).

Now, assume (*) holds for the formula $\psi(x_1, \ldots, x_n)$. We must prove that it holds for the formula

$$\phi(x_1, \ldots, x_{n-1}) = \exists x_n \, \psi(x_1, \ldots, x_n).$$

If $a_1, \ldots, a_{n-1} \in A_\xi$ and $\mathfrak{A}_\xi \models \phi[\![a_1, \ldots, a_{n-1}]\!]$, then there exists $a_n \in A_\xi$ such that

$$\mathfrak{A}_\xi \models \psi[\![a_1, \ldots, a_n]\!].$$

By the inductive hypothesis, $\mathfrak{A} \models \psi[\![a_1, \ldots, a_n]\!]$, so

$$\mathfrak{A} \models \phi[\![a_1, \ldots, a_{n-1}]\!].$$

Conversely, suppose $\mathfrak{A} \models \phi[\![a_1, \ldots, a_{n-1}]\!]$. Then, there exists some $a \in A$ such that

$$\mathfrak{A} \models \psi[\![a_1, \ldots, a_{n-1}, a]\!].$$

Since $a \in \bigcup A_\xi$, there exists $\eta < \alpha$ such that $a \in A_\eta$. Let ζ be the larger of ξ and η. Then,

$$a_1, \ldots, a_{n-1}, a \in A_\zeta.$$

By the inductive hypothesis, $\mathfrak{A}_\zeta \models \psi[\![a_1, \ldots, a_{n-1}, a]\!]$, so

$$\mathfrak{A}_\zeta \models \phi[\![a_1, \ldots, a_{n-1}]\!].$$

Since $\xi \leqslant \zeta$ and $\mathfrak{A}_\xi \prec \mathfrak{A}_\zeta$, it follows that

$$\mathfrak{A}_\xi \models \phi[\![a_1, \ldots, a_{n-1}]\!].$$

$\dashv$

Elementary chains are important and useful in the construction of models, especially saturated nonstandard universes.

Exercises

3.1 Prove Theorem 3.8.5.

3.2 Show that the structure $\mathfrak{B}_T$ defined in Theorem 3.9.1 is a Boolean algebra.

3.3 Prove Theorem 3.9.14.

3.4 Prove that if $\mathfrak{A}$ and $\mathfrak{B}$ are finite, then $\mathfrak{A} \cong \mathfrak{B}$ implies $\mathfrak{A} \equiv \mathfrak{B}$.

3.5 Prove Theorem 3.9.15 for atomic formulas, negations, and conjunctions.

References

Boolos, G. S., Burgess, J. P., & Jeffrey, R. (2002). *Computability and logic*. Cambridge University Press.

Maltsev, A. (1936). Untersuchungen aus dem Gebiete der mathema-tischen Logik. *Matematicheskii Sbornik, 1*, 323–336.

Chapter 4
Ultraproducts

Abstract Chapter 4 develops the ultraproduct construction and proceeds to elaborate the theory of ultraproducts up to the natural embedding. It places particular emphasis on the construction of saturated models, since saturated universes have convenient properties that enable the proper development of the topological spaces required by nonstandard economics. After constructing nonstandard universes in full detail and proving that they possess the desired properties relevant to the development of nonstandard analysis, the chapter thoroughly explores the nature and existence of saturated models and proves the equivalence of saturation with the existence of concurrent or finitely satisfiable binary relations, as well as families of sets with the finite intersection property.

Keywords Ultraproducts · Nonstandard universes · Saturated models

4.1 Preamble

Broadly speaking, the ultraproduct of a family $\{\mathfrak{A}_i\}_{i\in I}$ of structures of the same similarity type is a system $\mathfrak{A}$ whose universe is a partition of the Cartesian product $\prod_{i\in I} A_i$ modulo a certain equivalence relation. This partition is called the reduced product of the sets A_i, and its relations and functions are defined based on the relations and functions of the factor structures $\mathfrak{A}_i$. The interesting feature of the ultraproduct is that it inherits properties from the structures from which it was constructed.

Recall that $\prod_{i\in I} A_i$ (Definition 1.8.4) is the set of functions

$$\left\{ f \mid f : I \to \bigcup_{i\in I} A_i,\ f(i) \in A_i \right\}.$$

The reduced product of the sets A_i is defined via function equivalence modulo a filter $\mathscr{F}$. The idea is that functions f and g will be equivalent modulo $\mathscr{F}$ if and only if the set of indices i such that $f(i) = g(i)$ belongs to $\mathscr{F}$. We now formally

A. García de la Sienra, *The Logical Foundations of Nonstandard Analysis*, Synthese Library 530, https://doi.org/10.1007/978-3-032-05677-1_4

introduce this concept. From this point onward, the expression '(ultra)filter on I' is an abbreviation for '(ultra)filter on the Boolean algebra of all subsets of I'.

Definition 4.1.1 Let $\mathcal{F}$ be a filter on I. Given two functions $f, g \in \prod_{i \in I} A_i$, we say that f and g are *equivalent modulo* $\mathcal{F}$, or $\mathcal{F}$*-equivalent*, if and only if

$$\{i \in I \mid f(i) = g(i)\} \in \mathcal{F}.$$

We will express by $f \approx g$ that f and g are equivalent modulo $\mathcal{F}$. We will show that $\approx$ is an equivalence relation.

Theorem 4.1.2 *Let $\mathcal{F}$ be a filter, and let $\approx$ be the equivalence relation modulo $\mathcal{F}$. Then $\approx$ is indeed an equivalence relation.*

Proof Since, obviously,

$$\{i \in I \mid f(i) = f(i)\} = I \in \mathcal{F},$$

it follows immediately that $\approx$ is reflexive.

Suppose that $f \approx g$, meaning that

$$X = \{i \in I \mid f(i) = g(i)\} \in \mathcal{F}.$$

Clearly,

$$\{i \in I \mid g(i) = f(i)\} = X \in \mathcal{F},$$

which shows that $\approx$ is symmetric.

Finally, suppose that $f \approx g$ and $g \approx h$, and let

$$X = \{i \in I \mid f(i) = g(i)\} \quad \text{and} \quad Y = \{i \in I \mid g(i) = h(i)\}.$$

By hypothesis, $X, Y \in \mathcal{F}$, so $X \cap Y \in \mathcal{F}$. But, since

$$X \cap Y \subseteq \{i \in I \mid f(i) = h(i)\} = Z,$$

it follows that $Z \in \mathcal{F}$, proving that $\approx$ is transitive. $\dashv$

The equivalence class of a function f modulo $\mathcal{F}$, denoted $[f]$ or $\boldsymbol{f}$, always refers to the filter $\mathcal{F}$, which will be clear from the context. The quotient set $\prod_{i \in I} A_i / \approx$ is of particular interest and has a special name, introduced in the following definition.

Definición 4.1.3 For any filter $\mathcal{F}$, the *reduced Cartesian product* of the family of sets $\{A_i\}_{i \in I}$ modulo $\mathcal{F}$ is the quotient set $\prod_{i \in I} A_i / \approx$, denoted $\prod_{\mathcal{F}} A_i$. If $A_i = A$ for every $i \in I$, this reduced product is called the *reduced Cartesian power* and is denoted $\prod_{\mathcal{F}} A$. If $\mathcal{F}$ is an ultrafilter, the reduced power is called the *ultrapower* of A.

The *natural embedding* of A into the ultrapower $\prod_{\mathcal{F}} A$ is the function that assigns to each element $a \in A$ the equivalence class of the constant function $a = f(i)$ for all $i \in I$.

4.2 Ultraproduct of Models

If τ is a term, $\tau(x_1 \cdots x_n)$ means that the variables in τ belong to the set $\{x_1, \ldots, x_n\}$. Similarly, if ϕ is a formula, $\phi(x_1 \cdots x_n)$ means that its variables belong to the set $\{x_1, \ldots, x_n\}$, which may be empty. The result of substituting the variables $x_1, \ldots, x_n$ with elements $a_1, \ldots, a_n$ in the term $\tau(x_1 \ldots x_n)$ will be denoted $\tau[\![a_1 \ldots a_n]\!]$. Similarly, for formulas, we denote $\phi[\![a_1 \ldots a_n]\!]$.

Note that, for any $i \in I$ and function $f \in \prod_{i \in I} A_i$, we have $f(i) \in A_i$. Therefore, if F is an n-ary function term in the language $\mathcal{L}$ corresponding to the similarity type of the structures $\mathfrak{A}_i$, and G_i is its interpretation in structure $\mathfrak{A}_i$—i.e., it is a function $G_i : A_i^n \to A_i$—then G_i is defined for any n-tuple of the form

$$(f^1(i), \ldots, f^n(i))$$

and, moreover, $G_i(f^1(i), \ldots, f^n(i)) \in A_i$. In fact, for each $a \in A_i$, there exists (at least) one function $f \in \prod_{i \in I} A_i$ such that $a = f(i)$, since $\prod_{i \in I} A_i$ contains all possible functions from I to A_i. This implies that every n-tuple $(a_1, \ldots, a_n)$ of elements in A_i can be expressed in the form $(f^1(i), \ldots, f^n(i))$ for suitable functions $f^1, \ldots, f^n$.

Therefore, for functions $f^1, \ldots, f^n \in \prod_{i \in I} A_i$, the symbol

$$\langle G_i(f^1(i), \ldots, f^n(i)) \mid i \in I \rangle$$

denotes a sequence $\langle a_i \mid i \in I \rangle$, where $a_i = G_i(f^1(i), \ldots, f^n(i)) \in A_i$. This sequence is nothing more than a choice function—that is, the image of a function $f \in \prod_{i \in I} A_i$ with

$$f(i) = G_i(f^1(i), \ldots, f^n(i)).$$

We now explain how to construct the ultraproduct of structures $\{\mathfrak{A}_i\}_{i \in I}$ using the interpretation of the first-order language $\mathcal{L}$ common to all of them.

Definition 4.2.1 Let $\mathcal{F}$ be a proper filter on I, and for each $i \in I$, let $\mathfrak{A}_i$ be a model of $\mathcal{L}$. The *reduced product of models* $\mathfrak{A}_i$ (or *reduced power* if $\mathfrak{A}_i = \mathfrak{A}$ for all $i \in I$), denoted $\prod_{\mathcal{F}} \mathfrak{A}_i$ ($\prod_{\mathcal{F}} \mathfrak{A}$ if $\mathfrak{A}_i = \mathfrak{A}$ for every $i \in I$), is the model of $\mathcal{L}$ defined as follows:

(1) The universe of $\prod_{\mathcal{F}} \mathfrak{A}_i$ is the reduced Cartesian product $\prod_{\mathcal{F}} A_i$, where A_i is the universe of $\mathfrak{A}_i$.

(2) For each n-ary relation symbol P in $\mathcal{L}$, the interpretation of P in $\prod_{\mathcal{F}} \mathfrak{A}_i$ is the relation R such that

$$(f^1, \ldots, f^n) \in R \Leftrightarrow \{i \in I \mid \mathfrak{A}_i \models P(f^1(i), \ldots, f^n(i))\} \in \mathcal{F}.$$

(3) For each n-ary function symbol F in $\mathcal{L}$, its interpretation in $\prod_{\mathcal{F}} \mathfrak{A}_i$ is the function G defined as

$$G(f^1, \ldots, f^n) = [\langle G_i(f^1(i), \ldots, f^n(i)) \mid i \in I \rangle],$$

where G_i is the interpretation of F in $\mathfrak{A}_i$.

(4) A constant c in $\mathcal{L}$ is interpreted as the equivalence class

$$f = [\langle a_i \mid i \in I \rangle],$$

where a_i is the interpretation of c in $\mathfrak{A}_i$.

If $\mathcal{F}$ is an ultrafilter, the reduced product is called the *ultraproduct*, and the reduced power is called the *ultrapower*.

The relation R constructed in this way is never empty because the relations R_i are never empty. Moreover, for each $k = 1, \ldots, n$ and $i \in I$, if $(a_i^1, \ldots, a_i^n) \in R_i$, there exists a function $f^k \in \prod_{i \in I} A_i$ such that $a_i^k = f^k(i)$, as we observed earlier. Thus, we can find functions $f^1, \ldots, f^n$ such that, for each $i \in I$,

$$(f^1(i), \ldots, f^n(i)) \in R_i,$$

which implies that

$$\{i \in I \mid \mathfrak{A}_i \models P(f^1(i), \ldots, f^n(i))\} = I \in \mathcal{F},$$

and therefore $(f^1, \ldots, f^n) \in R$. The elements of R are precisely ordered n-tuples of equivalence classes modulo $\mathcal{F}$ of functions that agree on some element of the filter.

For each index set I, the filter $\mathcal{F} = \{I\}$ is the trivial filter. A particular case of the previous definition is the reduced product of models when the filter $\mathcal{F}$ is the trivial filter. In this case, if P, F, and c are relation, function, and constant symbols of $\mathcal{L}$, respectively, the interpretation of P in $\prod \mathfrak{A}_i$ is the relation R such that

$$(f^1, \ldots, f^n) \in R \quad \Leftrightarrow \quad \{i \in I \mid P(f^1(i), \ldots, f^n(i))\} = I.$$

The interpretations of F and c remain the same as in the general case. Thus, we can define the *direct product of models* as the reduced product in the special case where $\mathcal{F} = \{I\}$.

It is easy to see that $G(f^1, \ldots, f^n)$ is well-defined, in the sense that it does not depend on the chosen representatives of the equivalence classes. Indeed, if $f^1 \approx g^1, \ldots, f^n \approx g^n$, then

$$G(\boldsymbol{f}^1, \ldots, \boldsymbol{f}^n) = G(\boldsymbol{g}^1, \ldots, \boldsymbol{g}^n),$$

since the equivalence class $\boldsymbol{f}^k$ is, by hypothesis, identical to $\boldsymbol{g}^k$ for all $k = 1, \ldots, n$. This is expressed by saying that $\approx$ is a congruence relation.

Theorem of Expansion 4.2.2 *Let $\mathcal{L}'$ be an expansion of $\mathcal{L}$ and let I be a nonempty set. For each $i \in I$, let $\mathfrak{A}_i$ be a model for $\mathcal{L}$ and let $\mathfrak{A}'_i$ be an expansion of $\mathfrak{A}_i$ to $\mathcal{L}'$. Let $\mathcal{F}$ be a filter on I. Then, the reduced product $\prod_{\mathcal{F}} \mathfrak{A}'_i$ is an expansion of the reduced product $\prod_{\mathcal{F}} \mathfrak{A}_i$ for $\mathcal{L}'$.*

Proof For each $i \in I$, the models $\mathfrak{A}_i$ and $\mathfrak{A}'_i$ have the same universe, i.e., $A_i = A'_i$, so we have

$$\prod_{\mathcal{F}} A_i = \prod_{\mathcal{F}} A'_i.$$

Since $\mathfrak{A}'_i$ is an expansion of $\mathfrak{A}_i$, each symbol of $\mathcal{L}$ has the same interpretation in $\mathfrak{A}'_i$ as in $\mathfrak{A}_i$. According to the definition of reduced product of models, the interpretation of a symbol of $\mathcal{L}$ in $\prod_{\mathcal{F}} \mathfrak{A}_i$ depends only on its interpretation in the models $\mathfrak{A}_i$, the universe sets, and the filter $\mathcal{F}$. Consequently, each symbol of $\mathcal{L}$ has the same interpretation in $\prod_{\mathcal{F}} \mathfrak{A}'_i$ as in $\prod_{\mathcal{F}} \mathfrak{A}_i$, which implies that $\prod_{\mathcal{F}} \mathfrak{A}'_i$ is an expansion of $\prod_{\mathcal{F}} \mathfrak{A}_i$.$\dashv$

From now on, if τ is a term, the symbol '$\tau^{\mathfrak{A}}$' will denote the interpretation of the symbol τ in the model $\mathfrak{A}$.

Theorem 4.2.3 (Fundamental Theorem of Ultraproducts) *Let $\mathfrak{A}$ be the ultraproduct $\prod_{\mathcal{U}} \mathfrak{A}_i$ modulo the ultrafilter $\mathcal{U}$. Then:*

(1) *For each term $\tau(x_1, \ldots, x_n)$ in $\mathcal{L}$ and elements $\boldsymbol{f}^1, \ldots, \boldsymbol{f}^n \in \prod_{i \in I} A_i$, we have*

$$\tau^{\mathfrak{A}}[\![\boldsymbol{f}^1 \cdots \boldsymbol{f}^n]\!] = [\langle \tau^{\mathfrak{A}_i}[\![f^1(i) \cdots f^n(i)]\!] \mid i \in I \rangle].$$

(2) (Łoś's Theorem) *Given a formula $\phi(x_1, \ldots, x_n)$ in $\mathcal{L}$ and elements $\boldsymbol{f}^1, \ldots, \boldsymbol{f}^n \in A$, we have*

$$\mathfrak{A} \models \phi[\![\boldsymbol{f}^1 \cdots \boldsymbol{f}^n]\!] \Leftrightarrow \{i \in I \mid \mathfrak{A}_i \models \phi[\![f^1(i) \cdots f^n(i)]\!]\} \in \mathcal{U}.$$

(3) *For any sentence ϕ in $\mathcal{L}$,*

$$\mathfrak{A} \models \phi \Leftrightarrow \{i \in I \mid \mathfrak{A}_i \models \phi\} \in \mathcal{U}.$$

Proof The proof proceeds by induction on the complexity of the formulas.

First, if τ is an individual constant and $a_i = \tau^{\mathfrak{A}_i}$ is its interpretation in $\mathfrak{A}_i$, the interpretation of τ in $\mathfrak{A}$ is a function f such that

$$\tau^{\mathfrak{A}} = \boldsymbol{f} = [\langle \tau^{\mathfrak{A}_i} \mid i \in I \rangle].$$

If τ is a variable x, then for $n = 1$, we have

$$\tau[\![\boldsymbol{f}^1 \cdots \boldsymbol{f}^n]\!] = x[\![\boldsymbol{f}]\!] = \boldsymbol{f},$$

where $\boldsymbol{f} = \boldsymbol{f}^1$. Since $\boldsymbol{f}$ is an element of $\prod_{\mathscr{F}} A_i$, if we expand $\mathscr{L}$ with a constant c denoting a (i.e., $\boldsymbol{f}$), we obtain

$$\tau^{\mathfrak{A}} = [\langle a_i \mid i \in I \rangle] = [\langle \tau^{\mathfrak{A}_i} \mid i \in I \rangle] = [\langle x[\![a_i]\!] | \, i \in I \rangle] = [\langle \tau[\![a_i]\!] | \, i \in I \rangle],$$

where $a_i = f(i)$ is the interpretation of c in $\mathfrak{A}_i$.

Now, suppose

$$\tau(x_1, \ldots, x_n) = F(\tau_1(x_1, \ldots, x_n), \ldots, \tau_m(x_1, \ldots, x_n)),$$

where F is a function symbol in $\mathscr{L}$ and the terms $\tau_1, \ldots, \tau_m$ satisfy (1). By the definition of term interpretation,

$$\tau^{\mathfrak{A}}[\![\boldsymbol{f}^1 \cdots \boldsymbol{f}^n]\!] = G(\tau_1^{\mathfrak{A}}[\![\boldsymbol{f}^1 \cdots \boldsymbol{f}^n]\!], \ldots, \tau_m^{\mathfrak{A}}[\![\boldsymbol{f}^1 \cdots \boldsymbol{f}^n]\!]),$$

where G is the interpretation of F in $\mathfrak{A}$. By (1) for $\tau_1, \ldots, \tau_m$, we have

$$\tau_k^{\mathfrak{A}}[\![\boldsymbol{f}^1 \cdots \boldsymbol{f}^n]\!] = \boldsymbol{g}^k$$

with

$$\boldsymbol{g}^k = [\langle \tau_k^{\mathfrak{A}_i}[\![f^1(i) \cdots f^n(i)]\!] | \, i \in I \rangle].$$

Using the definition of reduced product of models,

$$G(\boldsymbol{g}^1, \ldots, \boldsymbol{g}^m) = [\langle G_i(g^1(i), \ldots, g^m(i)) \mid i \in I \rangle].$$

By the definition of term interpretation,

$$\tau^{\mathfrak{A}_i}[\![f^1(i), \ldots, f^n(i)]\!] = G_i(g^1(i), \ldots, g^m(i)).$$

Combining these results,

$$\tau^{\mathfrak{A}}[\![\boldsymbol{f}^1 \cdots \boldsymbol{f}^n]\!] = G(\boldsymbol{g}^1 \cdots \boldsymbol{g}^m) = [\langle \tau_k^{\mathfrak{A}_i}[\![f^1(i) \cdots f^n(i)]\!] | \, i \in I \rangle].$$

Thus, $\tau(x_1, \ldots, x_n)$ satisfies (1).

For the atomic formula $\phi(x_1, \ldots, x_n) = \tau_1 = \tau_2$, using (1),

$$\tau_1^{\mathfrak{A}}[\![f^1 \cdots f^n]\!] = \tau_2^{\mathfrak{A}}[\![f^1 \cdots f^n]\!]$$

is equivalent to

$$[\langle \tau_1^{\mathfrak{A}_i}[\![f^1(i) \cdots f^n(i)]\!] \mid i \in I \rangle] = [\langle \tau_2^{\mathfrak{A}_i}[\![f^1(i) \cdots f^n(i)]\!] \mid i \in I \rangle].$$

This is equivalent to

$$\{ i \in I \mid \mathfrak{A}_i \models \phi[\![f^1(i) \cdots f^n(i)]\!] \} \in \mathcal{U}.$$

Similarly, for the atomic formula $\phi(x_1, \ldots, x_n) = P(\tau_1, \ldots, \tau_n)$, using (1),

$$\mathfrak{A} \models \phi[\![f^1 \cdots f^n]\!] \leftrightarrow \{ i \in I \mid \mathfrak{A}_i \models P(f^1(i), \ldots, f^n(i)) \} \in \mathcal{U}.$$

By induction, the result extends to logical connectives and quantifiers. Thus, (2) holds for all formulas. Finally, (3) follows directly from (2).⊣

In the proof of point (2) of the previous theorem, in the case where ϕ is of the form $\neg\psi$, we can see why it is necessary for $\mathcal{U}$ to be an ultrafilter: the assertion is false for reduced products over filters that are not ultrafilters because we need to ensure that if a set does not belong to the filter, then its complement does. This is, in fact, the only point where $\mathcal{U}$ needs to be an ultrafilter.

These are the first results that impose conditions on the elements involved in the construction. As previously observed, Theorem 4.2.3 holds only for ultraproducts. This is not the only necessary restriction, as we will see later.

Let $\mathfrak{A}$ and $\mathfrak{B}$ be models of a language $\mathcal{L}$. We recall that $\mathfrak{A}$ and $\mathfrak{B}$ are elementarily equivalent, written $\mathfrak{A} \equiv \mathfrak{B}$, if they satisfy the same sentences. A function $\varphi : A \to B$ is called an elementary embedding of $\mathfrak{A}$ into $\mathfrak{B}$, written $\varphi : \mathfrak{A} \prec \mathfrak{B}$, if, for every formula $\phi(x_1 \cdots x_n)$ in $\mathcal{L}$ and every sequence $a_1, \ldots, a_n \in A$, we have

$$\mathfrak{A} \models \phi[\![a_1 \cdots a_n]\!] \quad \leftrightarrow \quad \mathfrak{B} \models \phi[\![\varphi(a_1) \cdots \varphi(a_n)]\!].$$

Recall that if there exists an elementary embedding of $\mathfrak{A}$ into $\mathfrak{B}$, we say that $\mathfrak{A}$ is elementarily embeddable in $\mathfrak{B}$, or alternatively, that $\mathfrak{A}$ is an elementary substructure of $\mathfrak{B}$, which we express as $\mathfrak{A} \prec \mathfrak{B}$.

Now, we will prove that the reduced product of models behaves well with respect to embeddings and homomorphisms.

Theorem 4.2.4 *Let $\mathcal{F}$ be a proper filter on I, and for each $i \in I$, let $\mathfrak{A}_i$ and $\mathfrak{B}_i$ be models of $\mathcal{L}$. Then:*

(1) *If, for each $i \in I$, $\mathfrak{A}_i$ is homomorphic to $\mathfrak{B}_i$, then $\prod_{\mathcal{F}} \mathfrak{A}_i$ is homomorphic to $\prod_{\mathcal{F}} \mathfrak{B}_i$.*
(2) *If, for each $i \in I$, $\mathfrak{A}_i$ is embeddable in $\mathfrak{B}_i$, then $\prod_{\mathcal{F}} \mathfrak{A}_i$ is embeddable in $\prod_{\mathcal{F}} \mathfrak{B}_i$.*

Proof Let $\mathfrak{A} = \prod_{\mathcal{F}} \mathfrak{A}_i$ and $\mathfrak{B} = \prod_{\mathcal{F}} \mathfrak{B}_i$. If φ_i is a homomorphism from $\mathfrak{A}_i$ to $\mathfrak{B}_i$ for each $i \in I$, we define the function

$$\varphi : \prod_{\mathcal{F}} A_i \to \prod_{\mathcal{F}} B_i$$

as the mapping that sends the element $\boldsymbol{f} = [\langle f(i) \mid i \in I \rangle]$ of $\prod_{\mathcal{F}} A_i$ to the element

$$\varphi(\boldsymbol{f}) = [\langle \varphi_i(f(i)) \mid i \in I \rangle]$$

of $\prod_{\mathcal{F}} B_i$. We will show that φ is a homomorphism from $\mathfrak{A}$ to $\mathfrak{B}$, i.e.,

$$(\boldsymbol{f}^1, \ldots, \boldsymbol{f}^n) \in P^{\mathfrak{A}} \Rightarrow (\varphi(\boldsymbol{f}^1), \ldots, \varphi(\boldsymbol{f}^n)) \in P^{\mathfrak{B}},$$

for any n-tuple $(\boldsymbol{f}^1, \ldots, \boldsymbol{f}^n)$ of elements of $\prod_{\mathcal{F}} A_i$ and any n-ary predicate P in the language of the structures.

By hypothesis, for each $i \in I$,

$$\mathfrak{A}_i \models P(f^1(i) \cdots f^n(i)) \Rightarrow \mathfrak{B}_i \models P(\varphi_i(f^1(i)) \cdots \varphi_i(f^n(i))).$$

Thus,

$$\{i \in I \mid \mathfrak{A}_i \models P(f^1(i) \cdots f^n(i))\} \subseteq$$
$$\subseteq \{i \in I \mid \mathfrak{B}_i \models P(\varphi_i(f^1(i)), \ldots, \varphi_i(f^n(i)))\}.$$

From this, it follows that

$$\{i \in I \mid \mathfrak{B}_i \models P(\varphi_i(f^1(i)), \ldots, \varphi_i(f^n(i)))\} \in \mathcal{F},$$

which concludes that

$$(\varphi(\boldsymbol{f}^1), \ldots, \varphi(\boldsymbol{f}^n)) \in P^{\mathfrak{B}}.$$

An analogous argument, whose details are left to the reader, proves the claim for functions and constants.

By definition, if $\mathfrak{A}_i$ is embeddable in $\mathfrak{B}_i$, then there exists a submodel $\mathfrak{C}_i$ of $\mathfrak{B}_i$ such that $\mathfrak{A}_i$ and $\mathfrak{C}_i$ are isomorphic. By the previous point, there is a homomorphism φ from $\prod_{\mathcal{F}} \mathfrak{A}_i$ to $\mathfrak{C} = \prod_{\mathcal{F}} \mathfrak{C}_i$. In fact, φ is an isomorphism.

First, φ is injective because

$$\varphi(\boldsymbol{f}) = \varphi(\boldsymbol{f}') \Rightarrow \{i \in I \mid \varphi_i(f(i)) = \varphi_i(f'(i))\} \in \mathcal{F}.$$

Since φ_i is injective by hypothesis,

$$\{i \in I \mid f(i) = f'(i)\} = \{i \in I \mid \varphi_i(f(i)) = \varphi_i(f'(i))\},$$

which implies that $\boldsymbol{f} = \boldsymbol{f}'$.

To show that φ is surjective, let $\boldsymbol{g}$ be an arbitrary element of $\prod_{\mathscr{F}} C_i$, so that g is a function $g : I \to \bigcup_{i \in I} C_i$ with $g(i) \in C_i$. We need to show that $\boldsymbol{g}$ is the equivalence class of a sequence of the form $\langle \varphi_i(f(i)) \mid i \in I \rangle$, where $\boldsymbol{f} \in \prod_{\mathscr{F}} A_i$.

Since φ_i is surjective for each i, there exists an $a_i \in A_i$ such that $\varphi_i(a_i) = g(i)$ for each $i \in I$. But we know that there exists a function $f \in \prod_{i \in I} A_i$ such that $f(i) = a_i$. It follows that

$$\varphi(\boldsymbol{f}) = [\langle \varphi_i(f(i)) \mid i \in I \rangle] = [\langle g(i) \mid i \in I \rangle] = \boldsymbol{g}.$$

Finally, to show that $\mathfrak{C}$ is a submodel of $\mathfrak{B}$, note that

$$\prod_{\mathscr{F}} A_i = \prod_{\mathscr{F}} C_i \subseteq \prod_{\mathscr{F}} B_i.$$

Thus, $\mathfrak{C}$ is indeed a submodel of $\mathfrak{B}$.⊣

By virtue of Theorem 4.2.3, the ultrapower $\prod_{\mathscr{F}} \mathfrak{A}$ satisfies ϕ if and only if $\mathfrak{A}$ satisfies ϕ. This leads to the following corollary, which forms the basis of the Transfer Principle.

Corollary 4.2.5 *If $\prod_{\mathscr{F}} \mathfrak{A}$ is an ultrapower of $\mathfrak{A}$, then $\prod_{\mathscr{F}} \mathfrak{A}$ and $\mathfrak{A}$ are elementarily equivalent; i.e.,*

$$\prod_{\mathscr{F}} \mathfrak{A} \equiv \mathfrak{A}.$$

Recall that we denote by $\mathfrak{A} \simeq \mathfrak{B}$ the assertion that $\mathfrak{A}$ and $\mathfrak{B}$ are isomorphic models. If two models are isomorphic, then they are elementarily equivalent, but the converse holds only when both structures are finite.

Now, we will prove that ultraproducts also inherit elementary equivalence if this relationship holds for each factor model.

Theorem 4.2.6 *Let $\mathscr{U}$ be an ultrafilter on I. Then:*

(1) If, for each $i \in I$, $\mathfrak{A}_i \simeq \mathfrak{B}_i$, then $\prod_{\mathscr{U}} \mathfrak{A}_i \equiv \prod_{\mathscr{U}} \mathfrak{B}_i$.
(2) If, for each $i \in I$, $\mathfrak{A}_i \prec \mathfrak{B}_i$, then $\prod_{\mathscr{U}} \mathfrak{A}_i \prec \prod_{\mathscr{U}} \mathfrak{B}_i$.

Proof

(1) This follows immediately from point (3) of Theorem 4.2.3. If ϕ is a sentence in $\mathscr{L}$, we have

$$\{i \in I \mid \mathfrak{A}_i \models \phi\} = \{i \in I \mid \mathfrak{B}_i \models \phi\}.$$

Thus,

$$\prod_{\mathcal{U}} \mathfrak{A}_i \models \phi \quad \Leftrightarrow \quad \prod_{\mathcal{U}} \mathfrak{B}_i \models \phi.$$

Therefore,

$$\prod_{\mathcal{U}} \mathfrak{A}_i \equiv \prod_{\mathcal{U}} \mathfrak{B}_i.$$

(2) This follows from the previous result and point (2) of Theorem 4.2.4, since $\mathfrak{A}_i \prec \mathfrak{B}_i$ implies $\mathfrak{A}_i \equiv \mathfrak{B}_i$. Since this holds for each $i \in I$, the ultraproducts are elementarily equivalent. Moreover, since $\mathfrak{A}$ is a substructure of $\mathfrak{B}$, the desired result follows. ⊣

In order to obtain a proper elementary extension using the ultraproduct construction, it is necessary to require that the ultrafilter be non-principal. This follows from Theorem 2.3.11, which states that a principal ultrafilter is generated by a singleton set. A property of ultraproducts constructed using principal ultrafilters is given below.

Theorem 4.2.7 *Let $\mathcal{U}$ be the principal ultrafilter generated by $\{j\} \subseteq I$. Then:*

$$\prod_{\mathcal{U}} \mathfrak{A}_i \equiv \mathfrak{A}_j.$$

Proof By definition, $f \approxeq g$ if and only if

$$\{i \in I \mid f(i) = g(i)\} \in \mathcal{U}.$$

Since $\mathcal{U}$ is the principal ultrafilter generated by $\{j\}$, it follows that $f \approxeq g$ whenever $f(j) = g(j)$. From Definition 4.2.1, it follows that the interpretation of function, relation, and constant symbols in $\prod_{\mathcal{U}} \mathfrak{A}_i$ depends only on their interpretation in $\mathfrak{A}_j$.⊣

A direct consequence of the previous theorem is that the ultraproduct must be taken over a non-principal ultrafilter; otherwise, the resulting ultraproduct is trivial. By Theorem 2.3.11, a non-principal ultrafilter does not contain finite sets, meaning that equivalence modulo $\mathcal{U}$ occurs only over infinite sets. The proof that a non-principal ultrafilter exists over any infinite index set was given in Theorem 2.3.13. Thus, in this sense, we say that equivalence modulo a non-principal ultrafilter occurs only over "large" sets.

4.3 Ultraproducts and Proper Elementary Extensions

Some applications of the Fundamental Theorem of Ultraproducts will be presented below. The first application provides another version of the compactness theorem via ultraproducts.

Theorem 4.3.1 (Compactness via Ultraproducts) *Let Σ be a consistent set of sentences in $\mathcal{L}$, let $I = S_\omega(\Sigma)$ be the set of all finite subsets of Σ, and let $\mathfrak{A}_i$ be a model of i for each $i \in I$. Then, there exists an ultrafilter $\mathcal{U}$ on I such that the ultraproduct $\prod_{\mathcal{U}} \mathfrak{A}_i$ is a model of Σ.*

Proof For each $\sigma \in \Sigma$, let $\hat{\sigma}$ be the set of all $i \in I$ such that $\sigma \in i$, so that $\hat{\sigma}$ consists of all finite subsets of Σ that contain σ. Let F be the family of all such sets $\hat{\sigma}$,

$$F = \{\hat{\sigma} \mid \sigma \in \Sigma\}.$$

We aim to extend F to a filter, which is possible because F posesses the finite intersection property. Indeed, given any finite list $\hat{\sigma}_1, \ldots, \hat{\sigma}_n$ of elements in F, the set $\Gamma = \{\sigma_1, \ldots, \sigma_n\}$ is contained in each of the sets $\hat{\sigma}_1, \ldots, \hat{\sigma}_n$, since $\sigma_j \in \Gamma \in I$ for each $j = 1, \ldots, n$. Therefore, the intersection

$$\hat{\sigma}_1 \cap \cdots \cap \hat{\sigma}_n$$

is nonempty.

By Theorem 2.3.4, F can be extended to a filter $\mathcal{F}$ on I, which can then be extended to an ultrafilter $\mathcal{U}$ by Theorem 2.3.6. Any ultrafilter on I that contains F will suffice.

If $i \in \hat{\sigma}$, then $\sigma \in i$, which implies that $\mathfrak{A}_i \models \sigma$. Hence, for each $\sigma \in \Sigma$,

$$\hat{\sigma} \subseteq \{i \in I \mid \mathfrak{A}_i \models \sigma\}.$$

Since $\hat{\sigma} \in \mathcal{U}$, it follows that

$$\{i \in I \mid \mathfrak{A}_i \models \sigma\} \in \mathcal{U}.$$

By the Fundamental Theorem of Ultraproducts,

$$\prod_{\mathcal{U}} \mathfrak{A}_i \models \sigma$$

for all $i \in I$, so $\prod_{\mathcal{U}} \mathfrak{A}_i$ is a model of Σ.⊣

4.4 The Natural Embedding

The application of Theorem 4.2.3 that we present here concerns the fact that every model $\mathfrak{A}$ can be elementarily and naturally embedded in its ultrapower. If appropriate precautions are taken to ensure that the ultrapower is non-trivial, then the ultrapower $\prod_{\mathcal{U}} \mathfrak{A}$ provides a new model that is an elementary extension of the factor models. That is, it has strictly greater cardinality than the original model while naturally containing $\mathfrak{A}$ as an elementary submodel. We will show that, for every model $\mathfrak{A}$, there exists a natural embedding of $\mathfrak{A}$ into the ultrapower $\prod_{\mathcal{U}} \mathfrak{A}$.

Definition 4.4.1 Let $\mathcal{F}$ be a proper filter on I, and let $\mathfrak{A}_i$ be a family of copies of the model $\mathfrak{A}$. The *natural embedding* of $\mathfrak{A}$ into $\prod_{\mathcal{F}} \mathfrak{A}_i$ is the function ι such that $\iota(a)$ is the equivalence class of the constant function with value $a = a_i$, that is:

$$\iota(a) = \langle a_i \mid i \in I \rangle_{\mathcal{F}}.$$

The restriction of $\prod_{\mathcal{F}} \mathfrak{A}$ to $\iota(A)$ is denoted by $\iota(\mathfrak{A})$.

We will show now that the natural embedding is actually an elementary embedding.

Theorem 4.4.2 *Let $\mathfrak{A}_i = \mathfrak{A}$ for all $i \in I$ be a model and let $\mathcal{U}$ be an ultrafilter. Then, the natural embedding ι of $\mathfrak{A}$ into its ultrapower $\prod_{\mathcal{U}} \mathfrak{A}_i$ is an elementary embedding.*

Proof Let $\phi(x_1, \ldots, x_n)$ be a formula in $\mathcal{L}$, and let $a_1, \ldots, a_n$ be elements of A. Then, by Theorem 4.2.3, we obtain the following chain of equivalences:

$$\prod_{\mathcal{U}} \mathfrak{A}_i \models \phi[\![\iota(a_1) \cdots \iota(a_n)]\!]$$

if and only if

$$\{i \in I \mid \mathfrak{A}_i \models \phi[\![a_1 \cdots a_n]\!]\} \in \mathcal{U}.$$

Since $\mathfrak{A}_i = \mathfrak{A}$ for all i, we have:

$$\mathfrak{A}_i \models \phi[\![a_1 \cdots a_n]\!]$$

which implies:

$$\mathfrak{A} \models \phi[\![a_1 \cdots a_n]\!].$$

Thus, ι is an elementary embedding.⊣

Next, we will see when the natural embedding is not surjective. This is crucial for proving the existence of a proper enlargement of the universe, where Robinsonian analysis can be developed—that is, an enlargement in which non-standard entities exist, such as nonzero infinitesimals. The argument is based on the concept of an ultrafilter that is complete with respect to a cardinal number, whose definition will be introduced shortly.

We will show that if the natural embedding

$$\iota : \mathfrak{A} \to \prod_{\mathcal{U}} \mathfrak{A}$$

is surjective, then the ultrafilter $\mathcal{U}$ is λ^+-complete, which in turn implies that $\mathcal{U}$ is principal. We conclude that if the ultrafilter $\mathcal{U}$ is non-principal, then the natural embedding is an enlargement—that is,

$$\left(\prod_{\mathcal{U}} A\right) \setminus A$$

is non-empty.

We denote by λ^+ the successor cardinal of λ; that is, the smallest cardinal greater than λ. For example, $(\lambda_\zeta)^+ = \lambda_{\zeta+1}$.

Definition 4.4.3 Let λ be an infinite cardinal, and let $\mathcal{F}$ be a filter on I. We say that $\mathcal{F}$ is λ^+-complete if and only if the intersection of any family of sets in $\mathcal{F}$ whose cardinality is less than λ^+ belongs to $\mathcal{F}$; that is,

$$G \subseteq \mathcal{F} \quad \text{and} \quad \#G < \lambda^+ \quad \Rightarrow \quad \bigcap G \in \mathcal{F}.$$

From this definition, it follows that every principal filter $\mathcal{F}$ on I is λ^+-complete, since it is generated by a singleton set $\{j\}$ (cf. Theorem 2.3.11). Thus,

$$\bigcap \mathcal{F} = \{j\},$$

which in turn implies that the intersection of any family of elements of $\mathcal{F}$ belongs to $\mathcal{F}$. The converse of this important result is expressed as the following theorem.

Theorem 4.4.4 *Let $\mathcal{F}$ be a filter on I. If $\mathcal{F}$ is λ^+-complete, then $\mathcal{F}$ is principal.*

Proof Let C be the set of complements of singleton sets that belong to $\mathcal{F}$, i.e.,

$$C = \{I \setminus \{i\} \mid I \setminus \{i\} \in \mathcal{F}\}.$$

We have

$$\#C \leq \lambda < \lambda^+.$$

By hypothesis, $\mathcal{F}$ is λ^+-complete and $C \subseteq \mathcal{F}$, so

$$\bigcap C \in \mathcal{F}.$$

On the other hand, if $X \in \mathcal{F}$, then $\bigcap C \in X$, since if $i \notin X$, then

$$X \subseteq I \setminus \{i\},$$

which implies that $I \setminus \{i\} \in C$ and that $i \notin \bigcap C$.

Thus, $\mathcal{F}$ is the principal filter generated by $\bigcap C$.⊣

In fact, by Theorem 2.3.11, we already know that every principal ultrafilter is generated by a singleton set. An example of a principal ultrafilter is as follows.

Let $\mathfrak{A}$ be the structure with universe $|\mathfrak{A}| = A = \{a, b, c\}$ and let $I = \{1, 2, 3\}$. The family

$$\mathcal{U} = \{\{2\}, \{1, 2\}, \{2, 3\}, I\}$$

is an ultrafilter on the Boolean algebra of all subsets of I, generated by $\{2\}$. The equivalence classes induced by this ultrafilter are:

$$f_1 = \{aaa, aab, aac, bab, bac, baa, cab, cac, caa\}$$

$$f_2 = \{aba, abb, abc, bba, bbb, bbc, cba, cbb, cbc\}$$

$$f_3 = \{aca, acb, acc, bca, bcb, bcc, cca, ccb, ccc\}$$

The reader may find amusing to verify that

$$\{i \in I \mid f_j(i) = f_k(i)\} \in \mathcal{U} \leftrightarrow j = k, \quad (j, k = 1, 2, 3).$$

After proving Theorem 2.3.12, we stated that an ultrafilter $\mathcal{U}$ on an infinite set I is called countably incomplete if there exists a countable partition $\{I_n \mid n \in \omega\}$ of I such that $I_n \notin \mathcal{U}$ for all $n \in \omega$. In terms of Definition 4.4.3, an ultrafilter $\mathcal{U}$ on ω is countably incomplete if and only if $\mathcal{U}$ is ω^+-incomplete. The following lemma can be seen as a generalization of Theorem 2.3.12 and characterizes λ^+-complete ultrafilters.

Lemma 4.4.5 *Let $\mathcal{F}$ be an ultrafilter on I. Then, $\mathcal{F}$ is λ^+-complete if and only if for every partition of I into fewer than λ^+ parts, one of them belongs to $\mathcal{F}$.*

Proof

($\Rightarrow$) Suppose that $\mathcal{F}$ is λ^+-complete, and let $\{X_\xi \mid \xi \in \eta\}$ be a partition of I where $\eta \leq \lambda$. By hypothesis, if all the sets $I \setminus X_\xi$ belonged to $\mathcal{F}$, we would have

$$\bigcap_{\xi<\eta}(I \setminus X_\xi) \in \mathcal{F}.$$

However, in reality,

$$\bigcap_{\xi<\eta}(I \setminus X_\xi) = I \setminus \bigcup_{\xi<\eta} X_\xi = \varnothing \notin \mathscr{F},$$

which implies that for some $\xi < \eta$, $I \setminus X_\xi \notin \mathscr{F}$, and hence $X_\xi \in \mathscr{F}$.

($\Leftarrow$) Suppose that for every partition of I into fewer than λ^+ parts, one of them belongs to $\mathscr{F}$. Let $\mathfrak{A} \subseteq \mathscr{F}$ such that $\#\mathfrak{A} = \eta < \lambda^+$. Enumerate $\mathfrak{A}$ as $\{X_\xi \mid \xi < \eta\}$. Define an injective function $f : I \to \eta + 1$ as follows:

$$f(i) = \begin{cases} \eta & \text{if } i \in \bigcap \mathfrak{A}, \\ \xi & \text{if } i \in I \setminus \bigcap \mathfrak{A}, \end{cases}$$

where ξ is the smallest $\xi < \eta$ such that $i \notin X_\xi$.
Let Y_η be the preimage of η under f, i.e.,

$$\bigcap \mathfrak{A} = f^{-1}(\eta).$$

For each $\xi < \eta$, let Y_ξ be the set of all i in $I \setminus \bigcap \mathfrak{A}$ where ξ is the smallest $\xi < \eta$ such that $i \notin X_\xi$. The sets Y_η and Y_ξ form a partition of I.
By hypothesis, there exists $\xi < \eta + 1$ such that $Y_\xi \in \mathscr{F}$. However, for $\xi < \eta$, $f(i) = \xi$ implies that $i \notin X_\xi$, so $Y_\xi \cap X_\xi = \varnothing$. Since $X_\xi \in \mathscr{F}$, it follows that $Y_\xi \notin \mathscr{F}$. Therefore, $Y_\eta \in \mathscr{F}$, but

$$Y_\eta = \bigcap \mathfrak{A},$$

so $\bigcap \mathfrak{A} \in \mathscr{F}$, proving that $\mathscr{F}$ is λ^+-complete.⊣

The following theorem provides necessary and sufficient conditions for the natural embedding to be surjective.

Theorem 4.4.6 *Let $\mathfrak{A}$ be a model of cardinality λ, and let $\mathscr{U}$ be an ultrafilter on I. Then, the natural embedding*

$$\iota : \mathfrak{A} \to \prod_{\mathscr{U}} \mathfrak{A}$$

is surjective if and only if $\mathscr{U}$ is λ^+-complete.

Proof Suppose that $\mathscr{U}$ is λ^+-complete, and let $[f] = \boldsymbol{f} \in \prod_{\mathscr{U}} A$, where f is a function from I to A. Since $|\mathfrak{A}| = \lambda$, the family of sets A_a of the form

$$\{f^{-1}(a) \mid a \in A\}$$

forms a partition of I into fewer than λ^+ parts. By Lemma 4.4.5, there exists $a \in A$ such that $A_a \in \mathcal{U}$. It follows that

$$f \approxeq \langle a \mid i \in I \rangle,$$

so that $f = \iota(a)$. This proves that $\iota(A) = \prod_{\mathcal{U}} A$.

Now, suppose that ι is surjective. Let $\{X_\xi \mid \xi \in \mu\}$ be a partition of I into $\mu < \lambda^+$ parts. Since $\mu \leq \lambda = |\mathfrak{A}|$, we can index the sets X_ξ using elements from A; that is,

$$I = \bigcup_{a \in B} X_a$$

with $B \subseteq A$. Define an injective function f from I to A by

$$f(i) = a \quad \text{if and only if} \quad i \in X_a.$$

Then $f \in \prod_{\mathcal{U}} A = \iota(A)$, so there exists some $a \in A$ such that $f = \iota(a)$. This means that $f^{-1}(a) \in \mathcal{U}$. But $f^{-1}(a) = X_a$, so $X_a \in \mathcal{U}$. By Lemma 4.4.5, $\mathcal{U}$ is λ^+-complete.⊣

We are now in a position to prove the main result we were aiming for.

Theorem 4.4.7 *Let $\mathcal{U}$ be a non-principal ultrafilter on a set I of cardinality λ. For every infinite set A, the natural embedding*

$$\iota : A \to \prod_{\mathcal{U}} A$$

is a proper embedding.

Proof If ι were not proper, then it would be surjective, meaning that $\mathcal{U}$ would be μ^+-complete for some $\mu < \lambda$. By Theorem 4.4.4, $\mathcal{U}$ would be principal. Thus, if $\mathcal{U}$ is non-principal, then the embedding ι is not surjective.⊣

4.5 Nonstandard Universes

Nonstandard analysis is conducted within a structure that we will call a 'nonstandard universe', which is a relationship between two classes that we will refer to as superstructures. This is why the ontology of nonstandard analysis requires the superstructure construction. A superstructure is a family of sets obtained by starting from a set of points conceived as urelements (*Grundobjekte*, in Ackermann's terminology), called the base set.

Broadly speaking, the superstructure is built upon the base set by taking, as the first stratum, the base set itself along with its power set. In general, stratum $n + 1$ is obtained from stratum $n \geqslant 0$ by taking the stratum n along with its power set. A more precise definition is given below.

Definition 4.5.1 Let $\mathbb{X}$ be a nonempty set. We define recursively $\mathfrak{v}_n(\mathbb{X})$, the universe of sets of rank n ($n < \omega$) over $\mathbb{X}$, as follows:

(1) The universe of sets of rank 0 is $\mathbb{X}$ itself; i.e.,

$$\mathfrak{v}_0(\mathbb{X}) = \mathbb{X}.$$

(2) For any n, the universe of sets of rank $n + 1$ is defined as

$$\mathfrak{v}_{n+1}(\mathbb{X}) = \mathfrak{v}_n(\mathbb{X}) \cup \mathcal{P}(\mathfrak{v}_n(\mathbb{X})),$$

where $\mathcal{P}(X)$ denotes the power set of X.

The union

$$\bigcup_{n<\omega} \mathfrak{v}_n(\mathbb{X})$$

is called a *superstructure over the set* $\mathbb{X}$, which is called the *base* set of the superstructure. The superstructure over $\mathbb{X}$ is denoted as $\mathfrak{v}(\mathbb{X})$.

The base set $\mathbb{X}$ is characterized by the conditions:

(1) $\varnothing \notin \mathbb{X}$, and
(2) For very $x \in \mathbb{X}$, $x \cap A = \varnothing$ for every $A \in \mathfrak{v}(\mathbb{X}) \setminus \mathbb{X}$

In other words, the elements of $\mathbb{X}$ are considered urelements (*Grundobjekte*) within superstructure $\mathfrak{v}(\mathbb{X})$ and are also called *individuals relative to* $\mathfrak{v}(\mathbb{X})$.

If $a \in \bigcup_{n<\omega} \mathfrak{v}_n(\mathbb{X})$, the rank of a is the smallest number n such that

$$a \in \mathfrak{v}_n(\mathbb{X}) \setminus \mathfrak{v}_{n-1}(\mathbb{X}).$$

If $\mathbb{X}$ is empty, then $\mathfrak{v}(\mathbb{X})$ is called a pure superstructure.

For nonstandard analysis, $\mathbb{X}$ is often chosen as the set $\mathbb{R}$ of real numbers, or a set that contains it.

A notable and relevant characteristic of superstructure $\mathfrak{v}(\mathbb{X})$ is transitivity.

Definition 4.5.2 We say that a set B is *transitive* if and only if, for every set A,

$$a \in A \in B \quad \Rightarrow \quad a \in B.$$

A set C is *strongly transitive* if the subsets of the elements of C are also elements of C, i.e.,

$$A \subseteq B \in C \quad \Rightarrow \quad A \in C.$$

It is possible to show that $\mathfrak{v}(\mathbb{X})$ is strongly transitive.

Lemma 4.5.3 *Superstructure $\mathfrak{v}(\mathbb{X})$ is transitive and, in fact, strongly transitive.*

Proof We first show that $\mathfrak{v}(\mathbb{X})$ is transitive. Suppose that $A \in \mathfrak{v}(\mathbb{X})$ and that $a \in A$. We need to show that $a \in \mathfrak{v}(\mathbb{X})$.

A cannot belong to $\mathbb{X}$ because the elements of $\mathbb{X}$ are not sets.

If A belongs to

$$\mathfrak{v}_1(\mathbb{X}) = \mathbb{X} \cup \mathcal{P}(\mathbb{X}),$$

then A is a subset of $\mathbb{X}$, meaning that a is an urelement. Therefore,

$$a \in \mathfrak{v}_1(\mathbb{X}) \subseteq \mathfrak{v}(\mathbb{X}).$$

Now, suppose that the proposition holds for sets in $\mathfrak{v}_n(\mathbb{X})$ and let $A \in \mathfrak{v}_{n+1}(\mathbb{X})$, so that either $A \in \mathfrak{v}_n(\mathbb{X})$, or A is a subset of $\mathfrak{v}_n(\mathbb{X})$.

In case 1, $a \in A$ implies $a \in \mathfrak{v}(\mathbb{X})$ by the inductive hypothesis. In case 2, $a \in A$ implies $a \in \mathfrak{v}_n(\mathbb{X}) \subset \mathfrak{v}(\mathbb{X})$.

Thus, $\mathfrak{v}(\mathbb{X})$ is transitive.

Next, we show that $\mathfrak{v}(\mathbb{X})$ is strongly transitive.

If B is a set in $\mathfrak{v}_1(\mathbb{X})$ and $A \subseteq B$, since B consists of urelements, the elements of A are also urelements, so

$$A \subseteq \mathbb{X}.$$

This implies that

$$A \in \mathcal{P}(\mathbb{X}) \subseteq \mathfrak{v}_1(\mathbb{X}) \subseteq \mathfrak{v}(\mathbb{X}).$$

Now, assume that the assertion holds for all sets B in $\mathfrak{v}_n(\mathbb{X})$. If $B \in \mathfrak{v}_{n+1}(\mathbb{X})$ and $A \subseteq B$, then:

Either $B \in \mathfrak{v}_n(\mathbb{X})$, in which case, by the inductive hypothesis, $A \in \mathfrak{v}(\mathbb{X})$.

Or $B \in \mathcal{P}(\mathfrak{v}_n(\mathbb{X}))$, meaning that

$$A \in \mathcal{P}(\mathfrak{v}_n(\mathbb{X})) \subseteq \mathfrak{v}_{n+1}(\mathbb{X}) \subset \mathfrak{v}(\mathbb{X}).$$

Thus, $\mathfrak{v}(\mathbb{X})$ is strongly transitive.$\dashv$

The previous argument shows that each level $\mathfrak{v}_n(\mathbb{X})$ is both transitive and strongly transitive. Other properties of superstructures will be presented in the following lemmas.

Lemma 4.5.4 *For each $n < \omega$:*

(1) $\mathfrak{v}_{n+1}(\mathbb{X}) = \mathbb{X} \cup \mathcal{P}[\mathfrak{v}_n(\mathbb{X})]$.
(2) $\mathfrak{v}_{n+1}(\mathbb{X}) \setminus \mathbb{X} = \mathcal{P}[\mathfrak{v}_n(\mathbb{X})]$.
(3) $\mathfrak{v}_n(\mathbb{X}) \subseteq \mathfrak{v}_{n+1}(\mathbb{X})$.

Proof For $n \geq 1$ and any set A,

$$\begin{aligned} A \in \mathcal{P}[\mathfrak{v}_{n-1}(\mathbb{X})] &\Rightarrow A \subseteq \mathfrak{v}_{n-1}(\mathbb{X}) \\ &\Rightarrow A \subseteq \mathfrak{v}_n(\mathbb{X}) \\ &\Rightarrow A \in \mathcal{P}[\mathfrak{v}_n(\mathbb{X})]. \end{aligned}$$

This chain of implications shows that

$$\mathcal{P}[\mathfrak{v}_{n-1}(\mathbb{X})] \subseteq \mathcal{P}[\mathfrak{v}_n(\mathbb{X})],$$

which leads to

$$\mathcal{P}[\mathfrak{v}_{n-1}(\mathbb{X})] \cup \mathcal{P}[\mathfrak{v}_n(\mathbb{X})] = \mathcal{P}[\mathfrak{v}_n(\mathbb{X})]. \tag{*}$$

For (1), observe that for $n = 0$,

$$\mathfrak{v}_{n+1}(\mathbb{X}) = \mathfrak{v}_1(\mathbb{X}) = \mathbb{X} \cup \mathcal{P}(\mathbb{X}) = \mathbb{X} \cup \mathcal{P}[\mathfrak{v}_0(\mathbb{X})] = \mathbb{X} \cup \mathcal{P}[\mathfrak{v}_n(\mathbb{X})].$$

Assuming the statement holds for $n > 0$, we obtain, using (*),

$$\begin{aligned} \mathfrak{v}_{n+1}(\mathbb{X}) &= \mathfrak{v}_n(\mathbb{X}) \cup \mathcal{P}[\mathfrak{v}_n(\mathbb{X})] \\ &= (\mathbb{X} \cup \mathcal{P}[\mathfrak{v}_{n-1}(\mathbb{X})]) \cup \mathcal{P}[\mathfrak{v}_n(\mathbb{X})] \\ &= \mathbb{X} \cup \mathcal{P}[\mathfrak{v}_n(\mathbb{X})]. \end{aligned}$$

For (2), note that $\mathbb{X}$ and $\mathcal{P}[\mathfrak{v}_n(\mathbb{X})]$ are disjoint for all n, so (2) follows immediately from (1).

For (3), for $n \geq 0$,

$$\begin{aligned} a \in \mathfrak{v}_n(\mathbb{X}) &\Rightarrow a \in \mathfrak{v}_n(\mathbb{X}) \cup \mathcal{P}[\mathfrak{v}_n(\mathbb{X})] \\ &\Rightarrow a \in \mathfrak{v}_{n+1}(\mathbb{X}). \dashv \end{aligned}$$

Superstructures have very nice closure properties, as shown in the following lemma.

Lemma 4.5.5 *The collection of sets in a superstructure is closed under the elementary set-theoretic operations. That is:*

(1) *If* $a_1, \ldots, a_n \in \mathfrak{v}_n(\mathbb{X})$, *then* $\{a_1, \ldots, a_n\} \in \mathfrak{v}_{n+1}(\mathbb{X}) \setminus \mathbb{X}$.
(2) *If* $a_1, \ldots, a_n \in \mathfrak{v}_n(\mathbb{X})$, *then* $\langle a_1, \ldots, a_n \rangle \in \mathfrak{v}_{n+2(n-1)}(\mathbb{X})$.

(3) *If* $a \in \mathfrak{v}_n(\mathbb{X}) \setminus \mathbb{X}$ *and* $b \subseteq a$, *then* $b \in \mathfrak{v}_n(\mathbb{X}) \setminus \mathbb{X}$.
(4) *If* $a, b \in \mathfrak{v}_n(\mathbb{X}) \setminus \mathbb{X}$, *then* $a \times b \in \mathfrak{v}_{n+3}(\mathbb{X}) \setminus \mathbb{X}$.
(5) *If* $a \in \mathfrak{v}_n(\mathbb{X})$, *then* $\mathcal{P}(\mathbb{X}) \in \mathfrak{v}_{n+1}(\mathbb{X}) \setminus \mathbb{X}$.

The proof is left as an exercise for the reader.

The practice of nonstandard analysis requires the use of *bounded* or *relativized quantifiers*. Expression $(\forall x \in A)\,\phi$ means

$$\forall x\,(x \in A \Rightarrow \phi).$$

Expression $(\exists x \in A)\,\phi$ means

$$\exists x\,(x \in A \wedge \phi).$$

A *bounded quantifier formula* is one obtained from atomic formulas using logical connectives and bounded quantifiers.

Analogous to the elementary embedding of a model into another, we define the concept of a bounded elementary embedding. A function

$$\varphi : A \to B$$

is a *bounded elementary embedding*, denoted

$$\varphi : \mathfrak{A} \prec_b \mathfrak{B},$$

if for every bounded quantifier formula $\phi(x_1 \cdots x_n)$ in $\mathcal{L}$ and every sequence $a_1, \cdots, a_n \in A$, we have:

$$\mathfrak{A} \models \phi[\![a_1 \cdots a_n]\!] \quad \Leftrightarrow \quad \mathfrak{B} \models \phi[\![\varphi(a_1) \cdots \varphi(a_n)]\!].$$

If such a function exists, we say that $\mathfrak{B}$ is a *bounded elementary extension* of $\mathfrak{A}$, or that $\mathfrak{A}$ is a bounded elementary submodel of $\mathfrak{B}$, and we write

$$\mathfrak{A} \prec_b \mathfrak{B}.$$

Notice that, since bounded quantifier formulas are abreviations of canonical ones, the existence of a bounded elementary extension follows from the Fundamental Theorem of Ultraproducts, Theorem 4.2.3.

A nonstandard universe is more complicated than a model since it actually consists of the existence of a monomorphism between two different models of the language $\mathcal{L}_\in$ (whose only symbol is the membership symbol for a set); namely, $\mathfrak{A} = \mathfrak{v}(\mathbb{X})$ and $\mathfrak{B} = \mathfrak{v}(\mathbb{Y})$, where $\mathbb{Y}$ is the reduced power of $\mathbb{X}$ modulo a certain ultrafilter $\mathcal{U}$. The monomorphism is an embedding, denoted by *, which maps $\mathfrak{v}(\mathbb{X})$ into a proper subclass of $\mathfrak{B}$.

Definition 4.5.6 A *nonstandard universe* is a triple:

$$\langle \mathfrak{v}(\mathbb{X}), \mathfrak{v}(\mathbb{Y}), {}^* \rangle$$

such that:

(1) $\mathbb{X}$ and $\mathbb{Y}$ are infinite base sets.
(2) Transfer Principle:

$$\langle \mathfrak{v}(\mathbb{X}), \in \rangle \prec_b \langle \mathfrak{v}(\mathbb{Y}), \in \rangle$$

is a bounded elementary embedding.
(3) ${}^*\mathbb{X} = \mathbb{Y}$.
(4) For every infinite subset A of $\mathbb{X}$,

$$\{{}^*a \mid a \in A\}$$

is a proper subset of ${}^*A$.

$*$ is called the star extension of A.

In fact, $*$ is a monomorphism that preserves relations and operations. As we shall see below, the existence of nonstandard universes follows from The Fundamental Theorem of Ultraproducts.

As a previous step to the proof of the existence of nonstandard universes, I will show now that the ultrapower of model $\mathfrak{A} = \langle \mathfrak{v}(\mathbb{X}), \in \rangle$ is model $\mathfrak{B} = \langle \mathfrak{v}(\mathbb{Y}), \in \rangle$, where $\mathbb{Y}$ is the reduced Cartesian product of $\mathbb{X}$ modulo a non-principal ultrafilter $\mathcal{U}$.

Theorem 4.5.7 *Let $\mathfrak{A}$ be the model $\langle \mathfrak{v}(\mathbb{X}), \in \rangle$, and let $\mathcal{U}$ be a non-principal ultrafilter. The ultrapower $\prod \mathfrak{A}$ is model $\mathfrak{B} = \langle \mathfrak{v}(\mathbb{Y}), \in \rangle$ where $\mathbb{Y}$ is the reduced Cartesian product of $\mathbb{X}$ modulo $\mathcal{U}$.*

Proof Let $\mathfrak{A}$ be the model $\langle \mathfrak{v}(\mathbb{X}), \in \rangle$, and let $\mathcal{U}$ be a non-principal ultrafilter. The universe of $\prod_{\mathcal{U}} \mathfrak{A}$ consists of the equivalence classes of sequences $\{a_i\}_{i \in I}$ in $[\mathfrak{v}(\mathbb{X})]^I$ modulo $\mathcal{U}$. That is, elements of the form:

$$[a_i] = \{\{b_i\} \in [\mathfrak{v}(\mathbb{X})]^I \mid \{i \in I \mid b_i = a_i\} \in \mathcal{U}\}.$$

For each $n \in \mathbb{N}$, we define the set W_n as the set of all sequences $\{a_i\}$ in $[\mathfrak{v}(\mathbb{X})]^I$ such that:

$$\{i \in I \mid a_i \in \mathfrak{v}_n(\mathbb{X})\} \in \mathcal{U}.$$

That is:

$$W_n = \{\{a_i\} \in [\mathfrak{v}(\mathbb{X})]^I \mid \{i \in I \mid a_i \in \mathfrak{v}_n(\mathbb{X})\} \in \mathcal{U}\}.$$

In fact, every sequence $\{a_i\}$ belongs to some W_n. Indeed, since $\mathfrak{v}(\mathbb{X})$ is strongly transitive and $\{a_i\} \subset \mathfrak{v}(\mathbb{X})$, we have:

$$\{a_i\} \in \mathfrak{v}(\mathbb{X}) = \bigcup_{i=1}^{\infty} \mathfrak{v}_n(\mathbb{X}).$$

Thus, there exists some $n \in \mathbb{N}$ such that $\{a_i\} \in \mathfrak{v}_n(\mathbb{X})$. The *rank* of $\{a_i\}$ *in* $\mathfrak{v}(\mathbb{X})$ is the smallest positive integer n such that $\{a_i\} \in \mathfrak{v}_n(\mathbb{X})$. If n is the rank of $\{a_i\}$ in $\mathfrak{v}_n(\mathbb{X})$, then the elements of $\{a_i\}$ belong to $V_{n-1}(X)$ and thus to $\mathfrak{v}_n(\mathbb{X})$ for all $i \in I$. Hence, we obtain:

$$\{i \in I \mid a_i \in \mathfrak{v}_n(\mathbb{X})\} = I \in \mathcal{U}.$$

That is, $\{a_i\} \in W_n$. Therefore:

$$W = \bigcup_{n=1}^{\infty} W_n = [\mathfrak{v}(\mathbb{X})]^I.$$

We can then define the *rank of* $\{a_i\}$ *in* W as the smallest positive integer n such that $\{a_i\} \in W_n(X)$.

Thus, W is the union of the increasing chain:

$$W_0 \subset W_1 \subset \cdots \subset W_n \cdots .$$

Clearly, $\{a_i\} \in [\mathfrak{v}(\mathbb{X})]^I$ has *rank* n if:

$$\{a_i\} \in W_n \setminus W_{n-1}.$$

Since $\mathcal{U}$ is an ultrafilter and:

$$\{i \in I \mid a_i \in \mathfrak{v}_n(\mathbb{X})\} \subset \{i \in I \mid a_i \in \mathfrak{v}_{n+1}(\mathbb{X})\},$$

we conclude that:

$$\{i \in I \mid a_i \in \mathfrak{v}_{n+1}(\mathbb{X})\} \in \mathcal{U}.$$

Hence, for all $m > n$:

$$\{i \in I \mid a_i \in \mathfrak{v}_m(\mathbb{X})\} \in \mathcal{U},$$

where n is the rank of $\{a_i\}$.

Since

$$\{i \in I \mid a_i \in \mathfrak{v}_0(\mathbb{X})\} = I,$$

whenever $\{a_i\}$ is a sequence of elements of $\mathbb{X}$, it follows that W_0 contains all sequences of elements of $\mathbb{X}$; that is $\mathbb{X}^I \subseteq W_0$.

If $\{a_i\} \in W_0$, the equivalence class $[a_i]$ of $\{a_i\} \in W_0$ is defined as:

$$[a_i] = \{\{b_i\} \in W \mid \{i \in I \mid b_i = a_i\} \in \mathcal{U}\}.$$

Set $\mathbb{Y}$ is then defined as the set of equivalence classes of sequences of rank 0:

$$\{[a_i] \mid \{a_i\} \in W_0\}.$$

For each constant sequence $\{a\} = \{a_i \mid a_i = a \text{ for all } i \in I\}$, there is a mapping ι that sends a to (a), and another mapping η that maps (a) to $[a]$. The composition $^* = \eta \circ \iota$ maps a to $[a]$ and, as we will show, is an isomorphic embedding of $\mathfrak{v}(\mathbb{X})$ into $\mathfrak{v}(\mathbb{Y})$.

Indeed, for $\{a_i\}$ in W_0, $[a_i]$ is a member of $\mathfrak{v}_0(^*\mathbb{Y}) = \mathbb{Y}$. Now suppose that $[a_i]$ is defined for all $\{a_i\} \in W_n$ and that it is an element of $\mathfrak{v}_n(^*\mathbb{Y})$. If $\{b_i\}$ is of rank $n+1$, we define $[b_i]$ as the set

$$\{[a_i] \mid \{a_i\} \in W_n \text{ and } \{i \in I \mid a_i \in b_i\} \in \mathcal{U}\}.$$

Observe that this definition specifies $[b_i]$ as a subset of $\mathfrak{v}_n(^*\mathbb{Y})$, and therefore as an element of $\mathfrak{v}_{n+1}(^*\mathbb{Y})$.

We will now show that if $\{a_i\}$ and $\{b_i\}$ are elements of W, then

(1) $[a_i] = [b_i]$ if and only if $\{i \in I \mid a_i = b_i\} \in \mathcal{U}$.
(2) $[a_i] \in [b_i]$ if and only if $\{i \in I \mid a_i \in b_i\} \in \mathcal{U}$.

To prove (1), suppose that $[a_i] = [b_i]$. This implies that $\{a_i\} \in [b_i]$ and therefore that $\{i \in I \mid a_i = b_i\} \in \mathcal{U}$.

Now suppose that $\{i \in I \mid a_i = b_i\} \in \mathcal{U}$ and that $[c_i] \in [a_i]$. It follows that $(c_i) \in W_n$ and that $\{i \in I \mid c_i \in a_i\} \in \mathcal{U}$. Now,

$$\{i \in I \mid c_i \in a_i\} \cap \{i \in I \mid a_i = b_i\} \subseteq \{i \in I \mid c_i \in b_i\}$$

and therefore $\{i \in I \mid c_i \in b_i\} \in \mathcal{U}$; that is, $[c_i] \in [b_i]$, from which it follows that $[a_i] \subseteq [b_i]$. A symmetric argument proves that $[b_i] \subseteq [a_i]$, and thus we conclude that $[a_i] = [b_i]$.

To demonstrate (2), suppose that $[a_i] \in [b_i]$ and, without loss of generality, that the rank of (b_i) is $n+1$. By definition of $[b_i]$, $\{a_i\} \in \mathfrak{v}_n(\mathbb{X})$ and $\{i \in I \mid a_i \in b_i\} \in \mathcal{U}$. It follows that $\{i \in I \mid a_i \in b_i\} \in \mathcal{U}$.

Now suppose that $\{i \in I \mid a_i \in b_i\} \in \mathcal{U}$. That the rank of (b_i) is $n+1$ means that

$$\{i \in I \mid b_i \text{ has rank } n+1\} \in \mathcal{U}.$$

The b_i of rank $n+1$ are subsets of $\mathfrak{v}_n(\mathbb{X})$ (by Lemma 4.5.4) and, therefore, the a_i that are elements of these b_i are also elements of $\mathfrak{v}_n(\mathbb{X})$. Now,

$$\begin{aligned} &\{i \in I \mid b_i \text{ has rank } n+1\} \cap \{i \in I \mid a_i \in b_i\} \\ &= \{i \in I \mid a_i \in b_i \text{ and } b_i \text{ has rank } n+1\} \\ &\subseteq \{i \in I \mid a_i \in \mathfrak{v}_n(\mathbb{X})\}, \end{aligned}$$

so that

$$\{i \in I \mid a_i \in \mathfrak{v}_n(\mathbb{X})\} \in \mathcal{U},$$

which means that $\{a_i\} \in \mathfrak{v}_n(\mathbb{X})$. We conclude that $[a_i] \in [b_i]$.

For a of rank $\geqslant 1$, we define

$$^*a = \{[b_i] \mid \{b_i\} \in W_n \text{ and } \{i \in I \mid b_i \in a\} \in \mathcal{U}\}.$$

Thus,

$$^*\mathfrak{v}_{n+1}(\mathbb{X}) = \{[b_i] \mid \{b_i\} \in W_n \text{ and } \{i \in I \mid b_i \in \mathfrak{v}_n(\mathbb{X})\} \in \mathcal{U}\}.$$

In particular,

$$\begin{aligned} ^*\mathbb{X} &= {}^*\mathfrak{v}_0(\mathbb{X}) \\ &= \{[b_i] \mid \{b_i\} \in W_0 \text{ and } \{i \in I \mid b_i \in \mathfrak{v}_0(\mathbb{X})\} \in \mathcal{U}\} \\ &\{[a_i] \mid \{a_i\} \in W_0\} \\ &= \mathbb{Y}. \end{aligned}$$

We conclude that $\mathfrak{v}(\mathbb{Y}) = \mathfrak{v}(^*\mathbb{X})$, and so

$$\langle \mathfrak{v}(\mathbb{Y}), \in\rangle = \prod_{\mathcal{U}} \mathfrak{A},$$

where $\mathbb{Y}$ is the reduced Cartesian power of $\mathbb{X}$ modulo $\mathcal{U}$.⊣

Theorem 4.5.8 *Superstructure* $\langle \mathfrak{v}(\mathbb{Y}), \in\rangle$ *is an enlargement of* $\langle \mathfrak{v}(\mathbb{X}), \in\rangle$.

Proof Let A be an infinite subset of $\mathbb{X}$. It suffices to show that the ultrapower $\prod_{\mathcal{U}} A$ is contained in $\mathbb{Y} = {}^*\mathbb{X}$ since, as the natural embedding $\iota : A \to \prod_{\mathcal{U}} A$ is a proper embedding, $\{^*a \mid a \in A\}$ is a proper subset of $^*A$. This follows from the fact that $\mathcal{U}$ is a non-principal ultrafilter, by an application of Theorem 4.4.7.

If $[a_i] \in \prod_{\mathcal{U}} A$, then $[a_i] \in W_0$ because $\{a_i\} \in \mathbb{X}^I$ and $\{i \mid a_i \in \mathfrak{v}_0(\mathbb{X})\} = I \in \mathcal{U}$. This follows from the assumption that $A \subseteq \mathbb{X} = \mathfrak{v}_0(\mathbb{X}) \subseteq \mathfrak{v}(\mathbb{X})$. Thus, $\prod_{\mathcal{U}} A \subseteq \mathbb{Y}$.⊣

Theorem 4.5.9 (Existence of a Nonstandard Universe) *Let $\mathfrak{v}(\mathbb{X})$ be a superstructure over an infinite base set $\mathbb{X}$. Then*

$$\langle \mathfrak{v}(\mathbb{X}), \mathfrak{v}(\mathbb{Y}), {}^* \rangle,$$

where $\langle \mathfrak{v}(\mathbb{Y}), \in \rangle = \prod_{\mathcal{U}} \mathfrak{A}$, is a nonstandard universe.

Proof Notice, first, that sequences $\{a_i\}$ are elements of the members $[a_i]$ of $\mathfrak{v}_0(\mathbb{X})$ and so, by regularity they cannot belong to sets containing the equivalence classes $[a_i]$, let alone sets above in the hierarchy. Hence, no sequence $\{a_i\}$ is in $\mathfrak{v}(\mathbb{Y})$. It follows that $[a_i] \cap B = \varnothing$ for every $B \in \mathfrak{v}(\mathbb{Y}) \setminus \mathbb{Y}$. Hence, the members $[a_i]$ of $\mathfrak{v}_0(\mathbb{X})$ are urelements relative to the sets in $\mathfrak{v}(\mathbb{Y})$. In other words, $\mathbb{Y}$ is a base set.

Since $\langle \mathfrak{v}(\mathbb{Y}), \in \rangle$ is an enlargement of $\langle \mathfrak{v}(\mathbb{X}), \in \rangle$, by Theorem 4.5.8 $\langle \mathfrak{v}(\mathbb{Y}), \in \rangle$ is a bounded elementary extension of $\langle \mathfrak{v}(\mathbb{X}), \in \rangle$ and

$${}^* : \langle \mathfrak{v}(\mathbb{X}), \in \rangle \to \langle \mathfrak{v}(\mathbb{Y}), \in \rangle$$

is a bounded elementary embedding.$\dashv$

4.6 α-Saturated Universes

The development of the topology necessary for nonstandard economic theory requires nonstandard universes with a special property known as α-saturation. This α has to be greater than the cardinal of the largest sets used in the development of the required topological spaces, which is usually that of the continuum. Hence, it will suffice to prove the existence of an α-saturated superstructure $\langle \mathfrak{v}(\mathbb{Y}), \in \rangle$ for an appropriate infinite cardinal α. In what follows, we will explain this property and show the existence of an enlargement of $\mathfrak{v}(\mathbb{X})$ that possesses it. The definition of α-saturation is given in the next definition.

Definition 4.6.1 Let α be an infinite cardinal number. A model $\mathfrak{A}$ is *α-saturated* if, and only if, for every subset $X \subset A$ with fewer than α elements, the expansion $(\mathfrak{A}, a)_{a \in X}$ realizes every type $\Sigma(v)$ of the language $\mathcal{L} \cup \{c_a \mid a \in X\}$ that is consistent with the theory of $(\mathfrak{A}, a)_{a \in X}$. A model $\mathfrak{A}$ is *saturated* if and only if it is $|A|$-saturated.

Notice that $\mathfrak{v}(\mathbb{Y})$ is sufficiently large for the development of mathematical analysis if $\mathbb{X} = \mathbb{R}$. Hence, α-saturation for $\alpha \geqslant |\mathfrak{v}(\mathbb{Y})|$ must suffice.

The former notion should not be confused with a weaker concept of saturation, which is saturation of a structure over another, as stipulated in the following definition.

Definition 4.6.2 Let $\mathfrak{B}$ be an elementary extension of $\mathfrak{A}$, and let α be a cardinal number. For each set $Y \subseteq A$, let $\mathcal{L}_Y$ denote the expansion of $\mathcal{L}$ obtained by adding a new constant symbol for each $y \in Y$. We say that $\mathfrak{B}$ is *α-saturated over* $\mathfrak{A}$ if and only if it satisfies the following condition:

For every set $Y \subseteq A$ with cardinality $|Y| < \alpha$, every set $\Sigma(x)$ of formulas of $\mathcal{L}_Y$ that is consistent with the theory of $(\mathfrak{A}, y)_{y \in Y}$ is satisfiable in $(\mathfrak{B}, y)_{y \in Y}$.

A model $\mathfrak{A}$ is said to be *α-saturated* if and only if $\mathfrak{A}$ is saturated over itself.

This notion should not be confused with sheer saturation. It turns out that the expansion $(\mathfrak{B}, y)_{y \in Y}$ of an elementary enlargement $\mathfrak{B}$ of a model $\mathfrak{A}$, modulo an α-regular ultrafilter, is α-saturated over $\mathfrak{A}$. Yet, what we need is to obtain an enlargement of superstructure $\mathfrak{v}(\mathbb{X})$ which is α-saturated over itself for some suitable α.

In order to achieve this goal, the notion of an α-regular model will be required. Its definition is given below. Then the strategy that we will adopt in order to prove the existence of α-saturated universes is the following. The definition of an α-regular filter is given in Definition 2.3.14, and the proof of the existence of such filters in Theorem 2.3.15. The existence of α-universal models are obtained as ultrapowers of a model $\mathfrak{A}$. An α-universal model is elementarily equivalent to all models elementarily embedded in it. Hence, starting with language $\mathcal{L}$ ($|\mathcal{L}| \leqslant \alpha$) for model $\mathfrak{A}$ and an α-regular ultrafilter $\mathcal{U}$, we produce an α-universal ultrapower modulo $\mathcal{U}$ of $\mathfrak{A}$, call it $\mathfrak{B}_1$, using Theorem 4.6.4. We then build, using the same α-regular ultrafilter $\mathcal{U}$, an elementary chain of models

$$\mathfrak{B}_1 \prec \mathfrak{B}_2 \prec \cdots \prec \mathfrak{B}_\xi \prec \cdots$$

starting with ultrapower $\mathfrak{B}_1 = \prod_{\mathcal{U}} \mathfrak{A}$ of $\mathfrak{A}$. By the same token (Theorem 4.6.4), all members of the chain are universal, and so, since they are elementarily equivalent, each one is elementarily embedded in the following ones, and so each $\mathfrak{B}_\xi$ is an elementary extension of $\mathfrak{A}$ of cardinality α^+. Finally, we define the desired extension as the union $\mathfrak{B} = \bigcup_{\eta < \alpha^+} \mathfrak{B}_\eta$, where $\mathfrak{B}_\eta = \bigcup_{\xi < \eta} \mathfrak{B}_\xi$, whenever η is a limit ordinal with $0 < \eta < \alpha^+$. The resulting model $\mathfrak{B}$ will be an elementary extension of $\mathfrak{A}$ of cardinality 2^α that is α^+-saturated.

The existence of α-regular ultrafilters is required also in order to prove the existence of saturated universes. A previous step consists of showing the existence of α-universal models. The definition of α^+-universality is as follows.

Definition 4.6.3 A model $\mathfrak{A}$ is said to be *α-universal* if and only if every model $\mathfrak{B}$ of cardinality less than α that is elementarily equivalent to $\mathfrak{A}$ is elementarily embedded in $\mathfrak{A}$:

$$\mathfrak{B} \equiv \mathfrak{A} \quad \text{and} \quad |B| < \alpha \quad \text{implies} \quad \mathfrak{B} \precsim \mathfrak{A}.$$

Examples of α-universal structures, where α is the cardinality of the continuum, include preference structures of cardinality smaller than the continuum. There is an interesting correlation between α-regularity and α-universality, which is stated in the following theorem.

Theorem 4.6.4 *Suppose that* $||\mathcal{L}|| \leq \alpha$ *and that* $\mathcal{U}$ *is an* α*-regular ultrafilter on* I*. Then, for every model* $\mathfrak{A}$ *with* $|A| = \alpha$*, the ultrapower* $\prod_{\mathcal{U}} \mathfrak{A}$ *is* α^+*-universal.*

Proof Let $\mathfrak{A}$ be any model of $\mathcal{L}$ and let $\mathcal{U}$ be an α-regular ultrafilter. Then, there exists a subset $E \subseteq \mathcal{U}$ with $|E| = \alpha$ such that each $i \in I$ belongs to only finitely many elements $e \in E$.

Let $\mathfrak{B}$ be a model such that $\mathfrak{B} \equiv \prod_{\mathcal{U}} \mathfrak{A}$ and $|B| < \alpha$. We need to show that $\mathfrak{B}$ is elementarily embedded in $\prod_{\mathcal{U}} \mathfrak{A}$. Since $\mathfrak{A} \equiv \prod_{\mathcal{U}} \mathfrak{A}$ (by Łoś's theorem), we have $\mathfrak{B} \equiv \mathfrak{A}$.

Let Γ_B be the elementary diagram of $\mathfrak{B}$ in the expanded language $\mathcal{L}_B$, which contains a constant c_b for each $b \in B$. The set Γ_B is the theory $\mathrm{Th}(\mathfrak{B}_B)$, i.e., the set of all statements in $\mathcal{L}_B$ that are true in $(\mathfrak{B}, b)_{b\in E}$. We have $|\Gamma_B| \leqslant \alpha$ because $||\mathcal{L}|| \leqslant \alpha$ and $|B| < \alpha$.

Since $|E| \leqslant \alpha$, there exists an injective function $h : \Gamma_B \to |E|$. Note that $h(\varphi)$ is an element of E and hence belongs to the ultrafilter $\mathcal{U}$.

Consider an element $i \in I$. Since there are only finitely many $e \in E$ to which i belongs, there are only finitely many sentences $\varphi \in \Gamma_B$ such that $i \in h(\varphi)$. Let $\varphi_1, \ldots, \varphi_n$ be such statements.

The sentence $\varphi_1 \wedge \cdots \wedge \varphi_n$ is true in $(\mathfrak{B}, b)_{b\in B}$ and is therefore consistent with the theory of $\mathfrak{A}$. It follows that there exists an expansion $(\mathfrak{A}, f_b(i))_{b\in B}$ of $\mathfrak{A}$ to $\mathcal{L}_B$ that is a model of $\varphi_1 \wedge \cdots \wedge \varphi_n$.

If $c_1, \ldots, c_n$ are the constants appearing in $\varphi_1 \wedge \cdots \wedge \varphi_n$ and $b_1, \ldots, b_n$ are their interpretations in $(\mathfrak{B}, b)_{b\in B}$, then the interpretations $a_j = f_{b_j}(i)$ $(j = 1, \ldots, n)$ of these constants satisfy the formula $\varphi_1 \wedge \cdots \wedge \varphi_n$; that is,

$$(\mathfrak{A}, f_b(i))_{b\in B} \models \varphi_1 \wedge \cdots \wedge \varphi_n.$$

We take such an expansion for each $i \in I$, thereby obtaining functions $f_b \in A^I$ such that, for every $i \in I$ and $\varphi \in \Gamma_B$,

$$\text{if} \quad i \in h(\varphi) \quad \text{then} \quad (\mathfrak{A}, f_b(i))_{b\in B} \models \varphi.$$

Then, by Łoś's theorem, since $h(\varphi) \in \mathcal{U}$ for every φ,

$$\prod_{\mathcal{U}} (\mathfrak{A}, f_b(i))_{b\in B} \models \varphi. \tag{4.1}$$

For each $b \in B$, let $a_b = [f_b(i)]$, where

$$[f_b(i)] = \{\{f_{b'}(i)\} \mid \{i \in I \mid f_{b'}(i) = f_b(i)\} \in \mathcal{U}\}.$$

By the definition of ultraproduct,

$$\prod_{\mathcal{U}} (\mathfrak{A}, f_b(i))_{b\in B} = \left(\prod_{\mathcal{U}} (\mathfrak{A}, a_b)_{b\in B} \right).$$

But (1) implies that $\prod_{\mathfrak{U}}(\mathfrak{A}, a_b)_{b\in B}$ is a model of Γ_B, from which we conclude that $\mathfrak{B} \precsim \prod_{\mathfrak{U}} \mathfrak{A}$. ⊣

Theorem 4.6.5 *Let $\mathfrak{A}$ be a model for a language $\mathscr{L}$ of power $\leqslant \alpha$. Then the ultrapower $\prod_{\mathfrak{U}} \mathfrak{A}$ of $\mathfrak{A}$ modulo an α-regular ultrafilter $\mathfrak{U}$ is an α^+-saturated model over $\mathfrak{A}$.*

Proof Let $\mathfrak{A}$ be a model for a language $\mathscr{L}$ of power $\leqslant \alpha$, let $\mathfrak{U}$ be an α-regular ultrafilter, and consider the ultrapower $\prod_{\mathfrak{U}} \mathfrak{A}$. Since $\mathfrak{A}$ is isomorphically embedded in $\mathfrak{A}$ by the natural embedding $\iota : \mathfrak{A} \to \prod_{\mathfrak{U}} \mathfrak{A}$, ι can be extended to an isomorphism $f : \mathfrak{B} \cong \prod_{\mathfrak{U}} \mathfrak{A}$, where $\mathscr{B}$ is an extension of $\mathfrak{A}$ (see Theorem 3.9.14).

Let $X \subseteq A$ with $|X| < \alpha^+$. Then, by Theorem 4.6.4, the ultrapower $\prod_{\mathfrak{U}}(\mathfrak{A}, x)_{x\in X}$ is α^+-universal. Since

$$\prod_{\mathfrak{U}}(\mathfrak{A}, x)_{x\in X} = (\prod_{\mathfrak{U}}(\mathfrak{A}, \iota(x))_{x\in X} \cong (\mathfrak{B}, x)_{x\in X},$$

it follows that $(\mathfrak{B}, x)_{x\in X}$ is α^+-universal.

Any set of formulas consistent with the theory of $(\mathfrak{A}, x)_{x\in X}$ is satisfiable in some model $(\mathfrak{C}, x)_{x\in X} \equiv (\mathfrak{A}, x)_{x\in X}$ of power $\leqslant \alpha$. Since $(\mathfrak{C}, x)_{x\in X}$ is elementarily embedded in $(\mathfrak{B}, x)_{x\in X}$, it follows that $\Sigma(x)$ is satisfiable in $(\mathfrak{B}, x)_{x\in X}$ and thus in $\prod_{\mathfrak{U}} \mathfrak{A}$. ⊣

Theorem 4.6.6 *Let $\mathfrak{A} = \langle \mathfrak{v}(\mathbb{X}), \in \rangle$ be a model of $\mathscr{L}$, where $||\mathscr{L}|| \leqslant \alpha$ and $\omega \leq |A| \leqslant \alpha^+$, where α^+ is the smallest cardinal greater than α and $\mathfrak{v}(\mathbb{X})$ is a superstructure over a nonempty set $\mathbb{X}$. Then there exists an α-saturated enlargement of $\mathfrak{A}$.*

Proof Let $\mathfrak{A} = \langle \mathfrak{v}(\mathbb{X}), \in \rangle$ be a model of $\mathscr{L}$, where $||\mathscr{L}|| \leqslant \alpha$ and $\omega \leq |A| \leqslant \alpha^+$, where α^+ is the smallest cardinal greater than α.[1] Also, let $\mathfrak{B}_1$ be the ultrapower $\prod_{\mathfrak{U}} \mathfrak{A}$ of $\mathfrak{A}$ modulo an α-regular ultrafilter $\mathfrak{U}$. By Theorem 4.6.5, $\mathfrak{B}_1$ is an elementary extension of $\mathfrak{A}$ of power α^+ such that, for every $Y \subset A$ of power α, $(\mathfrak{B}_1, y)_{y\in Y}$ realizes every type $\Sigma(v)$ of $(\mathfrak{A}, y)_{y\in Y}$.

Note that, by the same Theorem 4.6.5, $(\mathfrak{B}_{\xi+1}, a)_{a\in Y}$ realizes every type $\Sigma(v)$ of $(\mathfrak{B}_\xi, a)_{a\in Y}$ for every $Y \subset B_\xi$ of power α.

Thus, we construct an elementary chain $\mathfrak{B}_\xi$, starting from $\mathfrak{B}_1$, the ultrapower of the model $\mathfrak{A}$, where each $\mathfrak{B}_\xi$ is an elementary extension of $\mathfrak{B}_1$ and hence of $\mathfrak{A}$. The required elementary extension is the limit $\mathfrak{B}$ of the elementary chain

$$\mathfrak{B}_1 \prec \mathfrak{B}_2 \prec \cdots \prec \mathfrak{B}_\xi \prec \cdots$$

which is defined as the union

[1] That implies that α^+ is a regular cardinal, meaning that the supremum of any sequence of numbers smaller than α^+ is still less than α^+, so α^+ has cofinality greater than α.

$$\mathfrak{B} = \bigcup_{\eta<\alpha^+} \mathfrak{B}_\eta,$$

where $\mathfrak{B}_\eta = \bigcup_{\xi<\eta} \mathfrak{B}_\xi$ whenever η is a limit ordinal $0 < \eta < \alpha^+$.

We now show that this limit is an α^+-saturated elementary extension of $\mathfrak{A}$, meaning that $\mathfrak{B}$ is an elementary extension of $\mathfrak{A}$ of power α^+ that satisfies $\Sigma(b)$ for some $b \in B$. To this end, we adopt the following notation. We denote the universe $A = \mathfrak{v}(\mathbb{X})$ of $\mathfrak{A}$ by W^0, with $W^0 = \bigcup_{n<\omega} W_n{}^0$, where $W_n{}^0 = \mathfrak{v}_n(\mathbb{X})$. In general, given an α-regular ultrafilter $\mathcal{U}$, we define

$$W_0{}^{\xi+1} = \prod_{\mathcal{U}} W_0{}^\xi, \quad W_n{}^{\xi+1} = V_n(W_0{}^{\xi+1}), \quad W^{\xi+1} = \bigcup_{n<\omega} W_n{}^{\xi+1}.$$

We write the elementary embedding of $\mathfrak{B}_\xi$ into $\mathfrak{B}_{\xi+1}$ ($\xi < \alpha^+$) as $*^{\xi\xi+1}$. Since the relation $\prec$ is transitive, the composition of embeddings is well-defined, and therefore the embedding $*^{\xi\zeta}$ of $\mathfrak{B}_\xi$ into $\mathfrak{B}_\zeta$ ($\xi \leqslant \zeta$) is also well-defined.

We define the desired embedding $*$ as the elementary embedding $*^{0\alpha^+}$ of $\mathfrak{A}$ into the limit $\mathfrak{B}$, which is an α-saturated extension of $\mathfrak{A}$. If we denote $W_0 = W_0{}^{\alpha^+}$ by $\mathbb{Y}$, then $\mathbb{Y} = {}^*W_0{}^0 = {}^*\mathbb{X}$ and $\mathfrak{v}(\mathbb{Y})$ is a superstructure over $\mathbb{Y}$ with $W = W^{\alpha^+} \subseteq \mathfrak{v}(\mathbb{Y})$. The internal entities in $\mathfrak{v}(\mathbb{Y})$ coincide with the elements in W. ⊣

Equivalent formulations of α-saturation relevant to nonstandard analysis are given in the following theorems. First, an important definition.

Definition 4.6.7 Let $\mathfrak{A} = \langle \mathfrak{v}(\mathbb{Y}), \in \rangle$ be a superstructure and $R \in \mathfrak{v}(\mathbb{Y})$ an internal binary relation. R is *concurrent* or *finitely satisfiable* over a set B contained in the domain of R if, for each $n \in \mathbb{N}$ and each finite set $\{b_1, \ldots, b_n\} \subseteq B$, there exists a c in the image of R such that $(b_i, c) \in R$ for $1 \leq i \leq n$.

Theorem 4.6.8 *The nonstandard structure $\mathfrak{A} = \langle \mathfrak{v}(\mathbb{Y}), \in \rangle$ is α-saturated iff, for every internal binary relation $R \in \mathfrak{v}(\mathbb{Y})$ and every α-small subset B (internal or external) of the domain of R such that R is concurrent over B, there exists an element c in the image of R such that $(b, c) \in R$ for every $b \in B$.*

Proof Suppose that $\mathfrak{A} = \langle \mathfrak{v}(\mathbb{Y}), \in \rangle$ is α-saturated; that is, for every set $B \subset \mathfrak{v}(\mathbb{Y})$ of cardinality $|B| < \alpha$, every set $\Sigma(b)$ of formulas in $\mathcal{L}_B$ that is consistent with the theory of $(\mathfrak{A}, b)_{b\in B}$ is satisfiable in $(\mathfrak{A}, b)_{b\in B}$. Let R be an internal binary relation in $\mathfrak{v}(\mathbb{Y})$, and let B be an α-small subset of the domain of R such that R is concurrent over B. For each $b \in B$, let $\phi(b, x)$ be the formula Rbx, and let $\Sigma(x)$ be the set of such formulas. Since $\Sigma(x)$ is finitely satisfiable (as R is concurrent over B), each formula in $\Sigma(x)$ is consistent with the theory of $(\mathfrak{A}, b)_{b\in B}$. On the other hand, we observe that $\Sigma(x)$ contains fewer than α formulas in $\mathcal{L}_B$ (those for which $b \in B$). Hence, $\Sigma(x)$ is satisfiable; that is, for each $b \in B$, there exists a $c \in \mathfrak{v}(\mathbb{Y})$ such that Rbc.

The proof of the converse is obvious and is left to the reader.⊣

A further result relates α-saturation to the finite intersection property.

Theorem 4.6.9 *Nonstandard superstructure* $\mathfrak{A} = \langle \mathfrak{v}(\mathbb{Y}), \in \rangle$ *is* α*-saturated iff, for every internal set C and every* α*-small collection* $\mathcal{B}$ *(internal or external) consisting of internal subsets of C such that* $\mathcal{B}$ *has the finite intersection property, it follows that* $\bigcap B \neq \varnothing$.

Proof Suppose $\mathfrak{A} = \langle \mathfrak{v}(\mathbb{Y}), \in \rangle$ is α-saturated, let C be an internal set and $\mathcal{B}$ any α-small collection of internal subsets of C that has the FIP. Let R be the relation

$$R := \{\langle B, a \rangle \mid a \in B \in {}^*\mathcal{P}(C)\}.$$

In order to see that R is concurrent notice that, for each finite family $\{B_1, \ldots, B_n\}$ of elements of $\mathcal{B}$, there is an $a \in \cap B_i$, as $\mathcal{B}$ has the finite intersection property. Hence, for each finite set $\{B_1, \ldots, B_n\}$ there is an a such that $\langle B_i, a \rangle \in R$ for every $i = 1, \ldots, n$. This establishes that R is concurrent over $\mathcal{B}$. By the previous theorem (Theorem 4.6.8), there exists an element b in the range of R such that $\langle B, b \rangle \in R$ for every $B \in \mathcal{B}$—i.e. $b \in B$ for every $B \in \mathcal{B}$—and so $\bigcap B \neq \varnothing$.

The proof of the converse is obvious and is left to the reader.⊣

We have thus proven the existence of saturated nonstandard structures. These constitute the appropriate environment for the development of nonstandard analysis, particularly for the development of the topology. We will undertake this task in the following chapter.

Exercises

4.1 In Theorem 4.2.4 prove claim for functions and constants.

4.2 Check that equivalence classes induced by {2} satisfy

$$\{i \in I \mid f_j(i) = f_k(i)\} \in \mathcal{U} \leftrightarrow j = k, \quad (j, k = 1, 2, 3).$$

4.3 Prove Lemma 4.5.5.

4.4 Prove converse of Theorem 4.6.8.

4.5 Prove converse of Theorem 4.6.9.

Chapter 5
The Ontology of Calculus

Abstract Cantor declared himself an outspoken opponent of infinitesimal numbers and, in fact, offered an argument through which he aimed to demonstrate the impossibility of their existence—based precisely on set theory. His conclusion does not follow solely from the axioms of set theory, but rather requires an additional premise—called the "chain thesis"—which describes one of the properties commonly attributed to physical space. Paradoxically, despite such a forceful attempt at refutation, Cantor's theory provided the very framework needed to resolve, once and for all, the foundational problems of infinitesimal calculus. This chapter develops in detail the ontology of nonstandard calculus, showing the adequacy of nonstandard universes for analysis. It presents the transfer method and explains that it is only applicable to entities called "internal". The chapter also presents methods for dealing with them.

Keywords Ontology of calculus · Transfer method · Internal entities

5.1 Cantor's Rejection of Infinitesimals

Abraham Robinson (1967) equated Leibniz's conception of infinite entities with Hilbert's formalist view; the intuitionists' perspective with that of Aristotle; and the views of De l'Hospital and Fontenelle with that of Cantor.

> Cantor's belief in the actual existence of the infinities of Set Theory still predominates in the mathematical world today. His basic philosophy may be likened to that of de l'Hospital and Fontenelle although their infinite quantities were thought to be concrete and geometrical while Cantor's infinities are abstract and divorced from the physical world. Similarly, the intuitionists and other constructivists of our time may be regarded as the heirs to the Aristotelian traditions of basing Mathematics on the potential infinite. Finally, Leibniz' approach is akin to Hilbert's original formalism, for Leibniz, like Hilbert, regarded infinitary entities as ideal, or fictitious, additions to concrete Mathematics. Thus, we may conclude this talk with the observation that although the very subject matter of foundational research has changed radically over the last two hundred years, there is a remarkable permanency in the concern with the infinite in Mathematics and in the various philosophical attitudes which have been adopted towards this notion. (Robinson, 1967, 39–40.)

A. García de la Sienra, *The Logical Foundations of Nonstandard Analysis*,
Synthese Library 530, https://doi.org/10.1007/978-3-032-05677-1_5

However, Cantor declared himself an avowed enemy of infinitesimal numbers and, in fact, offered an argument (Cantor, 1887,1888) through which he sought to demonstrate their impossibility, precisely based on set theory. Moore (2002) provides a reconstruction of this argument, showing that its conclusion does not follow solely from the axioms of set theory but requires an additional premise—what Moore calls the "chain thesis"—which describes one of the properties usually attributed to physical space. Paradoxically, despite such a vigorous attempt at refutation, Cantor's theory ultimately provided the necessary framework to resolve, once and for all, the problem of the foundations of infinitesimal calculus.

5.2 The Adequacy of Superstructures for Analysis

The universe $\langle \mathfrak{v}(\mathbb{R}), \mathfrak{v}({}^*\mathbb{R}), {}^* \rangle$ is suitable for the development of nonstandard mathematical analysis, including theorems relevant to economic theory, such as the Hahn-Banach theorem and the separating hyperplane theorem. As we will see shortly, structures of the form $\langle \mathfrak{v}(\mathbb{X}), \in \rangle$ satisfy all the axioms of ZFC, except for the Schema of Replacement.

Theorem 5.2.1 $\mathfrak{A} = \langle \mathfrak{v}(\mathbb{X}), \in \rangle$ *is a model of the axioms of ZFC, except for the Schema of Replacement.*

Proof $\mathfrak{A}$ *satisfies the Axiom of Extensionality*; that is,

$$(\forall x \in \mathfrak{v}(\mathbb{X})) \quad ((x \in A \leftrightarrow x \in B) \rightarrow A = B),$$

for any sets A and B in $\mathfrak{v}(\mathbb{X})$. Suppose that $(\forall x \in \mathfrak{v}(\mathbb{X}))(x \in A \leftrightarrow x \in B)$ and that $x \in A$. Then $x \in A \in \mathfrak{v}(\mathbb{X})$, and by the transitivity of $\mathfrak{v}(\mathbb{X})$, we get $x \in \mathfrak{v}(\mathbb{X})$, so we can apply the above condition and conclude that $x \in B$; that is, $A \subseteq B$. Similarly, $B \subseteq A$, from which we conclude that $A = B$.

$\mathfrak{A}$ *Satisfies the Union Axiom* Let A be any set in $\mathfrak{v}(\mathbb{X})$. We want to show that there exists a set $B \in \mathfrak{v}(\mathbb{X})$ such that

$$x \in B \leftrightarrow \exists b\, x \in b \in A.$$

Since $A \in \mathfrak{v}(\mathbb{X})$, it follows that $A \subseteq \mathfrak{v}_n(\mathbb{X})$ for some $n < \lambda$. Because $\mathfrak{v}_n(\mathbb{X})$ is transitive for all $\alpha < \lambda$,

$$x \in \bigcup A \Rightarrow \exists b\, x \in b \in A \Rightarrow x \in \mathfrak{v}_\alpha(\mathbb{X}).$$

Therefore, $\bigcup A \subseteq \mathfrak{v}_n(\mathbb{X})$, from which we conclude that $\bigcup A \in \mathfrak{v}(\mathbb{X})$.

$\mathfrak{A}$ *Satisfies the Power Set Axiom* Let A be any set in $\mathfrak{v}(\mathbb{X})$. We want to prove that $\mathcal{P}(A) \in \mathfrak{v}(\mathbb{X})$. If the rank of A is α, then the rank of $\mathcal{P}(A)$ is $\alpha + 1$, from which it follows that $\mathcal{P}(A) \in \mathfrak{v}(\mathbb{X})$. Hence, we have shown the existence of a set $B \in \mathfrak{v}(\mathbb{X})$, namely $\mathcal{P}(A)$, such that for every $x \in \mathfrak{v}(\mathbb{X})$,

$$x \in B \leftrightarrow x \subseteq A \leftrightarrow (\forall z \in \mathfrak{v}(\mathbb{X}))(z \in x \Rightarrow z \in A).$$

This follows from the transitivity of $\mathfrak{v}(\mathbb{X})$, by which $z \in A$ implies $z \in \mathfrak{v}(\mathbb{X})$.

$\mathfrak{A}$ satisfies the Regularity Axiom. We want to show that if A is any nonempty set in $\mathfrak{v}(\mathbb{X})$, then there exists an element x of A that has nothing in common with A, i.e., $x \cap A = \varnothing$. Let x be an element of A with the smallest possible rank. By transitivity, $x \in \mathfrak{v}(\mathbb{X})$, and x satisfies the required condition.

$\mathfrak{A}$ satisfies the Pairing Axiom. This axiom was not included in our list of **ZFC** axioms because it is derivable from the Union Axiom.[1] Using the Schema of Existence of Sets, we prove the existence of an unordered pair for any two sets or urelements (Theorem 1.4.3). However, in **ZFC**, it is possible to derive the existence of unordered pairs solely from the Union Axiom (ZF2). Indeed, if $a, b \in \mathfrak{v}(\mathbb{X})$, then $\{a\}, \{b\} \in \mathfrak{v}(\mathbb{X})$, and the Union Axiom implies that $\{a, b\} = \{a\} \cup \{b\} \in \mathfrak{v}(\mathbb{X})$.

$\mathfrak{A}$ Satisfies the Axiom of Choice Let A be a family of pairwise disjoint nonempty sets. We need to show the existence of a function $f \equiv \mathfrak{v}(\mathbb{X})$, $f : A \to \bigcup A$, such that $f(B) \in B$ for all $B \in A$.

Consider the relation R whose domain is A and whose codomain is $\bigcup A$, defined such that the pair (B, x) belongs to R if and only if $x \in B$. For each $B \in A$, there are as many pairs of the form (B, x) satisfying this condition as there are elements in B. It follows that the relation F containing only one of these pairs for each B is a subset of R and therefore belongs to $\mathfrak{v}(\mathbb{X})$. Clearly, if $B \in A$ and (B, x) is the unique pair corresponding to B, then the function f defined by $f(B) = x$ is the required function.

$\mathfrak{A}$ Satisfies the Axiom of Infinity This holds whenever $\lambda \geq 2 \cdot \omega$, as in that case $V_{2\cdot\omega}(X)$ contains all the sets necessary for elementary mathematics (cf. Enderton, 1977, 252). If the base set X is infinite, λ can be ω. ⊣

A consequence of Theorem 5.2.1 is that $\mathfrak{v}(\mathbb{X})$ satisfies the material adequacy conditions required for a framework in which mathematical analysis can be developed. Indeed, we have the following result.

Theorem 5.2.2 $\mathfrak{v}(\mathbb{X})$ *satisfies the following conditions:*

(1) *If $a \in \mathfrak{v}(\mathbb{X})$, then $\{a\} \in \mathfrak{v}(\mathbb{X})$.*
(2) *$\mathfrak{v}(\mathbb{X})$ is closed under the formation of ordered n-tuples, Cartesian products, and subsets thereof.*
(3) *If the sets A and B belong to $\mathfrak{v}(\mathbb{X})$, then the set B^A of all functions from A to B also belongs to $\mathfrak{v}(\mathbb{X})$. Moreover, if $f \in B^A$, then f itself, the image of A under*

[1] Cf. Suppes (1969: 31).

f, the image of any proper subset of A under f, and the preimage of any subset of B under f all belong to $\mathfrak{v}(\mathbb{X})$.

(4) $\mathfrak{v}(\mathbb{X})$ *is closed under arbitrary intersections and unions of sets; that is, if the family of sets A is in* $\mathfrak{v}(\mathbb{X})$*, then so are* $\bigcup A$ *and* $\bigcap A$*. If A is in* $\mathfrak{v}(\mathbb{X})$*, then so are its power set and all of its subsets.*

(5) $\mathfrak{v}(\mathbb{X})$ *is strongly transitive; that is, for any set A in* $\mathfrak{v}(\mathbb{X})$*, there exists a transitive set* $B \in \mathfrak{v}(\mathbb{X})$ *such that* $A \subseteq B \subseteq \mathfrak{v}(\mathbb{X})$.

Proof

(1) Let a be an arbitrary element of $\mathfrak{v}(\mathbb{X})$, and let α be the rank of a, so that $a \in \mathfrak{v}_\alpha(\mathbb{X})$. Since $\{a\} \in \mathscr{P}[\mathfrak{v}_\alpha(\mathbb{X})] \subseteq \mathfrak{v}_{\alpha+1}(\mathbb{X}) \subseteq \mathfrak{v}(\mathbb{X})$, it follows that $\{a\} \in \mathfrak{v}_{\alpha+1}(\mathbb{X}) \subseteq \mathfrak{v}(\mathbb{X})$.
Thus, $\mathfrak{v}(\mathbb{X})$ satisfies the formula:

$$(\forall x)(\exists y)(x \in y \wedge (\forall z \in y)\, z = x).$$

(2) If a and b belong to $\mathfrak{v}(\mathbb{X})$, then by the Pairing Axiom, $\{a, b\} \in \mathfrak{v}(\mathbb{X})$. Let n be the maximum of rank a and rank b, so that $a, b \in \mathfrak{v}_\alpha(\mathbb{X})$. Then $\{a, b\}$ and $\{b\}$ belong to $\mathscr{P}(\mathfrak{v}_\alpha(\mathbb{X})) \subseteq \mathfrak{v}_{\alpha+1}(\mathbb{X})$, and

$$(a, b) = \{\{a, b\}, \{b\}\} \in \mathscr{P}(\mathfrak{v}_{\alpha+1}(\mathbb{X})) \subseteq \mathfrak{v}_{\alpha+2}(\mathbb{X}).$$

Thus, if A and B are sets such that $A, B \in \mathfrak{v}_{\alpha+1}(\mathbb{X})$, then since any elements $a \in A$ and $b \in B$ are in $\mathfrak{v}_\alpha(\mathbb{X})$, each pair (a, b) is in $\mathfrak{v}_{\alpha+2}$, and the set of all such pairs, namely $A \times B$, is in $\mathscr{P}(\mathfrak{v}_{\alpha+2}(\mathbb{X})) \subseteq \mathfrak{v}_{\alpha+3} \subseteq \mathfrak{v}(\mathbb{X})$.
Since any ordered n-tuple is defined recursively as an ordered pair, the Cartesian product of any finite family of sets in $\mathfrak{v}(\mathbb{X})$ is in $\mathfrak{v}(\mathbb{X})$. Any subset of a Cartesian product is in its power set and thus belongs to the next immediate stratum, hence to $\mathfrak{v}(\mathbb{X})$.

(3) If the sets A and B belong to $\mathfrak{v}(\mathbb{X})$, then the Cartesian product $A \times B$ also belongs to $\mathfrak{v}(\mathbb{X})$, and thus so do all relations $R \subseteq A \times B$. Since the domain of any such relation R is a subset of A, and A belongs to $\mathfrak{v}(\mathbb{X})$, it follows that the domain also belongs to $\mathfrak{v}(\mathbb{X})$. Similarly, the codomain of R, being a subset of B, belongs to $\mathfrak{v}(\mathbb{X})$. Any function $f \in B^A$ is a relation $R \subseteq A \times B$, so it, along with its domain and codomain, belongs to $\mathfrak{v}(\mathbb{X})$. Finally, since the image of any function is a subset of its codomain, it follows that the image also belongs to $\mathfrak{v}(\mathbb{X})$.

(4) Since $\mathfrak{v}(\mathbb{X})$ satisfies the Union Axiom and the Power Set Axiom, the arbitrary union of any element of $\mathfrak{v}(\mathbb{X})$, as well as its power set and all its subsets, are in $\mathfrak{v}(\mathbb{X})$. It remains to show that if A is in $\mathfrak{v}(\mathbb{X})$, then so is $\bigcap A$. If A is a family of sets in $\mathfrak{v}(\mathbb{X})$, its elements b belong to $\mathfrak{v}(\mathbb{X})$, and therefore so do the elements of b. Since the intersection of the b is just a subset of $\bigcap A$, it follows that $\bigcap A$ is in A_{n+1}, where n is the rank of A.

(5) We have already shown that $\mathfrak{v}(\mathbb{X})$ is strongly transitive. $\dashv$

5.3 The Method of Transference

The cultivation of nonstandard analysis—or Robinsonian analysis, as some prefer to call it—requires a certain set-theoretic framework whose existence we have already proved. Recall that this consists of two superstructures, $\mathfrak{v}(\mathbb{R})$ and $\mathfrak{v}({}^*\mathbb{R})$, where $\mathfrak{A} = \langle \mathfrak{v}(\mathbb{R}), \in \rangle$ and $\mathfrak{B} = \langle \mathfrak{v}({}^*\mathbb{R}), \in \rangle$ are structures corresponding to language $\mathscr{L}_\in$, and of an elementary embedding * of $\mathfrak{A}$ into $\mathfrak{B}$.

Actually, we wish to consider an expansion of language $\mathscr{L}_\in$ for these structures that includes an individual constant term τ for each element a of $\mathfrak{A}$, which we will occasionally denote simply as a for brevity. The (expanded) language $\mathscr{L}_\in$, interpreted in $\mathfrak{A}$, will be denoted $\mathscr{L}_{\mathfrak{A}}$; the same language interpreted in $\mathfrak{B}$ will be denoted $\mathscr{L}_{\mathfrak{B}}$. Each constant τ of the expanded language naturally has an interpretation in $\mathfrak{B}$, which will be precisely the image $\tau^{\mathfrak{B}}$ of $\tau^{\mathfrak{A}}$ under the transfer mapping defined below.

Definition 5.3.1 Let $\mathfrak{A} = \langle \mathfrak{v}(\mathbb{R}), \in \rangle$ and $\mathfrak{B} = \langle \mathfrak{v}({}^*\mathbb{R}), \in \rangle$ be models of the language $\mathscr{L}_\in$. Function * is a proper elementary embedding of $\mathfrak{A}$ into $\mathfrak{B}$ that satisfies the following conditions:

(1) The image of the interpretation $P^{\mathfrak{A}}$ of the predicate P in $\mathscr{L}$ is

$$^*P^{\mathfrak{A}} = P^{\mathfrak{B}},$$

the interpretation of P in the structure $\mathfrak{B}$.

(2) The image of the interpretation $F^{\mathfrak{A}}$ of the function symbol F in $\mathscr{L}$ is

$$^*F^{\mathfrak{A}} = F^{\mathfrak{B}},$$

the interpretation of F in the structure $\mathfrak{B}$.

(3) The image of the interpretation $c^{\mathfrak{A}}$ of the constant symbol c in $\mathscr{L}$ is

$$^*c^{\mathfrak{A}} = c^{\mathfrak{B}},$$

the interpretation of c in the structure $\mathfrak{B}$.

Symbol '$^*P$' denotes the relation $P^{\mathfrak{B}}$; '$^*F$' denotes the function $F^{\mathfrak{B}}$, and '$^*c$' denotes $c^{\mathfrak{B}}$. We use the symbol $\rightsquigarrow$ for the semantic designation relation. The function * is called the *transfer mapping*.

The transfer method is based on the fact that * is a bounded elementary embedding. Since the elementary embedding * is proper, the image $^*A$ of any infinite set in $\mathfrak{v}(\mathbb{R})$ contains elements that are not the image of any element of $\mathfrak{v}(\mathbb{R})$ under *. For this reason, the transfer mapping is called an *enlargement*. It is called a transfer mapping because it "transfers" true statements in $\mathfrak{A}$ to the interpretation of the same statements in $\mathfrak{B}$.

Definition 5.3.2 Let $\mathscr{L}_{\in}$ be the language common to $\mathfrak{A}$ and $\mathfrak{B}$, $\mathscr{L}_{\mathfrak{A}}$ the language $\mathscr{L}$ interpreted in $\mathfrak{A}$, and $\mathscr{L}_{\mathfrak{B}}$ the same language interpreted in $\mathfrak{B}$. If $\phi_{\mathfrak{A}}$ is a statement in $\mathscr{L}_{\mathfrak{A}}$ that is true in $\mathfrak{A}$, and $\phi_{\mathfrak{B}}$ is its interpretation in $\mathfrak{B}$, then by the *transfer principle*, we understand that a statement is true in its respective structure if and only if the corresponding statement is true in the other structure:

$$\mathfrak{A} \models \phi_{\mathfrak{A}} \quad \text{if and only if} \quad \mathfrak{B} \models \phi_{\mathfrak{B}}.$$

Based on the transfer principle, we have the *transfer rule*: if a statement ϕ is true in $\mathfrak{A}$, then its transferred version $\phi_{\mathfrak{B}}$ is true in $\mathfrak{B}$.

In more informal language, the subscript in $\phi_{\mathfrak{A}}$ is usually omitted, and $\phi_{\mathfrak{B}}$ is simply denoted as $^*\phi$. The transfer mapping has the properties stated in the following theorem.

Theorem 5.3.3 *The transfer mapping* * *has the following properties:*

(1) * *is injective; that is,* $a = b$ *if and only if* $^*a =^* b$.
(2) $a \in A$ *if and only if* $^*a \in^* A$.
(3) $A \subseteq B$ *if and only if* $^*A \subseteq^* B$.
(4) *If* $A \subseteq X$, *then* $^*A \subseteq^* X$.
(5) $^*(A \cap B) =^* A \cap^* B$.
(6) $^*(A \cup B) =^* A \cup^* B$.
(7) $^*(A \setminus B) =^* A \setminus^* B$.
(8) $^*\{a_1, \ldots, a_m\} = \{^*a_1, \ldots, ^*a_m\}$, *so that* $^*A = \{^*a \mid a \in A\}$ *if* A *is finite.*
(9) *If* A *is transitive in* $\mathfrak{v}(\mathbb{X})$, *then* $^*A$ *is transitive in* $\mathfrak{v}(\mathbb{Y})$.
(10) $^*\mathcal{P}(A) \subseteq \mathcal{P}(^*A)$.
(11) $^*\langle a_1, \ldots, a_m\rangle = \langle ^*a_1, \ldots, ^*a_m\rangle$.
(12) $\langle a_1, \ldots, a_m\rangle \in R$ *if and only if* $\langle ^*a_1, \ldots, ^*a_m\rangle \in^* R$.
(13) *If* $R \in \mathfrak{v}(\mathbb{X})$ *is an* m*-ary relation, then so is* $^*R$.
(14) $^*(A_1 \times \cdots \times A_n) = (^*A_1 \times \cdots \times^* A_n)$, *so that* $^*(A^n) = (^*A)^n$ *for* $n \in \mathbb{N}$.
(15) *If* $f : A \to B$ *belongs to* $\mathfrak{v}(\mathbb{X})$, *then* $^*f :^* A \to^* B$ *and*

$$^*[f(a)] =^* f(^*a).$$

Proof

(1) Suppose that $a = b$, where $a, b \in \mathfrak{v}(\mathbb{X})$. Then, we have the following facts: (i) There is a constant c in $\mathscr{L}$ to which $\mathfrak{A}$ assigns the object a, which is the same as b. Consequently, $\mathfrak{B}$ assigns $^*a$, that is, $^*b$, to c. In other words,

$$^*a = c^{\mathfrak{B}} =^* b.$$

(2) Suppose that $a \in A$; that is,

$$\mathfrak{A} \models Pc_1c_2,$$

where P denotes the membership relation in the structure $\mathfrak{A}$, c_1 designates a, and c_2 designates A. Then, by transfer,

$$\mathfrak{B} \models Pc_1c_2,$$

where now P denotes the membership relation in $\mathfrak{B}$, c_1 designates $^*a$, and c_2 designates $^*A$. Thus, $^*a \in {}^*A$ holds in $\mathfrak{B}$.

(3) The proof of this proposition presents a difficulty that is common in applications of the Transfer Principle: the symbol $\subseteq$ does not belong to the language $\mathcal{L}$, which contains only a binary predicate (for the relation $\in$) and individual constants. Therefore, one must appeal to the definitions of informal symbols given in Chap. 1, in this case, the definition of subclass (Definition 1.2.1). If c_1 designates A and c_2 designates B in $\mathfrak{A}$, the assumption that $A \subseteq B$ in $\mathfrak{A}$ translates as

$$\mathfrak{A} \models \forall x(x \in c_1 \rightarrow x \in c_2).$$

By transfer, we obtain

$$\mathfrak{B} \models \forall x(x \in c_1 \rightarrow x \in c_2),$$

where now c_1 designates $^*A$ and c_2 designates $^*B$ in $\mathfrak{B}$, which informally means that every element of $^*A$ is an element of $^*B$; that is, $^*A \subseteq^* B$.

(4) This assertion presents a difficulty similar to the previous one, but now the non-primitive symbol is that of intersection. If c_1, c_2, c_3 are constants in $\mathcal{L}$ such that c_1 designates A, c_2 designates B, and c_3 designates $A \cap B$ in $\mathfrak{A}$, then we have

$$\mathfrak{A} \models \forall x(x \in c_3 \leftrightarrow (x \in c_1 \wedge x \in c_2)).$$

Thus,

$$\mathfrak{B} \models \forall x(x \in c_3 \leftrightarrow (x \in c_1 \wedge x \in c_2)),$$

where now c_1 designates $^*A$, c_2 designates $^*B$, and c_3 designates $^*(A \cap B)$ in $\mathfrak{B}$, which means that

$$^*(A \cap B) =^* A \cap^* B.$$

(8) Let $c_1, \ldots, c_m, c_{m+1}$ be constants of $\mathcal{L}$ such that c_1 designates $a_1, \ldots, c_m$ designates a_m, and c_{m+1} designates $\{a_1, \ldots, a_m\}$ in $\mathfrak{A}$. We then have

$$\mathfrak{A} \models \forall x(x \in c_{m+1} \leftrightarrow (x = c_1 \vee \cdots \vee x = c_m)).$$

By transfer,

$$\mathfrak{B} \models \forall x(x \in c_{m+1} \leftrightarrow (x = c_1 \vee \cdots \vee x = c_m)),$$

where now c_1 designates $^*a_1$, ..., c_m designates $^*a_m$, and c_{m+1} designates $^*\{a_1, \ldots, a_m\}$. In informal terms,

$$^*\{a_1, \ldots, a_m\} = \{^*a_1, \ldots, ^*a_m\}.$$

(10) Let c be a constant of $\mathscr{L}$ such that c designates A in $\mathfrak{A}$. If A is transitive in $\mathfrak{v}(\mathbb{X})$, then we have

$$\mathfrak{A} \models \forall x \forall y((x \in y \wedge y \in c) \rightarrow x \in c).$$

By transfer,

$$\mathfrak{B} \models \forall x \forall y((x \in y \wedge y \in c) \rightarrow x \in c),$$

which translates to $^*A$ being transitive in $\mathfrak{v}(\mathbb{Y})$. Similarly, if A is strongly transitive, then

$$\mathfrak{A} \models \forall x \forall y((x \subseteq y \wedge y \in c) \rightarrow x \in c),$$

and by transfer,

$$\mathfrak{B} \models \forall x \forall y((x \subseteq y \wedge y \in c) \rightarrow x \in c),$$

where now c designates $^*A$.

(11) Let c_1 and c_2 be constants of $\mathscr{L}$, such that, in $\mathfrak{A}$, c_1 designates $\mathscr{P}(A)$ and c_2 designates A. Consequently, we have

$$\mathfrak{A} \models \forall x(x \in c_1 \rightarrow \forall y(y \in x \rightarrow y \in c_2)).$$

By transfer, it follows that

$$\mathfrak{B} \models \forall x(x \in c_1 \rightarrow \forall y(y \in x \rightarrow y \in c_2)),$$

where now c_1 designates $^*\mathscr{P}(A)$ and c_2 designates $^*A$. In informal terms, everything in $^*\mathscr{P}(A)$ is a subset of $^*A$, and hence it is in $\mathscr{P}(^*A)$. This establishes that

$$^*\mathscr{P}(A) \subseteq \mathscr{P}(^*A).$$

Following the method illustrated here, proving the remaining propositions is an amusing exercise left to the reader. $\dashv$

5.4 Internal Entities

The transfer method applies only to certain sets of $\mathfrak{v}(\mathbb{Y})$ called internal sets. For this reason, when applying these concepts in the sciences, it is important to distinguish within $\mathfrak{v}(\mathbb{Y})$—the ontology of infinitesimal calculus—internal from external elements.

An element of $\mathfrak{v}(\mathbb{Y})$ is *internal* if it is standard—that is, if it is an image $^*a$ of an element a in $\mathfrak{v}(\mathbb{X})$—or if it belongs to a standard set, meaning a set of the form $^*A$, where $A \in \mathfrak{v}(\mathbb{X})$. In other words, the internal elements of $\mathfrak{v}(\mathbb{Y})$ are either standard elements or elements of standard elements of $\mathfrak{v}(\mathbb{Y})$. Naturally, an element of $\mathfrak{v}(\mathbb{Y})$ is called *external* if it is not internal. The set of all internal elements of $\mathfrak{v}(\mathbb{Y})$ will be denoted by $^*\mathfrak{v}(\mathbb{X})$.

We define the *standard image* of set $A \in \mathfrak{v}(\mathbb{Y})$ as the set $\mathbf{ime}\, A$ of all standard elements of $^*A$. Notice that $\mathbf{ime}\, A$ is a proper subset of $^*A$ whenever A is infinite. However, if $A = \{a_1, \ldots, a_n\}$, then $\mathbf{ime}\, A$ coincides with $^*A$, since:

$$^*A = \{^*b \mid b \in \{a_1, \ldots, a_n\}\} = \{^*a_1, \ldots, ^*a_n\} = \mathbf{ime}\, A.$$

Incidentally, this shows that standard elements are internal. However, the set $\mathbf{ime}\, A$ itself is external whenever A is infinite.

Next, we will examine what kinds of sets are internal, provide some criteria for identifying them, and outline some methods for constructing them.

Theorem 5.4.1 $^*\mathcal{P}(A)$ *is the set of all internal subsets of* $^*A$*, meaning:*

$$^*\mathcal{P}(A) = \mathcal{P}(^*A) \cap^* \mathfrak{v}(\mathbb{X}) = \{B \subseteq^* A \mid B \textit{ is internal}\}.$$

Proof From Theorem 5.3.3(11), statement (11), we know that $^*\mathcal{P}(A) \subseteq \mathcal{P}(^*A)$. Now suppose that $B \in \mathcal{P}(^*A)$ is an internal set, meaning there exists some $C \in \mathfrak{v}(\mathbb{X})$ such that $B \in^* C$. The following sentence:

$$\forall x(x \in c_1 \to \forall y(y \in x \to (y \in c_2 \to x \in c_3)))$$

is true in $\mathfrak{A}$ if c_1 designates C, c_2 designates A, and c_3 designates $\mathcal{P}(A)$, since it expresses that every element x of C whose elements all belong to A—that is, every subset of A—belongs to $\mathcal{P}(A)$. But then, by the transfer principle, we obtain:

$$\mathfrak{B} \models \forall x(x \in c_1 \to \forall y(y \in x \to (y \in c_2 \to x \in c_3))),$$

where now c_1 designates $^*C$, c_2 designates $^*A$, and c_3 designates $^*\mathcal{P}(A)$. This means that every element of $^*C$ that is a subset of $^*A$ belongs to $^*\mathcal{P}(A)$, which in particular implies that $B \in {}^*\mathcal{P}(A)$. $\dashv$

It is convenient to simplify the notation and transition to a more informal language, even though statements can always be formally expressed in $\mathcal{L}_{\in}$. In particular, recall that statements of the form:

$$\forall x (x \in A \to \dots)$$

can be written using a bounded quantifier:

$$(\forall x \in A) \dots$$

Similarly, we can abbreviate complex expressions using defined terms; for example:

$$A \subseteq B$$

instead of:

$$\forall x (x \in A \to x \in B).$$

The results expressed in Theorem 5.3.3 justify these simplifications.

For instance, the Minimum Number Principle can be expressed as follows:

$$(\forall x \in \mathcal{P}(\mathbb{N}))(x \neq \varnothing \to (\exists y \in x)(\forall z \in x) y \leq z)).$$

By transfer, we obtain:

$$(\forall x \in {}^*\mathcal{P}(\mathbb{N}))(x \neq \varnothing \to (\exists y \in x)(\forall z \in x)\, y \leq z),$$

which means, by virtue of Theorem 5.4.1, that every internal subset of ${}^*\mathbb{N}$ has a first element. Since ${}^*\mathbb{N} \setminus \mathbb{N}$ has no first element, it cannot be internal.

The symbol $\leqslant$ does not belong to the lexicon of $\mathcal{L}_{\in}$, but we can introduce it via a definition, because all relations between real numbers belong to $\mathfrak{v}(\mathbb{X})$. In the next chapter, we will introduce the non-Archimedean ordered field of what we will call hyperreal numbers, which are the elements of ${}^*\mathbb{R}$. For example, we can define the symbol $\leqslant$ as follows. Let $c_1 = \mathbf{dom}\, R$ and $c_2 = \mathbf{im}\, R$. Then, the formula:

$$x \in c_1 \wedge y \in c_2$$

is satisfied in $\mathfrak{A}$ with the pair $r, r' \in \mathfrak{v}(\mathbb{X})$ if and only if r, r' are real numbers and $r \leq r'$:

$$\mathfrak{A} \models x \in c_1 \wedge y \in c_2 \quad \Leftrightarrow \quad r \leqslant r' \text{ and } r, r' \in \mathbb{R}.$$

Theorem 5.4.2 *Every internal set belongs to a standard transitive set. Every element of an internal set is internal and* ${}^*\mathfrak{v}(\mathbb{X})$ *is strongly transitive.*

Proof From Lemma 4.5.3, we know that both the strata $\mathfrak{v}_n(\mathbb{X})$ and $\mathfrak{v}(\mathbb{X})$ itself are transitive and strongly transitive. By Theorem 5.4.2, sets ${}^*\mathfrak{v}_n(\mathbb{X})$ and ${}^*\mathfrak{v}(\mathbb{X})$ are also transitive and strongly transitive.

To show that every internal set belongs to a standard transitive set, note that every set B in $\mathfrak{v}(\mathbb{X})$ is a subset of some stratum $\mathfrak{v}_n(\mathbb{X})$. Now, if A is an internal set, there exists $B \in \mathfrak{v}(\mathbb{X})$ such that $A \in {}^*B$, meaning $A \in {}^*B \in {}^*\mathfrak{v}_n(\mathbb{X})$, and therefore, $A \in {}^*\mathfrak{v}_n(\mathbb{X})$.

If a is an element of an internal set B, since $a \in B \in {}^*\mathfrak{v}_n(\mathbb{X})$ for some n, we obtain $a \in {}^*\mathfrak{v}_n(\mathbb{X})$, which means that a is internal. ⊣

Some properties of internal sets are displayed in the following theorem.

Theorem 5.4.3 *Internal sets possess the following properties:*

(1) *If A, B, and C are internal sets, then so are $A \cap B$, $A \cup B$, $A \setminus B$, and $A \times B$.*
(2) *If the elements $a_1, \ldots, a_n$ are internal, then so are $\{a_1, \ldots, a_n\}$ and $\langle a_1, \ldots, a_n\rangle$.*
(3) *If the binary relation R is internal, then its domain* **dom** *R, its range* **im** *R, its inverse R^{-1}, and the image of an internal set A under R, defined as*

$$\{b \in \mathfrak{v}(\mathbb{Y}) \mid (a, b) \in R \text{ for some } a \in A\},$$

are also internal.
(4) *If the binary relations R and S are internal, then so is their composition $R \circ S$.*
(5) *If the function f is internal and $a \in$* **dom** *f, then $f(a)$ is internal. If $A \subseteq$* **dom** *f is internal, then $f(A)$ is internal. Moreover, the preimage $f^{-1}(B)$ of an internal subset B of* **im** *f is also internal.*

Proof

(1) Since $\mathfrak{v}(\mathbb{X})$ is closed under set-theoretic operations, the following statements hold:

$$(\forall x \in \mathfrak{v}(\mathbb{X}))(\forall y \in \mathfrak{v}(\mathbb{X}))(x \cap y \in \mathfrak{v}(\mathbb{X})),$$
$$(\forall x \in \mathfrak{v}(\mathbb{X}))(\forall y \in \mathfrak{v}(\mathbb{X}))(x \cup y \in \mathfrak{v}(\mathbb{X})),$$
$$(\forall x \in \mathfrak{v}(\mathbb{X}))(\forall y \in \mathfrak{v}(\mathbb{X}))(x \setminus y \in \mathfrak{v}(\mathbb{X})).$$

By transfer, we obtain:

$$\forall x \in {}^*\mathfrak{v}(\mathbb{X}))(\forall y \in {}^*\mathfrak{v}(\mathbb{X}))(x \cap y \in {}^*\mathfrak{v}(\mathbb{X})),$$
$$(\forall x \in {}^*\mathfrak{v}(\mathbb{X}))(\forall y \in {}^*\mathfrak{v}(\mathbb{X}))(x \cup y \in {}^*\mathfrak{v}(\mathbb{X})),$$
$$(\forall x \in {}^*\mathfrak{v}(\mathbb{X}))(\forall y \in {}^*\mathfrak{v}(\mathbb{X}))(x \setminus y \in {}^*\mathfrak{v}(\mathbb{X})).$$

By universal specification, if A and B are internal, then so are $A \cap B$, $A \cup B$, and $A \setminus B$.

(2) Suppose that the elements $a_1, \ldots, a_n$ are internal, and let

$$m = \max\{\text{rank of } a_1, \ldots, \text{rank of } a_n\}.$$

Then $a_1, \ldots, a_n \in {}^*\mathfrak{v}_m(\mathbb{X})$, so

$$\{a_1, \ldots, a_n\} \in {}^*\mathfrak{v}_m(\mathbb{X}).$$

Similarly, by Lemma 4.5.5,

$$\langle a_1, \ldots, a_n \rangle \in {}^*\mathfrak{v}_{m+2(n-1)}(\mathbb{X}).$$

(3) The domain and range of a binary relation in $\mathfrak{v}(\mathbb{X})$ belong to $\mathfrak{v}(\mathbb{X})$ due to the transitivity of $\mathfrak{v}(\mathbb{X})$. Hence, we have:

$$(\forall x \in \mathfrak{v}(\mathbb{X}))(x \text{ is a binary relation } \Rightarrow$$
$$(\exists y, z \in \mathfrak{v}(\mathbb{X}))(y = \mathbf{dom}\, x \wedge z = \mathbf{im}\, x)).$$

By transfer:

$$(\forall x \in {}^*\mathfrak{v}(\mathbb{X}))(x \text{ is a binary relation } \Rightarrow$$
$$(\exists y, z \in {}^*\mathfrak{v}(\mathbb{X}))(y = \mathbf{dom}\, x \wedge z = \mathbf{im}\, x)).$$

Thus, if R is an internal binary relation, then **dom** R and **im** R are internal. The inverse of a binary relation R in $\mathfrak{v}(\mathbb{X})$ belongs to $\mathcal{P}(\mathbf{im}\, R \times \mathbf{dom}\, R) \in \mathfrak{v}(\mathbb{X})$, so:

$$(\forall x, y \in \mathfrak{v}(\mathbb{X}))(x \text{ is a binary relation } \wedge$$
$$y \text{ is the inverse of } x \Rightarrow y \in \mathcal{P}(\mathbf{im}\, y \times \mathbf{dom}\, x)).$$

By transfer and substitution, we obtain $R^{-1} \in {}^*\mathcal{P}(\mathbf{im}\, y \times \mathbf{dom}\, x)$, meaning R^{-1} is internal.

For an internal set A, define

$$B = \{b \in \mathfrak{v}(\mathbb{Y}) \mid (a, b) \in R \text{ for some } a \in A\},$$

which simplifies to $B = \mathbf{im}\, R \cap A$; so, B is internal.

Similarly, since the composition of binary relations is defined in $\mathfrak{v}(\mathbb{X})$, we conclude that $R \circ S$ is internal. ⊣

The Internal Set Definition Principle (ISDP) provides a method for defining internal sets. It ensures that a family of elements of an internal set, defined in $\mathbb{Y}$ by a formula containing only internal constants, is also internal. In its general formulation it adopts the following form.

Theorem 5.4.4 *If $\phi(x)$ is a formula interpreted in $\mathfrak{B}$, whose only free variable is x and whose constants denote internal elements, then for any internal set A, the set*

$$\{a \in A \mid \phi(a)\}$$

is internal.

Proof Let $\phi(x)$ be a formula interpreted in $\mathfrak{B}$, where x is the only free variable, and the constants in the formula denote internal elements. Let A be an internal set in ${}^*\nu(\mathbb{X})$. Denote by $c_1, \ldots, c_n$ the constants appearing in ϕ, and let $a_1, \ldots, a_n$ be their respective interpretations in $\mathfrak{B}$, which are internal elements in ${}^*\nu(\mathbb{X})$.

Replacing these constants with variables $x_1, \ldots, x_n$ in ϕ, we observe that $A, a_1, \ldots, a_n$ belong to an internal set ${}^*T$, which is transitive (Theorem 5.4.2). Thus, we have:

$$\mathfrak{A} \models (\forall x_1, \ldots, x_n, y \in T)(\exists u \in \mathcal{P}(T)(\forall x \in T)(x \in u \\ \leftrightarrow (x \in y \wedge \phi(x, x_1, \ldots, x_n))).$$

Indeed, regardless of the choice of elements $x_1, \ldots, x_n, y$ in T, it is always possible to collect into a set all elements of T that are in y and satisfy the condition $\phi(x, x_1, \ldots, x_n)$. By transfer, we obtain:

$$\mathfrak{B} \models (\forall x_1, \ldots, x_n, y \in^* T)(\exists u \in \mathcal{P}(^*T))(\forall x \in^* T)(x \in u \\ \leftrightarrow (x \in y \wedge \phi(x, x_1, \ldots, x_n))).$$

By universal specification, substituting $x_1, \ldots, x_n$ with the original constants $c_1, \ldots, c_n$, we conclude:

$$\mathfrak{B} \models (\exists u \in \mathcal{P}(^*T))(\forall x \in^* T)(x \in u \leftrightarrow (x \in A \wedge \phi(x))).$$

Thus, in $\mathfrak{B}$, there exists an internal set containing exactly those elements of A that satisfy the condition $\phi(x)$. ⊣

An example of an application of this principle is the following. If $\phi(x)$ is the formula:

$$\forall y \forall z \forall w((y \in x \wedge z \in x \wedge w \in^* \mathbb{R} \wedge 0 \leq w \leq 1) \rightarrow wy + (1 - w)z \in x)$$

and A is the set $\mathcal{P}(^*\mathbb{R}^n)$, then the set

$$\{a \in A \mid \phi(a)\}$$

is the class of all convex subsets of $^*\mathbb{R}^n$ (called 'Q-convex' in the literature). Since the only constant involved is $^*\mathbb{R}$, which is internal as it is the image of a standard object under the transfer mapping, the principle allows us to infer that these subsets are internal.

To exhibit an external element of $\nu(\mathbb{Y})$, we will now introduce the Internal Induction Principle, namely:

Theorem 5.4.5 *If A is an internal subset of $^*\mathbb{N}$ that contains 1 and is closed under the successor function $S(n) = n + 1$, then $A = {}^*\mathbb{N}$.*

Demostración In $\mathfrak{v}(\mathbb{X})$, the usual principle of mathematical induction states:

$$(\forall x \in \mathcal{P}\mathbb{N})(1 \in x \wedge (\forall y(y \in x \Rightarrow S(y) \in \mathbb{N})) \Rightarrow x = \mathbb{N}).$$

By transfer:

$$(\forall x \in {}^*\mathcal{P}\mathbb{N})(1 \in x \wedge (\forall y(y \in x \Rightarrow S(y) \in {}^*\mathbb{N})) \Rightarrow x = {}^*\mathbb{N}).$$

If A is an internal subset of $^*\mathbb{N}$, then $A \in {}^*\mathcal{P}\mathbb{N}$. Substituting x with A, we obtain:

$$(1 \in A \wedge (\forall y(y \in A \Rightarrow S(y) \in {}^*\mathbb{N}))) \Rightarrow A =^* \mathbb{N}. \dashv$$

By Theorem 5.4.5, $\mathbb{N}$ cannot be internal, since otherwise, $^*\mathbb{N}$ would be identical to $\mathbb{N}$.

We have in general the following facts.

Lemma 5.4.6 *Let A be an infinite set, let $f : \mathbb{N} \to A$ be an injective function, and let*

$$F = \{f(n) \mid n \in \mathbb{N}\}$$

be the image of $\mathbb{N}$ under f. Then:

$$^*f^{-1}(\mathbf{ime}\, F) = \mathbb{N}.$$

Proof The set $\mathbf{ime}\, F \subset^* F$ is simply the embedding of F in $\mathfrak{v}(\mathbb{Y})$, while $^*f$ is the enlarged function

$$^*f : {}^*\mathbb{N} \to {}^*A.$$

In $\mathfrak{v}(\mathbb{X})$, we have $f^{-1}(F) = \mathbb{N}$, and by transfer in $\mathfrak{v}(\mathbb{Y})$, we obtain

$$^*f^{-1}({}^*F) =^* \mathbb{N}.$$

Since $^*f(n) = f(n)$ is standard whenever $n \in \mathbb{N}$, the inverse images $^*f^{-1}(^*f(n))$ of the elements of **ime** F are precisely the numbers in $\mathbb{N}$.

Theorem 5.4.7 *The standard image* **ime** A *of any infinite set* $A \in \mathfrak{v}(\mathbb{X})$ *is an external element of* $\mathfrak{v}(\mathbb{Y})$.

Proof If **ime** F were internal, then since $^*F$ is standard and therefore internal, the intersection

$$\mathbf{ime}\, F \cap^* F = \{^*a \mid a \in F\} = \mathbf{ime}\, F$$

would also be internal. However, this is impossible because then

$$\mathbb{N} = {}^*f^{-1}(\mathbf{ime}\, F)$$

would also be internal, which contradicts our previous results. ⊣

With this, we conclude our foundational discussion of nonstandard analysis and proceed to its further development.

Exercises

5.1 Prove propositions (12)–(16) of Theorem 5.3.3.

References

Cantor, G. (1887, 1888). Mitteilungen zur Lehre vom Transfiniten. *Zeitschrift für Philosophie und philosophische Kritik*, *91*, 811–825; 18–88; *92*, 240–265.

Cantor, G. (1977). *Elements of set theory*. Academic Press.

Moore, M. E. (2002). A Cantorian argument against infinitesimals. *Synthese*, *133*, 305–330.

Robinson, A. (1967). *The metaphysics of the calculus*. Lakatos (1967), pp. 28–40. Reimpreso en Hintikka (1969), pp. 153–163.

Suppes, P. (1969). *Axiomatic set theory*. Princeton: D. Van Nostrand.

Chapter 6
The Hyperreal Numbers

Abstract Chapter 6 develops in detail the theory of hyperreal numbers; i.e., the field of real numbers together with infinitesimals and their algebraic transformations. It applies, in this special case, the principles of extension and transfer from the ordinary field of real numbers to that of the hyperreals. The chapter provides a detailed analysis of the algebraic structure of the set of limited (or bounded) numbers and outlines the rules of operation for hyperreal numbers in general. It concludes with a study of hypernatural, hyperinteger, and hyperrational numbers, which are important in the formulation of nonstandard economic theory.

Keywords Hyperreal Numbers · Principle of extension · Algebra of hyperreal numbers

6.1 Concurrent Relations

Just as the structure $\mathfrak{v}(\mathbb{R})$ contains the real numbers and all functions and relations between them—and much more than that—which allows for the development of classical mathematical analysis within it, structure $\mathfrak{v}({}^*\mathbb{R})$—or rather its internal part ${}^*\mathfrak{v}(\mathbb{R})$—allows us to develop nonstandard analysis and its applications. We shall suppose, from now on, that the structure $\mathfrak{A} = \langle \mathfrak{v}({}^*\mathbb{R}), \in \rangle$ is α-saturated for $\alpha \geqslant \#[\mathfrak{v}({}^*\mathbb{R})]$.

We begin by noting that we can refer to any object—number, relation, or function—in $\mathfrak{v}(\mathbb{R})$ by introducing a constant to name it and display its properties. For example, we know that $\mathbb{N} \in \mathfrak{v}(\mathbb{R})$, and that is why we can introduce term '$\mathbb{N}$' as a constant to refer to this important set. Then, we can use the transfer mapping * to refer to the extension of the object in $\mathfrak{v}({}^*\mathbb{R})$, for example, ${}^*\mathbb{N}$.

Theorem 3.9.13 allowed us to demonstrate the existence of an infinitesimal number in ${}^*\mathbb{R}$. We shall see below that the existence of infinitesimals entails the existence of a large set of numbers that we will call 'hyperreals'. We will denote such a set by the symbol $\mathbb{H}$. In this chapter we will explore set $\mathbb{H}$. We will classify its elements and show that, with the operations ${}^*{+}$, ${}^*{\cdot}$, the elements 0 and 1, and the relation ${}^*{\leqslant}$, set $\mathbb{H}$ forms a non-Archimedean ordered field.

A. García de la Sienra, *The Logical Foundations of Nonstandard Analysis*, Synthese Library 530, https://doi.org/10.1007/978-3-032-05677-1_6

A concept that will be useful in exploring set $\mathbb{H}$ is that of a concurrent binary relation. In its most general terms, it can be defined as follows:

Definition 6.1.1 We say that a binary relation $R \in \mathfrak{v}(\mathbb{R})$ is *concurrent* or *finitely satisfiable* if, for any subset $\{a_1, \ldots, a_n\}$ of the domain of R, there exists an element b such that $a_i R b$ for all $i \in \{1, \ldots, n\}$. The entity c is called *a bound* for R if $\langle a, c\rangle \in {}^*R$ for every a for which there exists a b such that $\langle a, b\rangle \in R$.

An example of a concurrent relation among real numbers is relation $<$ since, for any numbers $a_1, \ldots, a_n, b_1, \ldots, b_n$ such that $a_1 < b_1, \ldots, a_n < b_n$, any number $b > b_1, \ldots, b_n$ is such that $a_1 < b, \ldots, a_n < b$. A bound for $<$ would be a hyperreal number c greater than a for every real number a that is less than some real number b; that is, for every real number. Such a number c would, of course, be unlimited, meaning greater in absolute value than any real number. What is interesting is that it can be proven that every concurrent binary relation among real numbers possesses a bound in $\mathbb{H}$. This is the meaning of the following theorem.

Theorem 6.1.2 *Every concurrent binary relation R in $\mathfrak{v}(\mathbb{X})$ has a bound in ${}^*\mathbb{R}$.*

Proof Let R be a concurrent relation in $\mathfrak{v}(\mathbb{R})$. For each $x \in \mathbf{dom}\, R$, let B_x be the set of all y in the image of x under R,

$$B_x = \{y \in \mathbf{im}\, R \mid \langle x, y\rangle \in R\},$$

and let $\mathscr{A} = \{B_x \mid x \in \mathbf{dom}\, R\}$ be the set of all B_x.

Let $\mathscr{B} = \{B_{r_1}, \ldots, B_{r_n}\}$ be any finite subset of $\mathscr{A}$, and let $y_1, \ldots, y_n$ be such that $\langle x_1, y_1\rangle, \ldots, \langle x_n, y_n\rangle \in R$, so that $y_i \in B_{x_i}$ for $i = 1, \ldots, n$. Since R is concurrent, there exists a y such that $\langle x_1, y\rangle, \ldots, \langle x_n, y\rangle \in R$; that is, $y \in B_{x_i}$ for each $i = 1, \ldots, n$. This proves that $\mathscr{A}$ has the finite intersection property (FIP), and therefore, since $\langle \mathfrak{v}(\mathbb{Y}), \in\rangle$ is α-saturated, by Theorem 4.6.9 there exists an element c in ${}^*\mathfrak{v}(\mathbb{X})$ such that

$$c \in \bigcap\{{}^*B_x \mid x \in \mathbf{dom}\, R\}.$$

In other words, $\langle x, c\rangle \in {}^*\mathfrak{v}(\mathbb{X})$ for all ${}^*x \in \mathbf{ime}\,(\mathbf{dom}\, R)$.⊣

Theorem 6.1.2 allows us to explore the universe of hyperreals and establish their properties. Hyperreal numbers are also called nonstandard numbers. Notice that, according to this terminology, standard numbers are nonstandard, so it is preferable to refer to the elements of ${}^*\mathbb{R}$ as hyperreals.

The most remarkable result of the theory of nonstandard universes is the existence of a field where infinitesimal numbers reside—exactly the kind envisioned by Leibniz and criticized by Berkeley. Indeed, consider the relation R defined on $\mathbb{R}^+$ by

$$\langle x, y\rangle \in R \quad \text{if and only if} \quad 0 < x < y.$$

This relation is concurrent, since $0 < x_i < y_i$ implies the existence of some x such that $0 < x < y_i$, from which it follows that there exists a number $c \in {}^*R$ such that

$$0 < c < y \quad \text{for all } y \in \mathbb{R}^+.$$

6.2 The Ordered Field of Hyperreals

The existence of infinitesimals implies the existence of other numbers that are not real. First, it implies the existence of numbers greater (or smaller) than any real number. Indeed, if r is a positive real number, the statements of the form

$$\forall x((0 < x \wedge x < r) \Rightarrow (0 < r^{-1} \wedge r^{-1} < x^{-1}))$$

are true in $\mathfrak{v}(\mathbb{R})$ for each real r; therefore, by transfer, they are also true in $\mathfrak{v}({}^*\mathbb{R})$. Thus, if we universally specify by taking a positive infinitesimal ε as a substitution instance, we can infer that $r < \varepsilon^{-1}$ for every real $r > 0$. Numbers that are greater than any real number will be called 'unlimited'. More generally, we have the following definition.

Definition 6.2.1 Let x be an element of $\mathbb{H}$.

(1) x is *standard* or *real* if $x \in \mathbf{ime}\,\mathbb{R}$.
(2) x is a *positive infinitesimal* if $0 < x < y$ for every positive standard number y.
(3) x is a *negative infinitesimal* if $y < x < 0$ for every negative standard number y.
(4) x is *infinitesimal* if it is either negative infinitesimal, positive infinitesimal, or zero.
(5) x is *unlimited* if x is not limited; that is, x is greater than all standard numbers.
(6) x is *appreciable* if it is limited but not infinitesimal; that is, if $r < |x| < s$ for some $r, s \in \mathbb{R}$ with $r > 0$.

As we mentioned earlier, the hyperreal numbers form a pleasant and familiar algebraic structure.

Theorem 6.2.2 *The structure*

$$\mathfrak{H} = \langle \mathbb{H}, {}^*+, {}^*\cdot, {}^*0, {}^*1, {}^*< \rangle$$

is a non-Archimedean ordered field.

Proof An *ordered field* is any structure

$$\langle K, +, \cdot, 0, 1, < \rangle$$

that satisfies the following axioms:

(1) Associativity of addition:

$$\forall x \forall y \forall z \;\; x + (y + z) = (x + y) + z.$$

(2) Existence of identity element for addition:

$$\forall x \;\; (x + 0 = x \wedge 0 + x = x).$$

(3) Existence of additive inverses:

$$\forall x \exists y \;\; (x + y = 0 \wedge y + x = 0).$$

(4) Associativity of multiplication:

$$\forall x \forall y \forall z \;\; x \cdot (y \cdot z) = (x \cdot y) \cdot z.$$

(5) Commutativity of multiplication:

$$\forall x \forall y \;\; x \cdot y = y \cdot x.$$

(6) Existence of identity element for multiplication:

$$\forall x \;\; (x \cdot 1 = x \wedge 1 \cdot x = x).$$

(7) Distributivity of multiplication over addition:

$$\forall x \forall y \forall z \;\; x \cdot (y + z) = (x \cdot y) + (x \cdot z).$$

(8) Existence of multiplicative inverses for nonzero elements:

$$\forall x (x \neq 0 \Rightarrow \exists y \;\; x \cdot y = 1).$$

(9) $<$ is a strict simple ordering of set K.

(10) Order preservation under addition:

$$\forall x \forall y (x < y \Rightarrow x + z < y + z).$$

(11) Order preservation under multiplication:

$$\forall x \forall y ((0 < x \wedge 0 < y) \Rightarrow 0 < x \cdot y).$$

Since these sentences are satisfied by the ordered field of real numbers, it follows immediately by the transfer principle that they are also satisfied by the ordered field of hyperreals, where the quantifiers now range over the set $\mathbb{H}$.⊣

A field is said to be *Archimedean* if it satisfies the additional axiom:
11. Archimedean property:

$$\forall x \forall y (x < y \Rightarrow \exists z (Nz \wedge y < zx)),$$

where Nz means that z is a positive integer. Naturally, (11) fails in $\mathbb{H}$ if N is interpreted as the predicate "is a standard positive integer". However, this is not the case when N is properly interpreted in $\mathfrak{v}({}^*\mathbb{R})$, because in that case, N represents the set of positive hyperintegers!

6.3 The Principles of Extension and Transfer in $\mathbb{R}$

In Chap. 5, we demonstrated the principles of extension and transfer in a general setting for all elements of $\mathfrak{v}(\mathbb{X})$. Here, we restrict these principles to real numbers and relations defined on them.

Definition 6.3.1 (Extension Principle) For every relation R defined on $\mathbb{R}^n$, there exists a relation ${}^*R$ defined on ${}^*\mathbb{R}^n$ such that $R \subseteq {}^*R$. The relation ${}^*R$ is called the natural extension of R.

In particular, for every function $f : X \to Y$, defined on a domain $X \subseteq \mathbb{R}^m$ with an image $Y \subseteq \mathbb{R}^n$ (where m, n are positive integers), there exists a function ${}^*f : {}^*X \to {}^*Y$. The function ${}^*f$ is called the natural extension of f.

We have already seen the transfer method. When applied to real numbers and their supersets, it takes the following form:

Definition 6.3.2 (Transfer Method) *Any statement in the language that holds for one or more relations or functions defined on any subset of* $\mathbb{R}^n$ *also holds for the natural extensions of these relations or functions in* $\mathfrak{H}$.

In order to formulate the third principle, we need to develop a deeper understanding of the set of hyperreal numbers $\mathbb{H}$. The most remarkable fact is that $\mathbb{H}$ contains a subset $\mathbb{I}$ consisting of infinitesimal numbers that are distinct from zero.

We denote that a number x is infinitesimally close to y with the notation $x \approx y$. It follows that $x \approx 0$ if and only if x is infinitesimal.

It is easy to see that appreciable numbers exist that are not real. If ε is a positive infinitesimal and r is a positive real number, then $r + \varepsilon$ is limited and belongs to $\mathbb{H}$, but it is not infinitesimal because $0 < r < r + \varepsilon < s$ for any real number s greater than r.

In general, limited numbers are of the form $r + \varepsilon$, with $\varepsilon \in \mathbb{I}$. Appreciable numbers have the same form, but with $r \neq 0$.

Since $\mathfrak{H}$ is an ordered field, we can classify the elements of $\mathbb{H}$ as follows:

(1) *Standard numbers*: These are of the form $^*r$, where r is a real number. The standard numbers correspond to the real numbers embedded in $\mathfrak{v}(^*\mathbb{R})$. When referring to r in $\mathbb{H}$, we mean $^*r$, since standard numbers form an Archimedean ordered field isomorphic to the real number field. For the same reason, we identify **im** $\mathbb{R}$ with $\mathbb{R}$.
(2) *limited numbers*: The sum of a real number and an infinitesimal, $r + \varepsilon$, is not standard but is limited, meaning it is constrained by real numbers: $s < r+\varepsilon < t$, for $s, t \in \mathbb{R}$.
(3) *unlimited numbers*: These are precisely the multiplicative inverses of infinitesimal numbers.

The set of all hyperreal numbers of the form $r + \varepsilon$, where r is standard and ε is infinitesimal, is called the monad or halo of r. We will refer to it as the 'halo' of r and denote it by $\text{hal}(r)$.

Every limited number a belongs to the halo of some real number, meaning it always has a standard part.

Theorem 6.3.3 *For every limited hyperreal number a, there exists a unique real number r such that $r \approx a$.*

Proof First, we observe that the sum of infinitesimal numbers is also infinitesimal. Indeed, let ε and δ be any infinitesimal numbers, and let r be an arbitrary real number. Then,

$$|\varepsilon| < r/2 \quad \text{and} \quad |\delta| < r/2.$$

Therefore,

$$|\varepsilon + \delta| \leqslant |\varepsilon| + |\delta| < r.$$

If there were more than one real number in the halo of a, say $r \approx a$ and $s \approx a$ for real numbers r, s, then the differences $r - a$ and $a - s$ would be infinitesimal. But then

$$r - s = (r - a) + (a - s) \neq 0$$

would be infinitesimal, which is impossible because the difference between two real numbers is only infinitesimal if it is exactly zero.$\dashv$

Definition 6.3.4 We introduce the following terminology:

(1) The *shadow* (or standard part) of a limited number a is its unique standard part, denoted by $\text{sh}(a)$. In some texts, it is also denoted as $\text{st}(a)$.
(2) A hyperreal number a is *infinitely close* to a hyperreal number b if $a - b$ is infinitesimal. In this case, we write $a \approx b$.

(3) A hyperreal number a is at a *limited distance* from a hyperreal number b if $a-b$ is limited. In this case, we write $a \sim b$.
(4) The *halo* (or *monad*) of a hyperreal number a is the set of all hyperreal numbers that are infinitely close to a:

$$\text{hal}(a) = \{b \in \mathbb{H} \mid b \approx a\}.$$

The halo is also called the monad of a and is sometimes denoted as $\mu(a)$.
(5) The *galaxy* of a hyperreal number a is the set of all hyperreal numbers that are at a limited distance from a:

$$\text{gal}(a) = \{b \in \mathbb{H} \mid b \sim a\}.$$

The reader can easily verify that both $\approx$ and $\sim$ are equivalence relations. The halo of 0 is the set $\mathbb{I}$ of all infinitesimal numbers. The galaxy of 0 is the set $\mathbb{L}$ of all limited numbers. The galaxy of an unlimited number a consists exclusively of unlimited numbers, though not all unlimited numbers belong to it, since some unlimited numbers are at unlimited distances from each other.

There is a subtle yet fundamental aspect regarding infinitesimal numbers that mathematicians have long recognized: infinitesimals exist in different orders of magnitude.

In standard analysis, particularly in the works of Russian mathematicians (see Barenenkov et al., 1977: 31–34), an *infinitesimal* is defined as a function of a real variable whose limit is zero as the independent variable approaches a given value. This leads to the following precise definitions.

Definition 6.3.5 An *infinitesimal* is a function $f : \mathbb{R} \to \mathbb{R}$ such that:

$$\lim_{x \to a} f(x) = 0$$

for some real number a. If f and g are both infinitesimals, we say that:

(1) f and g are of the same order if and only if:

$$\lim_{x \to a} \frac{f(x)}{g(x)} = c,$$

where c is a nonzero constant.
(2) f is of higher order with respect to g if and only if:

$$\lim_{x \to a} \frac{f(x)}{g(x)} = 0.$$

(3) f and g are equivalent in order if and only if:

$$\lim_{x \to a} \frac{f(x)}{g(x)} = 1.$$

This definition can be reformulated, *mutatis mutandis*, to describe the order relations between nonstandard infinitesimal numbers.

Definition 6.3.6 (Orders of Infinitesimals) Let $\varepsilon, \delta \in \mathbb{H}$ be nonzero infinitesimal numbers. We say that:

(1) ε and δ are of the same order if and only if:

$$\frac{\varepsilon}{\delta} \approx c,$$

where c is a nonzero standard number.

(2) ε is of higher order with respect to δ if and only if:

$$\frac{\varepsilon}{\delta} \approx 0.$$

(3) ε and δ are equivalent in order if and only if:

$$\frac{\varepsilon}{\delta} \approx 1.$$

Since unlimited numbers are simply the multiplicative inverses of nonzero infinitesimals, their order relations can be defined analogously.

Definition 6.3.7 (Orders of Unlimited Numbers) Let H, K be unlimited numbers. We say that:

(1) H and K are of the same order if and only if:

$$H/K \approx c,$$

where c is a nonzero standard number.

(2) H is of higher order with respect to K (or equivalently, K is of lower order than H) if and only if:

$$H/K \text{ is unlimited} \quad \text{or, equivalently,} \quad K/H \text{ is infinitesimal.}$$

(3) H and K are equivalent in order if and only if:

$$H/K \approx 1.$$

The following theorem establishes an interesting connection between the galaxies of unlimited numbers and their order relations.

Theorem 6.3.8 *Let H be an unlimited hyperreal number. The galaxy of H is the set of hyperreal numbers that have an equivalent order of magnitude to that of H.*

Proof If H is unlimited and $a \in \text{gal}(H)$, then there exists a limited hyperreal b such that:

$$a = H + b.$$

Thus,

$$\frac{H+b}{H} = 1 + \varepsilon \approx 1,$$

where $\varepsilon = b/H$ is infinitesimal.

Conversely, if H and a are of the same order:

$$\frac{a}{H} \approx 1,$$

then a is nearly equal to H, meaning:

$$a = H + \varepsilon, \quad \varepsilon \in \mathbb{I}.$$

This confirms that $a \in \text{gal}(H)$.⊣

The reader can verify that hyperreal numbers obey the following fundamental arithmetic properties.

Theorem 6.3.9 (Hyperreal Arithmetic Rules) *Let $\varepsilon, \delta \in \mathbb{I}$ (infinitesimals), b, c be appreciable hyperreals, and H, K be unlimited hyperreal numbers. Then:*

(1) *Real Numbers*

- *The only real infinitesimal is 0.*
- *Every real number is limited.*

(2) *Negatives*

- $-\varepsilon$ *is infinitesimal.*
- $-b$ *is appreciable.*
- $-H$ *is unlimited.*

(3) *Reciprocals*

- *If $\varepsilon \neq 0$, then $1/\varepsilon$ is unlimited.*
- $1/b$ *is appreciable.*
- $1/H$ *is infinitesimal.*

(4) *Sums*

- $\varepsilon + \delta$ *is infinitesimal.*

- *$b + \varepsilon$ is appreciable.*
- *$b + c$ is limited (possibly infinitesimal).*
- *$H + \varepsilon$ and $H + b$ are unlimited.*

(5) *Products*

- *$\delta \cdot \varepsilon$ and $b \cdot \varepsilon$ are infinitesimal.*
- *$b \cdot c$ is appreciable.*
- *$H \cdot b$ and $H \cdot K$ are unlimited.*

(6) *Quotients*

- *ε/b, ε/H, and b/H are infinitesimal.*
- *b/c is appreciable.*
- *b/ε, H/ε, and H/b are unlimited, provided $\varepsilon \neq 0$.*

(7) *Roots*

- *If $\varepsilon > 0$, then $\sqrt[n]{\varepsilon}$ is infinitesimal.*
- *If $b > 0$, then $\sqrt[n]{b}$ is appreciable.*
- *If $H > 0$, then $\sqrt[n]{H}$ is unlimited.*

(8) *Indeterminate Forms*

- ε/δ
- H/K
- $H\varepsilon$
- $H + K$.

6.4 Algebraic Structure of Limited Numbers

The set of halos of limited numbers has an algebraic structure isomorphic to the real number field. This allows us to think of halos as if they were real numbers.

Theorem 6.4.1 (Algebraic Structure of Limited Numbers) *The set $\mathbb{L}$ of limited numbers, equipped with addition and multiplication, forms a commutative ring $\mathfrak{L}$. The set $\mathbb{I}$ of infinitesimals is a bilateral ideal of $\mathfrak{L}$, and the collection of halos corresponds to the cosets of $\mathfrak{L}$. The canonical function $\varphi : \mathbb{L}/\mathbb{I} \to \mathbb{R}$, defined by:*

$$\varphi(\mathrm{hal}(a)) = \mathrm{sh}(a),$$

is an isomorphism.

Proof We first observe that $\mathbb{L}$ is a commutative ring, as $\langle \mathbb{L}, +, 0 \rangle$ is an abelian group under addition and multiplication is associative, commutative, and distributive with 1 as the multiplicative identity. Function $\mathrm{sh} : \mathbb{L} \to \mathbb{R}$ is a ring homomorphism with kernel $\mathbb{I}$. The cosets of $\mathbb{I}$ (i.e., the equivalence classes of elements differing by an infinitesimal) correspond precisely to the halos of limited numbers. These cosets take the form:

$$a + \mathbb{I} = \{a + \varepsilon \mid a \in \mathbb{L},\ \varepsilon \in \mathbb{I}\}.$$

where a is a fixed element.

Thus, the halos form a ring when we define addition and multiplication as follows:

$$(a + \mathbb{I}) + (b + \mathbb{I}) = (a + b) + \mathbb{I},$$

$$(a + \mathbb{I})(b + \mathbb{I}) = ab + \mathbb{I}.$$

The canonical function $\varphi : \mathbb{L}/\mathbb{I} \to \mathbb{R}$, defined by:

$$\varphi(\mathrm{hal}(a)) = \mathrm{sh}(a),$$

is an isomorphism. Therefore, the function mapping $\mathrm{hal}(a)$ to $\mathrm{sh}(a)$ establishes an isomorphism between $\mathfrak{L}$ and the ring of real numbers.[1]

Since $\mathbb{I}$ is an additive subgroup of $\langle \mathbb{L}, +, 0\rangle$ and satisfies:

$$a\mathbb{I} = \mathbb{I}a \subseteq \mathbb{I}, \quad \forall a \in \mathbb{L},$$

it follows that $\mathbb{I}$ is an ideal of the ring $\mathfrak{L}$.⊣

Given their importance in the application of nonstandard analysis, it is crucial to examine the properties of the nonstandard extensions of the special sets $\mathbb{N}$ (natural numbers), $\mathbb{Z}$ (integers), and $\mathbb{Q}$ (rationals), all of which contain unlimited elements. By the transfer principle, these extensions inherit the same first-order properties as their standard counterparts. For example, we have already seen that the hypernatural numbers $^*\mathbb{N}$ satisfy the Principle of Least Number and the Internal Induction Principle.

6.5 The Hyperintegers

Some authors identify the set of natural numbers $\mathbb{N}$ with the set of positive integers, but this is not possible if $\mathbb{N}$ is identified with ω, the first limit ordinal. In the construction of ω, the first element is $\varnothing$ (see the proof of Theorems 1.6.11 and 1.4.12). Consequently, as established in Chap. 1, we include 0 as the first element of $\mathbb{N}$.

[1] See Fraleigh (2003, 239–240).

We observe that the set of hypernaturals ${}^*\mathbb{N}$ contains unlimited elements, since the following statement holds in $\mathfrak{v}(\mathbb{R})$:

$$(\forall x \in \mathbb{R})(\exists y \in \mathbb{N}) \quad y > x.$$

By transfer, this statement remains true in $\mathfrak{v}({}^*\mathbb{R})$:

$$(\forall x \in {}^*\mathbb{R})(\exists y \in {}^*\mathbb{N}) \quad y > x.$$

If we take an unlimited positive hypernatural H, we see that there exists a hypernatural number greater than H.

The set of hyperintegers ${}^*\mathbb{Z}$ is defined as:

$$\mathbb{N} \cup \{x \in \mathbb{H} \mid x = -n \text{ for some } n \in {}^*\mathbb{N}\}.$$

Since $\mathfrak{v}(\mathbb{R})$ satisfies the formula:

$$(\forall x \in \mathbb{Q})(\exists y \in \mathbb{Z})(\exists z \in \mathbb{Z}) \left(z \neq 0 \wedge x = \frac{y}{z}\right),$$

by transfer, the analogous statement holds in $\mathfrak{v}({}^*\mathbb{R})$:

$$(\forall x \in {}^*\mathbb{Q})(\exists y \in {}^*\mathbb{Z})(\exists z \in {}^*\mathbb{Z}) \left(z \neq 0 \wedge x = \frac{y}{z}\right).$$

Thus, hyperrationals can be defined as quotients of two hyperintegers.

It is convenient to collect the most relevant properties of ${}^*\mathbb{N}$ in a theorem. By definition, for hypernaturals K, H, we write $\gamma(K) < \gamma(H)$ if and only if $K < H$ and $K \not\sim H$.

Theorem 6.5.1 (Properties of Hypernatural Numbers) *The set of hypernatural numbers ${}^*\mathbb{N}$ satisfies the following properties:*

(1) ${}^*\mathbb{N}$ *is closed under addition and multiplication.*
(2) *The only limited elements of* ${}^*\mathbb{N}$ *are the members of* $\mathbb{N}$.
(3) *For a fixed* $K \in {}^*\mathbb{N}$, *the hypernatural galaxy of* K *is defined as:*

$$\gamma(K) = \{K \pm n \mid n \in \mathbb{N}\}.$$

The elements of $\gamma(K)$ *form a copy of the set* $\mathbb{Z}$ *of all integers.*
(4) $\gamma(K) = \text{gal}(K) \cap {}^*\mathbb{N}$.
(5) *If* $H \in {}^*\mathbb{N}$, *then* $\gamma(K) = \gamma(H)$ *if and only if* $K \sim H$.
(6) *The standard natural numbers satisfy:*

$$\mathbb{N} = \text{gal}(0) \cap {}^*\mathbb{N}.$$

(7) *There is no maximum or minimum hypernatural galaxy.*
(8) *The set of hypernatural galaxies is dense with respect to the order relation* $<$.

Proof

1. Closure under addition and multiplication follows from the transfer principle applied to the standard property:

$$\forall x \forall y((x \in \mathbb{N} \wedge y \in \mathbb{N}) \rightarrow (x + y \in \mathbb{N} \wedge x \cdot y \in \mathbb{N})).$$

   By transfer, this holds in ${}^*\mathbb{N}$.
2. Limited elements of ${}^*\mathbb{N}$ are precisely natural numbers: If $x \in {}^*\mathbb{N}$ is limited, then there exists $n \in \mathbb{N}$ such that $x \leqslant n$. The following formula holds in $\mathfrak{v}(\mathbb{X})$:

$$\forall x(x \leqslant n \rightarrow (x = 0 \vee x = 1 \vee \cdots \vee x = n)).$$

   By transfer, it holds in $\mathfrak{v}({}^*\mathbb{R})$, ensuring that any limited hypernatural must be standard.
3. The galaxy $\gamma(K)$ forms a copy of $\mathbb{Z}$: The mapping $\varphi(p) = K + p$ embeds $\mathbb{Z}$ into $\gamma(K)$ in a surjective manner.
4. Characterization of $\gamma(K)$ as $\text{gal}(K) \cap {}^*\mathbb{N}$: The statement:

$$\forall x \forall y((x \in \mathbb{N} \wedge y \in \mathbb{N} \wedge x < y) \rightarrow y - x \in \mathbb{N})$$

   holds in $\mathfrak{v}(\mathbb{R})$, so by transfer, it holds in $\mathfrak{v}({}^*\mathbb{R})$. Since $n < K$ for any $n \in \mathbb{N}$ and any unlimited hypernatural K, we conclude that $K - n \in {}^*\mathbb{N}$ and $K + n \in {}^*\mathbb{N}$. Since $\gamma(K) \subseteq \text{gal}(K)$, it follows that $\gamma(K) = \text{gal}(K) \cap {}^*\mathbb{N}$.
5. Equivalent galaxies: If $\gamma(K) = \gamma(H)$, then K and H are at a limited distance. Conversely, if $H \sim K$, then $H - K$ is a limited hyperinteger, meaning H and K belong to the same galaxy.
6. Natural numbers as the galaxy of 0: Since $\mathbb{N} \subseteq \text{gal}(0)$ and $\mathbb{N} \subseteq {}^*\mathbb{N}$, we have $\mathbb{N} \subseteq \text{gal}(0) \cap {}^*\mathbb{N}$. Conversely, any limited hypernatural in gal(0) must be natural.
7. No maximum or minimum galaxy: If K is any hypernatural, then $H = 2K > K$ and $H - K$ is unlimited, ensuring that no maximum galaxy exists. Similarly, one can show that no minimum galaxy exists.
8. Density of hypernatural galaxies: Given any two hypernaturals H, K with $\gamma(H) < \gamma(K)$, we find that:

$$\gamma(H) < \gamma(H + K/2) < \gamma(K).$$

   Since $K/2 \in {}^*\mathbb{N}$ or $K + 1/2 \in {}^*\mathbb{N}$ (by transfer), we conclude that hypernatural galaxies form a dense order. $\dashv$

This foundational study of hyperintegers and their structure within nonstandard analysis lays the groundwork for further exploration of infinitesimal calculus, to which we turn now.

Exercises

6.1 Prove that $\approx$ and $\sim$ of Definition 6.3.4 are equivalence relations.

6.2 Show that the hyperreal numbers obey rules of Theorem 6.3.9.

References

Barenenkov, G., Demidovich, B., Efimenko, V., Kogan, S., Lunts, G., Porshneva, E., Sichova, E., Frolov, S., Shostak, R., & Yampolski, A. (1977). *Problemas y ejercicios de análisis matemático*. MIR.

Fraleigh, J. B. (2003). *A first course in abstract algebra*. Addison Wesley.

Chapter 7
Infinitesimal Calculus

Abstract Chapter 7 does not aim to be a comprehensive treatment of the discipline of differential and integral calculus, but rather to clarify the fundamental concepts and develop the main techniques. After introducing the usual topology of real vector spaces, it presents their nonstandard extension and proves basic theorems about continuous functions. It then establishes the rules of differential calculus, defines the integral as a sum of infinitesimal rectangles, and develops some key integrals such as those of the exponential, logarithmic, and basic trigonometric functions. The final part is devoted to intuitively clarifying the meaning of partial derivatives, differentiation in several variables, and the concept of the total differential, which acquires an intuitive meaning when viewed from the nonstandard perspective.

Keywords Nonstandard differential and integral calculus · Nonstandard multivariate calculus

7.1 Introduction

This chapter is not intended to be a regular treatise on calculus, much less a comprehensive one. Readers interested in following a typical course on nonstandard calculus are referred to Keisler (2000). My goal here is merely to introduce some fundamental key concepts, such as continuous functions, derivatives, integrals (of a continuous function over a closed interval), and total differentials in multivariable calculus. I prove all the basic differentiation rules and define some fundamental derivatives, specifically the logarithmic, exponential, and basic trigonometric derivatives. I have found that these notions are the most commonly required in mathematical economics courses and that calculus can be developed with relative ease based on them.

We will begin with notions of the usual topology of $\mathbb{R}^n$, which we will later relate to its extensions in ${}^*\mathbb{R}^n$ (in a later chapter, we will address topology in more general terms).

A. García de la Sienra, *The Logical Foundations of Nonstandard Analysis*, Synthese Library 530, https://doi.org/10.1007/978-3-032-05677-1_7

7.2 The Topology of $\mathbb{R}^n$

The cornerstone of the usual topology of $\mathbb{R}^n$ is the concept of an open ball, as the family of open balls forms a basis for this topology. If ϱ is the Euclidean metric on $\mathbb{R}^n$, the *open ball* of radius $\varepsilon > 0$ centered at the point $\boldsymbol{p} \in \mathbb{R}^n$ is the set

$$B_\varepsilon(\boldsymbol{p}) = \{\boldsymbol{q} \in \mathbb{R}^n \mid \varrho(\boldsymbol{p}, \boldsymbol{q}) < \varepsilon\}.$$

An *interior point* of a subset $S \subseteq \mathbb{R}^n$ is a point $\boldsymbol{p} \in S$ such that $B_\varepsilon(\boldsymbol{p}) \subseteq S$ for some $\varepsilon > 0$. A set is *open* in $\mathbb{R}^n$ if all of its elements are interior points. The family $\mathcal{T}$ of all open sets in $\mathbb{R}^n$ is called the *usual topology* of $\mathbb{R}^n$. A *neighborhood* of $\boldsymbol{p}$ is any set that contains an open set to which point $\boldsymbol{p}$ belongs.

Point $\boldsymbol{p}$ of a subset $S \subseteq \mathbb{R}^n$ is a *boundary point* iff any neighborhood of $\boldsymbol{p}$ contains elements both of S and its complement $\mathbb{R}^n \setminus S$. It is easy to see that, in order to show that a point $\boldsymbol{p}$ is a boundary point of a set S, it is sufficient to prove that every open ball $B_\varepsilon(\boldsymbol{p})$ contains points from both S and its complement $\mathbb{R}^n \setminus S$. A set is *closed* if it is the complement of an open set. It is easy to see that a set is closed if and only if it contains all its boundary points. A subset S of $\mathbb{R}^n$ or $^*\mathbb{R}^n$ is *bounded* if and only if the set of distances

$$\{\varrho(\boldsymbol{p}, \boldsymbol{q}) \mid \boldsymbol{p}, \boldsymbol{q} \in S\} \quad \text{or} \quad \{^*\varrho(\boldsymbol{p}, \boldsymbol{q}) \mid \boldsymbol{p}, \boldsymbol{q} \in S\},$$

where $^*\varrho$ is the extension of the Euclidean metric, has an upper bound in $\mathbb{R}$.

An *infinitesimal vector* is a vector whose norm is infinitesimal. The reader may verify that this is the case if and only if all its coordinates are infinitesimal. Since **ime** $(\mathbb{R}^n)$ is isomorphic to $\mathbb{R}^n$, we identify the image $^*\boldsymbol{p}$ of any real vector $\boldsymbol{p}$ with $\boldsymbol{p}$ itself. Thus, we will say that $S \subseteq^* S$ for sets $S \subseteq \mathbb{R}^n$.

By definition, the open balls $B_\varepsilon(\boldsymbol{p})$ satisfy the following formula:

$$(\forall \varepsilon \in \mathbb{R})(\forall \boldsymbol{q} \in \mathbb{R}^n)\,(\varepsilon > 0\forall \to (\boldsymbol{q} \in B_\varepsilon(\boldsymbol{p}) \leftrightarrow \varrho(\boldsymbol{p}, \boldsymbol{q}) < \varepsilon)). \tag{1}$$

By transfer, its nonstandard counterpart is obtained:

$$(\forall \varepsilon \in^* \mathbb{R})(\forall \boldsymbol{q} \in^* \mathbb{R}^n)\,(\varepsilon > 0\forall \to (\boldsymbol{q} \in B_\varepsilon(\boldsymbol{p}) \leftrightarrow^* \varrho(\boldsymbol{p}, \boldsymbol{q}) < \varepsilon)). \tag{2}$$

Balls $B_\varepsilon(\boldsymbol{p})$, defined via the extension $^*\varrho$ of the Euclidean metric, form the basis of a topology on $^*\mathbb{R}^n$ called the *S-topology*.[1] It can be verified that these sets are nonempty for $\varepsilon \approx 0$ because

$$(\forall \boldsymbol{p} \in {}^*\mathbb{R}^n)(\forall \varepsilon \in {}^*\mathbb{R})\; B_\varepsilon(\boldsymbol{p}) \neq \varnothing$$

---

[1] Cf. Robinson (1974, 107).

is the transferred statement of

$$(\forall \boldsymbol{p} \in \mathbb{R}^n)(\forall \varepsilon \in \mathbb{R})(\varepsilon > 0 \forall B_\varepsilon(\boldsymbol{p}) \neq \varnothing).$$

Moreover, it is easy to see that, for every infinitesimal vector $\boldsymbol{\varepsilon}$, the number $\boldsymbol{p} + \boldsymbol{\varepsilon}$ belongs to $B_\varepsilon(\boldsymbol{p})$, since ${}^*\varrho(\boldsymbol{p}, \boldsymbol{p} + \boldsymbol{\varepsilon}) \approx 0$.

Note that $B_\varepsilon(\boldsymbol{p})$ is defined for any point $\boldsymbol{p}$ in space, not just for standard points. The *monad* or *halo* of the point $\boldsymbol{p}$, denoted $\mathrm{hal}(\boldsymbol{p})$, is defined as the set of all points $\boldsymbol{q}$ such that the distance ${}^*\varrho(\boldsymbol{p}, \boldsymbol{q})$ is infinitesimal; in this case, we also write $\boldsymbol{p} \approx \boldsymbol{q}$. A vector $\boldsymbol{q}$ is called *limited* or *finite* if there exists a standard vector $\boldsymbol{p}$ such that ${}^*\varrho(\boldsymbol{p}, \boldsymbol{q})$ is infinitesimal. Thus, the open balls of infinitesimal radius around points in ${}^*\mathbb{R}^n$ are contained within their halos; notice that this implies that halos themselves are open. A vector $\boldsymbol{q}$ is called *nearstandard* if $\boldsymbol{q}$ is in the halo of a standard point $\boldsymbol{p}$. Obviously, every limited point $\boldsymbol{p}$ is nearstandard and has a unique shadow $\mathrm{sh}(\boldsymbol{p})$.

We now state the following interesting result.

Theorem 7.2.1 *A point $\boldsymbol{p} \in S \subseteq \mathbb{R}^n$ is in the boundary of S iff* $\mathrm{hal}(\boldsymbol{p}) \cap^* S \neq \varnothing \neq \mathrm{hal}(\boldsymbol{p}) \cap {}^*\mathbb{R}^n \setminus^* S$.

Proof Suppose $\boldsymbol{p} \in S \subseteq \mathbb{R}^n$ is in the boundary of S. By transfer, V is a neighborhood of $\boldsymbol{p}$ in ${}^*S$ iff V intersects both S and its complement in ${}^*\mathbb{R}^n$. Since the halo of any point is open, $\mathrm{hal}(\boldsymbol{p})$ is a neighborhood of $\boldsymbol{p}$.

Suppose now that $\mathrm{hal}(\boldsymbol{p})$ has points in common both with ${}^*S$ and its complement, and let V be any neighborhood of $\boldsymbol{p}$ in $\mathbb{R}^n$. Then $\boldsymbol{p} \in \mathrm{hal}(\boldsymbol{p}) \subseteq {}^*V$ and so ${}^*V$ has points in common both with ${}^*S$ and its complement. By transfer, the same holds for V in $\mathbb{R}^n$. $\dashv$

The former fact entitles us to prove the following interesting result.

Theorem 7.2.2 *A set $S \subseteq \mathbb{R}^n$ is closed and bounded if and only if all points of ${}^*S$ are nearstandard.*

Proof Since S is bounded, there exists a real number r satisfying the formula

$$(\forall \boldsymbol{p} \in S)(\forall \boldsymbol{q} \in S)\, \varrho(\boldsymbol{p}, \boldsymbol{q}) < r. \tag{1}$$

By transfer, since ${}^*r = r$,

$$(\forall \boldsymbol{p} \in^* S)(\forall \boldsymbol{q} \in^* S)\, {}^*\varrho(\boldsymbol{p}, \boldsymbol{q}) < r. \tag{2}$$

Let $\boldsymbol{q}$ be an arbitrary point in ${}^*S$ and let $\boldsymbol{p}$ be a standard point in the same set. By (2), ${}^*\varrho(\boldsymbol{p}, \boldsymbol{q})$ is finite, and therefore $\boldsymbol{q}$ is limited, meaning it is nearstandard.

Now, suppose that all points of ${}^*S$ are nearstandard. In particular, for standard points $\boldsymbol{p}$ and $\boldsymbol{q}$, we have

$$\varrho(\boldsymbol{p}, \boldsymbol{q}) = {}^*\varrho(\boldsymbol{p}, \boldsymbol{q}) < r, \tag{3}$$

which shows that S is bounded.

Now, let $\boldsymbol{p}$ be a standard point on the boundary of S, meaning there are points in $\mathrm{hal}(\boldsymbol{p})$ that belong both to ${}^*S$ and to its complement ${}^*\mathbb{R}^n \setminus {}^*S$. If $\boldsymbol{p}$ did not belong to S, then it would not belong to ${}^*S$ either, and thus it would be an interior point of ${}^*\mathbb{R}^n \setminus {}^*S$, implying that

$$\mathrm{hal}(\boldsymbol{p}) \subseteq^* \mathbb{R}^n \setminus^* S.$$

However, this is impossible because some points of $\mathrm{hal}(\boldsymbol{p})$ are in ${}^*S$. We conclude that S is closed and bounded. ⊣

In $\mathbb{R}^n$, due to a theorem due to Heine-Borel (see Theorem 8.3.2), sets that are closed and bounded are called compact, and we will follow that practice here, in what follows.

7.3 Continuous Functions

The topological notions introduced in the previous section allow us to introduce a nonstandard definition of continuity for a function. This definition is particularly elegant and intuitive. The intuition behind the continuity of a function is that infinitesimal changes in the independent variable correspond to infinitesimal changes in the dependent variable.

Definition 7.3.1 Let ϱ_m be the Euclidean metric in $\mathbb{R}^m$ and ϱ_n the Euclidean metric in $\mathbb{R}^n$. If $D \subseteq \mathbb{R}^m$ is the domain of the function $f : D \to \mathbb{R}^n$ and $\boldsymbol{p}$ is an interior point of D, we say that f is *continuous* at $\boldsymbol{p}$ if and only if ${}^*\varrho_n(f(\boldsymbol{p}), f(\boldsymbol{p}+\boldsymbol{\varepsilon})) \approx 0$ for every infinitesimal vector $\boldsymbol{\varepsilon}$. The function is *continuous on D* if f is continuous at every point of D.

A very important result in economic theory is expressed in the following theorem, which guarantees the existence of minima and maxima of continuous functions defined over compact domains.

Theorem 7.3.2 *If D is a compact subset of $\mathbb{R}^n$ and $f : D \to \mathbb{R}$ is continuous, then $f(D)$ is compact.*

Proof Suppose that $D \subseteq \mathbb{R}^n$ is compact, so that all points of ${}^*D$ are nearstandard. Suppose that $\boldsymbol{r}$ is an arbitrary point of ${}^*f({}^*D)$. We will show that $\boldsymbol{r}$ is nearstandard.

Let $\boldsymbol{q} \in^* D$ such that $\boldsymbol{r} = {}^*f(\boldsymbol{q})$, and let $\boldsymbol{p}$ be the standard point in ${}^*D$ such that $\boldsymbol{q} \in \mathrm{hal}(\boldsymbol{p})$. Since $\boldsymbol{q} \approx \boldsymbol{p}$ and f is continuous, we have ${}^*f(\boldsymbol{q}) \approx^* f(\boldsymbol{p})$, and therefore $\boldsymbol{r} \in \mathrm{hal}[{}^*f(\boldsymbol{p})]$. Since ${}^*f(\boldsymbol{p}) = f(\boldsymbol{p})$ is standard, $\boldsymbol{r}$ is nearstandard. By Theorem 7.2.2, $f(D)$ is compact. ⊣

Corollary 7.3.3 *If $D \subseteq \mathbb{R}^n$ is compact and $f : D \to \mathbb{R}$ is continuous, then f has a global maximum and a global minimum on D.*

Proof It is enough to observe that $f(D)$ has an infimum and a supremum, which must belong to $f(D)$ since this set is closed. ⊣

The following theorem states that the nonstandard definition of a continuous function is equivalent to the more common definition in standard analysis.

Theorem 7.3.4 (**$\varepsilon - \delta$ Continuity**) *A function $f : D \to \mathbb{R}$ is continuous at the point $\boldsymbol{x}_0 \in D \subseteq \mathbb{R}^n$ if and only if for every real $\varepsilon > 0$, there exists a real $\delta > 0$ such that*

$$|f(\boldsymbol{x}) - f(\boldsymbol{x}_0)| < \varepsilon$$

whenever

$$\varrho(\boldsymbol{x}, \boldsymbol{x}_0) < \delta.$$

Proof Assume that f is continuous at $\boldsymbol{x}_0$, and let

$$A = \{\boldsymbol{y} \in f(D) \mid |f(\boldsymbol{x}_0) - \boldsymbol{y}| < \varepsilon\}.$$

Since A is bounded, the argument from Theorem 7.3.2 implies that the set $f^{-1}(A) \subseteq D$ is bounded, and therefore so is

$$B = \{\varrho(\boldsymbol{x}, \boldsymbol{x}_0) \mid \boldsymbol{x} \in f^{-1}(A)\}.$$

Thus, we can choose $\delta = \sup(B)$ and verify that

$$\varrho(\boldsymbol{x}, \boldsymbol{x}_0) < \delta \Rightarrow |f(\boldsymbol{x}) - f(\boldsymbol{x}_0)| < \varepsilon.$$

Conversely, assume that for every real $\varepsilon > 0$, there exists a real $\delta > 0$ such that $|f(\boldsymbol{x}) - f(\boldsymbol{x}_0)| < \varepsilon$ whenever $\varrho(\boldsymbol{x}, \boldsymbol{x}_0) < \delta$. Let $\hat{\boldsymbol{x}}$ be an arbitrary point in ${}^*D$ infinitely close to $\boldsymbol{x}_0$, such that ${}^*\varrho(\hat{\boldsymbol{x}}, \boldsymbol{x}_0) < \lambda$ for some infinitesimal $\lambda > 0$. For any real $\varepsilon > 0$, there exists a real $\delta > 0$ such that

$$\varrho(\boldsymbol{x}, \boldsymbol{x}_0) < \delta \Rightarrow |f(\boldsymbol{x}) - f(\boldsymbol{x}_0)| < \varepsilon.$$

Since ${}^*\varrho(\hat{\boldsymbol{x}}, \boldsymbol{x}_0) < \lambda < \delta$ for all such δ, it follows that ${}^*\varrho(\hat{\boldsymbol{x}}, \boldsymbol{x}_0) < \delta$ for all real $\varepsilon > 0$. Therefore,

$$|f(\hat{\boldsymbol{x}}) - f(\boldsymbol{x}_0)| < \varepsilon$$

for every real $\varepsilon > 0$. It follows that $|f(\hat{\boldsymbol{x}}) - f(\boldsymbol{x}_0)|$ is infinitesimal (positive or zero), meaning that $f(\hat{\boldsymbol{x}}) \approx f(\boldsymbol{x}_0)$. That is, f is continuous at $\boldsymbol{x}_0$. ⊣

An interesting property of continuous functions on $\mathbb{R}$ is the Darboux's property. A function $f : [a, b] \to \mathbb{R}$ is said to have *Darboux's property* if and only if for every $y \in [f(a), f(b)]$, there exists some $x \in [a, b]$ such that $y = f(x)$.

Intuitively, continuous functions possess this property because continuity implies the absence of jumps. Recall that a sequence of points in $\mathbb{R}^n$ is a function s from $\mathbb{N}$ to $\mathbb{R}^n$. The term $s(k)$ is denoted as s^k.

Theorem 7.3.5 (Darboux's Property) *If $f : [a, b] \to \mathbb{R}$ is a continuous function, then it has the Darboux's property.*

Proof Let y be an arbitrary point in $[f(a), f(b)]$, and define the sets:

$$A_y = \{f(y) \in \mathrm{Im}(f) \mid f(x) \leq y\}$$
$$A^y = \{f(y) \in \mathrm{Im}(f) \mid f(x) \geq y\}.$$

If $y = \sup A_y$, take a sequence $f(x^k)$ converging to y. Since the domain of f is bounded, we can assume, without loss of generality, that x^k converges to some point x^0, which belongs to $[a, b]$ because this interval is closed. Thus,

$$y = \lim_{x^k \to x^0} f(x^k) = f(x^0).$$

A similar argument proves the same result if $y = \inf A^y$.

We will now show that the case where $\sup A_y < \inf A^y$ is impossible. If this were true, there would exist points x^* and x_* in $[a, b]$ such that

$$f(x^*) = \sup A_y, \quad f(x_*) = \inf A^y.$$

Now, if $x^* \leq x_*$, define:

$$A_x = \{x \in [a, b] \mid x \leq x^*\}$$
$$A^x = \{x \in [a, b] \mid x \geq x_*\}.$$

(If $x^* \geq x_*$, reverse the inequalities.) If $A_x \cup A^x$ were different from $[a, b]$, there would be elements in this interval for which f is not defined, which contradicts our hypothesis. ⊣

7.4 Single-Variable Calculus: The Differential

A concept that is often difficult for beginners in calculus is the differential of a function. We will see that nonstandard analysis provides an intuitive explanation of this notion.

Definition 7.4.1 Let $f : D \to \mathbb{R}$ be a function defined on a domain $D \subseteq \mathbb{R}$, and let p be an interior point of D. We say that S is the *slope* of f at p if and only if

$$S = \text{sh}\left(\frac{{}^*f(p+\Delta p) - {}^*f(p)}{\Delta p}\right)$$

for every nonzero infinitesimal Δp.

The quotient

$$\frac{{}^*f(p+\Delta p) - {}^*f(p)}{\Delta p}$$

is called the *Newton quotient*. If ${}^*f(p)$ is not defined, or ${}^*f(p+\Delta p)$ is not defined for some nonzero infinitesimal Δp, or if the Newton quotient is unlimited, or if it has different shadows for different infinitesimals $\Delta p \neq 0$, then we say that the slope of f at p does not exist.

A well known result can be obtained in a more intuitive way as follows.

Definition 7.4.2 (Derivative of a Function) Let $f : D \to \mathbb{R}$ be a function defined on the domain $D \subseteq \mathbb{R}$. The *derivative* of f is the function f' whose value at $p \in D$ is the slope of f at p, provided that such a slope exists.

One consequence of derivability is continuity. That can be easily seen as follows.

Theorem 7.4.3 *If f is a differentiable function on the domain $D \subseteq \mathbb{R}$, then f is continuous on D.*

Proof Let x be an arbitrary point in D. Since f is differentiable at x, the number

$$\text{sh}\left(\frac{{}^*f(x+\varepsilon) - {}^*f(x)}{\varepsilon}\right)$$

is well-defined and unique for any nonzero infinitesimal ε. It follows that

$$\frac{{}^*f(x+\varepsilon) - {}^*f(x)}{\varepsilon}$$

is a bounded hyperreal number a. Therefore,

$$^*f(x+\varepsilon) -^* f(x) = \varepsilon a$$

is infinitesimal. This proves that f is continuous at x. ⊣

We now transition from classical calculus concepts to more intuitive notions in nonstandard analysis.

Definition 7.4.4 Let f be a differentiable function with real values defined on a domain $D \subseteq \mathbb{R}$. The *graph* of f is the curve:

$$C = \{(x, f(x)) \mid x \in D\}.$$

If $(p, f(p)) \in C$, the *tangent line* to the curve at the point $(p, f(p))$ is the straight line that: Passes through the point $(p, f(p))$, and has the same slope as f at p. If we define:

$$\alpha = f'(p), \quad \beta = f(p) - f'(p) \cdot p,$$

then the equation of the tangent line is:

$$L(x) = \alpha x + \beta.$$

Since $\alpha = f'(p)$, the tangent line has the same slope as f and it passes through the point $(p, f(p))$:

$$\begin{aligned} L(p) &= \alpha p + \beta \\ &= f'(p) \cdot p + f(p) - f'(p) \cdot p \\ &= f(p). \end{aligned}$$

Function $L(x)$ is clearly parallel to the line:

$$y = \alpha x = \hat{L}(x).$$

If we start at the point $(p, f(p))$ on the curve and move an infinitesimal distance $\Delta p \approx 0$ from p along the x-axis, this motion corresponds to a displacement on the curve C from $(p, f(p))$ to $(p + \Delta p, f(p + \Delta p))$. At the same time, moving along the tangent line at $(p, f(p))$ takes us from $(p, L(p)) = (p, f(p))$ to $(p + \Delta p, L(p + \Delta p))$ (Fig. 7.1).

The key observation is that this motion on the tangent line closely approximates the motion along the curve because the distance between points $(p + \Delta p, f(p + \Delta p))$ and $(p + \Delta p, L(p + \Delta p))$ is infinitesimal. In other words, if $\Delta y = f(p + \Delta p) - f(p)$, and $\mathrm{d}y = L(p + \Delta p) - L(p)$, then $\Delta y \approx \mathrm{d}y$, since both are infinitesimal. To explore the exact relationship between Δy and $\mathrm{d}y$, we prove the following theorem.

Theorem 7.4.5 (Increment Theorem for Single Variable) *Let f be a differentiable function at p and let Δp be an infinitesimal. Then Δy is infinitesimal and*

$$\Delta y = f'(p)\Delta p + \Delta p \cdot \eta(\Delta p),$$

for some infinitesimal $\eta(\Delta p)$ that depends on p and Δp.

Proof Since f is continuous at p, Δy is infinitesimal. If $\Delta p = 0$, then $\Delta y = f'(p)\Delta p = 0$, and we can simply take $\eta(\Delta p) = 0$.

If $\Delta p \neq 0$, we note that, by definition,

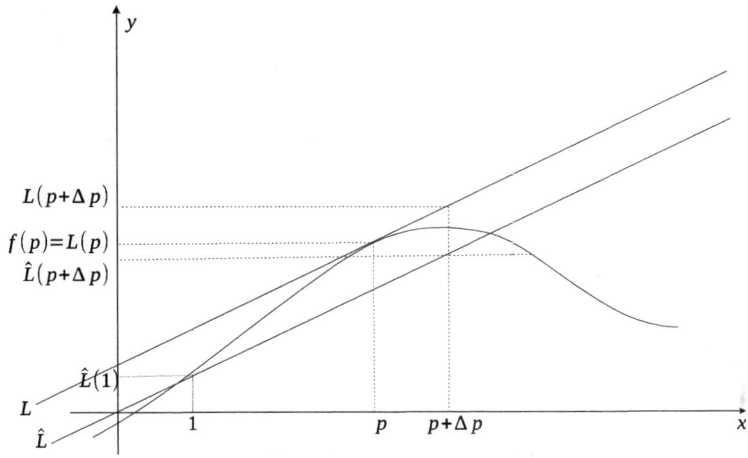

Fig. 7.1 Tangent line and its translation to the origin

$$\frac{\Delta y}{\Delta p} \approx f'(p).$$

Thus, there exists an infinitesimal $\eta(\Delta p)$ such that

$$\frac{\Delta y}{\Delta p} = f'(p) + \eta(\Delta p).$$

Multiplying both sides by Δp, we get:

$$\Delta y = f'(p)\Delta p + \Delta p \cdot \eta(\Delta p). \dashv$$

The magnitude $f'(p) \cdot \Delta p$ is called the *differential* of y.

It is the distance between the point $\hat{L}(p)$ and the point $\hat{L}(p+\Delta p)$, which is equal to the distance between the points $L(p)$ and $L(p+\Delta p)$. Indeed,

$$\begin{aligned} f'(p) \cdot \Delta p &= \hat{L}(\Delta p) \\ &= \hat{L}(p) + \hat{L}(\Delta p) - \hat{L}(p) \\ &= \hat{L}(p + \Delta p) - \hat{L}(p) \\ &= (\hat{L}(p + \Delta p) + \beta) - (\hat{L}(p) + \beta) \\ &= L(p + \Delta p) - L(p) \\ &= \mathrm{d}y. \end{aligned}$$

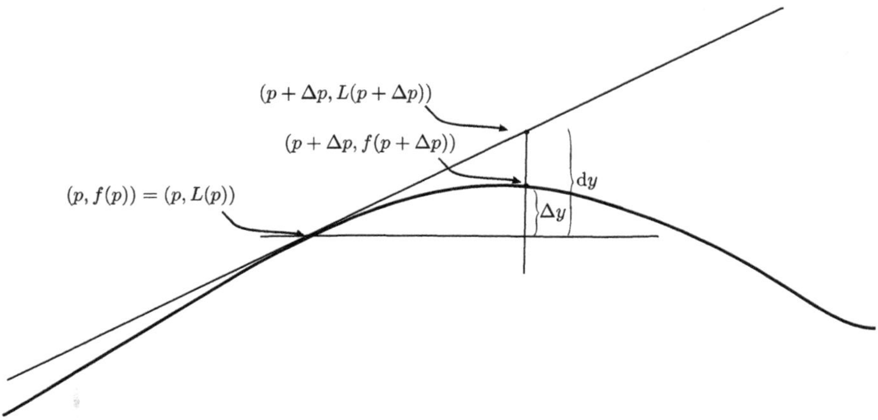

Fig. 7.2 The differential $\mathrm{d}y$ is an infinitesimal hyperreal number

Thus, the exact relationship between Δy and $\mathrm{d}y$ is given by the equation

$$\Delta y = \mathrm{d}y + \Delta p \cdot \eta(\Delta p).$$

Since $f'(p) \cdot \Delta p = \hat{L}(\Delta p) = \alpha \Delta p$ is infinitesimal, the function $\hat{L}$ is said to closely approximate the difference Δy near 0 (Fig. 7.2).

In general, the differential of x, written as $\mathrm{d}x$, satisfies:

$$\mathrm{d}y = f'(x)\mathrm{d}x.$$

This provides another way to write the derivative of f at x:

$$f'(x) = \frac{\mathrm{d}y}{\mathrm{d}x}.$$

This is the famous Leibniz notation, whose original meaning is now clear: the derivative of f at p is the value at 1 of function

$$\hat{L}(x) = f'(p) \cdot x,$$

which is obtained by translating the tangent line at $(p, f(p))$ to the origin. This is expected, since the slope of the tangent line at $(p, f(p))$ is precisely:

$$\frac{\hat{L}(1)}{1} = f'(p). \dashv$$

7.5 Differentiation Rules

Particularly beautiful and elegant is the nonstandard derivation of differentiation rules. In what follows, we assume that f is a differentiable function in the interior of its domain.

Functions of the form $f(x) = ax + b$ are called "linear" because their graph is a straight line. However, they are not strictly linear, as they do not satisfy the correct definition of a linear function, which requires the satisfaction of the following condition:

$$f(ax + by) = af(x) + bf(y).$$

Nevertheless, the derivative of a function $f(x) = ax + b$ is the same as that of its translated to the origin, function $g(x) = ax$.

Theorem 7.5.1 *If $f(x) = ax + b$ then*

$$\frac{\mathrm{d}}{\mathrm{d}x}(ax + b) = a \qquad a \in \mathbb{R}.$$

Proof

$$\begin{aligned}\frac{\Delta y}{\Delta x} &= \frac{[a(x + \Delta x) + b] - ax - b}{\Delta x}\\ &= \frac{a\Delta x}{\Delta x}\\ &= a.\end{aligned}$$

Hence,

$$\mathrm{sh}\left(\frac{\Delta y}{\Delta x}\right) = a. \dashv$$

The following five theorems are self-explanatory and straightforward.

Theorem 7.5.2 *If f y g are differentiable in x,*

$$\frac{\mathrm{d}}{\mathrm{d}x}[f(x) + g(x)] = \frac{\mathrm{d}}{\mathrm{d}x}f(x) + \frac{\mathrm{d}}{\mathrm{d}x}g(x).$$

Proof

$$\frac{\Delta y}{\Delta x} = \frac{f(x + \Delta x) + g(x + \Delta x) - f(x) - g(x)}{\Delta x}$$

$$= \frac{f(x+\Delta x)-f(x)}{\Delta x}+\frac{g(x+\Delta x)-g(x)}{\Delta x}$$

Hence,

$$\begin{aligned}\mathrm{sh}\left(\frac{\Delta y}{\Delta x}\right) &= \mathrm{sh}\left(\frac{f(x+\Delta x)-f(x)}{\Delta x}\right)+\mathrm{sh}\left(\frac{g(x+\Delta x)-g(x)}{\Delta x}\right)\\ &= f'(x)+g'(x).\dashv\end{aligned}$$

Theorem 7.5.3 *If f y g are differentiable in x,*

$$\frac{\mathrm{d}}{\mathrm{d}x}[f(x)\cdot g(x)] = f'(x)\cdot g(x)+f(x)\cdot g'(x).$$

Proof

$$\begin{aligned}\frac{\Delta y}{\Delta x} &= \frac{f(x+\Delta x)\cdot g(x+\Delta x)-f(x)\cdot g(x)}{\Delta x}\\ &= \frac{(f(x)+\Delta f)\cdot(g(x)+\Delta g)-f(x)\cdot g(x)}{\Delta x}\\ &= \frac{f(x)\cdot g(x)+\Delta f\cdot g(x)+f(x)\cdot\Delta g}{\Delta x}+\\ &\quad+\frac{\Delta f\cdot\Delta g-f(x)\cdot g(x)}{\Delta x}\\ &= \frac{f(x)\cdot\Delta g+g(x)\cdot\Delta f+\Delta f\cdot\Delta g}{\Delta x}.\end{aligned}$$

Therefore,

$$\begin{aligned}\mathrm{sh}\left(\frac{\Delta y}{\Delta x}\right) &= \mathrm{sh}\left(f(x)\frac{\Delta g}{\Delta x}+g(x)\frac{\Delta f}{\Delta x}+\frac{\Delta f\cdot\Delta g}{\Delta x}\right)\\ &= f(x)\cdot g'+g(x)\cdot f'(x)+\mathrm{sh}(\Delta f)\cdot\mathrm{sh}\left(\frac{\Delta g}{\Delta x}\right)\\ &= f(x)\cdot g'(x)+g(x)\cdot f'(x).\dashv\end{aligned}$$

Theorem 7.5.4 *If $f(x)$ is differentiable, $f(x)\neq 0$ and $h(x)=1/f(x)$, then $h(x)$ is differentiable and*

$$h'(x) = -\frac{f'(x)}{[f(x)]^2}.$$

Proof

$$\begin{aligned}
\frac{\Delta h}{\Delta x} &= \frac{1/f(x+\Delta x) - 1/f(x)}{\Delta x} \\
&= \frac{1/f(x+\Delta x) + -1/f(x)}{\Delta x} \\
&= \frac{f(x) - f(x+\Delta x)/f(x)\cdot f(x+\Delta x)}{\Delta x} \\
&= \frac{-(f(x+\Delta x) - f(x))}{\Delta x \cdot f(x)\cdot f(x+\Delta x)} \\
&= \frac{-(f(x+\Delta x) - f(x)}{\Delta x} \cdot \frac{1}{f(x+\Delta x)\cdot f(x)}.
\end{aligned}$$

By the continuity of f, $f(x+\Delta x) \approx f(x)$, and so

$$\begin{aligned}
\operatorname{sh}\left(\frac{\Delta h}{\Delta x}\right) &= -f'(x)\cdot \frac{1}{\operatorname{sh}(f(x+\Delta x))\cdot \operatorname{sh} f(x)} \\
&= -\frac{f'(x)}{[f(x)]^2}. \dashv
\end{aligned}$$

Theorem 7.5.5 *If $f(x)$ and $g(x)$ are differentiable, and $g(x) \neq 0$, $h(x) = f(x)/g(x)$ es diferenciable y*

$$h'(x) = \frac{f'(x)}{g(x) + f(x)\cdot g'(x)}[g(x)]^2.$$

Proof Exercise. ⊣

Theorem 7.5.6 *If $f(x) = x^n$, where n is a positive integer,*

$$\frac{\mathrm{d}}{\mathrm{d}x}x^n = nx^{n-1}.$$

Proof The proof is by induction on n. For $n = 1$, the assertion reduces to

$$\frac{\mathrm{d}}{\mathrm{d}x}x^1 = 1 = 1\cdot x^{1-1},$$

which is true because $x^0 = 1$ for all $x \in \mathbb{R}$.

Assume as the induction hypothesis that the assertion holds for $k \geq 1$, so that the derivative of x^k is kx^{k-1}, and let $f(x) = x^k$, $g(x) = x$. Then, the derivative of x^{k+1} is

$$\begin{aligned} f'(x) \cdot g(x) + f(x) \cdot g'(x) &= kx^{k-1}x + x^k \cdot 1 \\ &= kx^k + x^k \\ &= (k+1)x^k. \end{aligned}$$

Therefore, the assertion holds for every positive integer n. ⊣

Actually, Theorem 7.5.6 still holds if n is replaced with any fraction. The proof of this fact is left to the reader.

Theorem 7.5.7 *If* $f(x) = x^{nm^{-1}}$,

$$\frac{\mathrm{d}}{\mathrm{d}x} f(x) = nm^{-1} x^{nm^{-2}}.$$

Proof Exercise. ⊣

The next theorem is the famous chain rule.

Theorem 7.5.8 *If f is differentiable with respect to x and g is differentiable with respect to $y = f(x)$, then $h(x) = g(f(x))$ is differentiable with respect to x and*

$$h'(x) = g'(f(x)) \cdot f'(x).$$

Proof Let Δx be a nonzero infinitesimal, let

$$\Delta f = f(x + \Delta x) - f(x), \qquad \Delta g = (f(x) + \Delta f) - g(f(x)),$$

and let

$$\begin{aligned} \Delta h &= h(x + \Delta x) - h(x) \\ &= g(f(x + \Delta x)) - g(f(x)) \\ &= g(\Delta f + f(x)) - g(f(x)) \\ &= \Delta g. \end{aligned}$$

Since f es continuous $\Delta f \approx 0$ and there are two cases.

(1) $\Delta f = 0$. Then $\Delta h = 0$ and

$$\begin{aligned} \mathrm{sh}\left(\frac{\Delta h}{\Delta x}\right) &= 0 \\ &= g' f(x) \cdot 0 \\ &= g' f(x) \cdot \mathrm{sh}\left(\frac{\Delta f}{\Delta x}\right) \end{aligned}$$

$$= g' f(x) \cdot f'(x).$$

(2) $\Delta f \neq 0$. In this case, since $\Delta h = \Delta g$,

$$\begin{aligned} \text{sh}\left(\frac{\Delta h}{\Delta x}\right) &= \text{sh}\left(\frac{\Delta g}{\Delta x}\right) \\ &= \text{sh}\left(\frac{\Delta g}{\Delta f} \cdot \frac{\Delta f}{\Delta x}\right) \\ &= \text{sh}\left(\frac{\Delta g}{\Delta f}\right) \cdot \text{sh}\left(\frac{\Delta f}{\Delta x}\right) \\ &= g' f(x) \cdot f'(x) \end{aligned}$$

If f es differentiable at x and g is differentiable at $f(x)$, $h(x) = g(f(x))$ is differentiable at x and

$$h'(x) = g'(f(x))f'(x)$$

We have that $\Delta g = (f(x) + \Delta f) - g(f(x))$. Let Δx be such that $\Delta F = f(x + \Delta x) - f(x)$ and $\Delta h = h(x + \Delta x) - h(x)$. Then $g(f(x) + \Delta f) = g(f(x + \Delta x))$ y

$$\begin{aligned} \Delta g &= g(f(x + \Delta x) - g(f(x)) \\ &= g \cdot f(x + \Delta x) - f(x) \\ &= h(x + \Delta x) - h(x) \\ &= \Delta h \end{aligned}$$

The doctrine of trigonometric derivatives follows from the next simple notions (along with trigonometric identities). but that of exponential and logarithmic functions requires considerations of integral calculus, which are introduced in the next section. ⊣

Theorem 7.5.9 *If* $\theta \approx 0$ *then* $\sin\theta \approx 0$.

Proof Length of arc $\overset{\frown}{AB}$ subtended by an angle $\theta < \pi/2$ (of θ radians) is equal to θ by definition and is always greater than the length of the opposite leg AC (see Fig. 7.3), which is equal to $y = \sin\theta$. Thus, we have

$$0 < \sin\theta < \theta.$$

Therefore, when $\theta \approx 0$, we have $\sin\theta \approx 0$. ⊣

Theorem 7.5.10 *If* $\theta \approx 0$ *then* $\cos\theta \approx 1$.

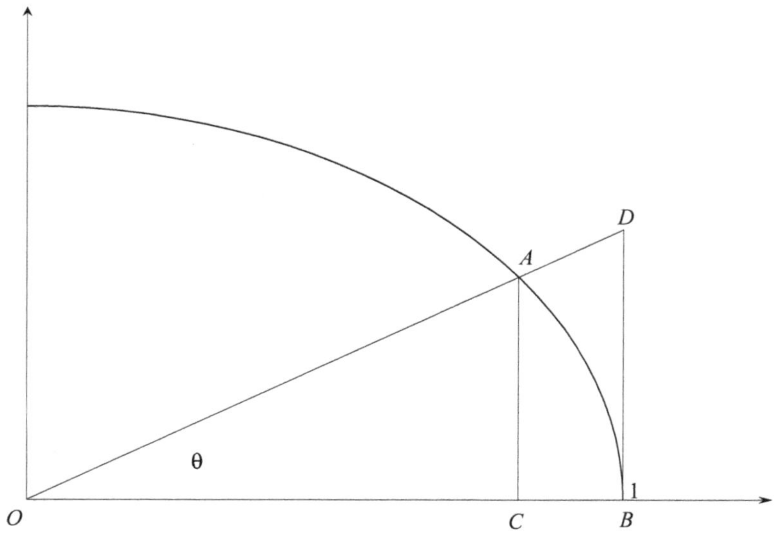

Fig. 7.3 Length of arc $\overset{\frown}{AB}$ is greater than that of the opposite leg $\overline{AC}$

Proof

$$\begin{aligned}\text{sh}\,(\cos\theta) &= \text{sh}\left(\sqrt{1-\sin^2\theta}\right)\\ &= \sqrt{\text{sh}(1-\sin^2\theta)}\\ &= \sqrt{1-[\text{sh}(\sin\theta)]^2}.\end{aligned}$$

And so, when $\theta \approx 0$, $\text{sh}\,(\cos\theta) = 1$. ⊣

If the reader has the list of trigonometric identities at hand, they will be able to deduce the derivatives of all trigonometric functions with relative ease from these two basic propositions.

7.6 The Integral

Given a continuous function $f : [a, b] \to \mathbb{R}$, with $f(x) \geq 0$ for all x, the integral of f from a to b is defined as "the area under the curve" of the function's graph, as illustrated in Fig. 7.4.

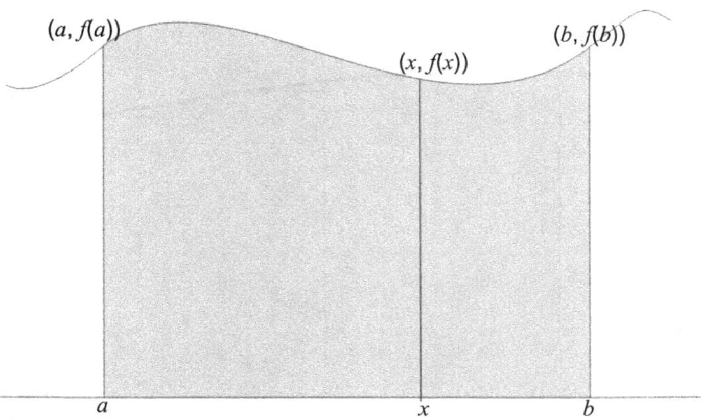

Fig. 7.4 Area under the curve

It is clear that we can approximate the area under the curve using rectangles, as illustrated in Fig. 7.5. By taking an interval Δx smaller than $b - a$, we can approximate the area under the curve as the sum of all rectangles with a base of Δx and a height of $f(x)$, plus, if necessary, an additional rectangle with a base of q if Δx does not exactly divide the segment $\overline{ab}$. More precisely, we consider points of the form $a + i\Delta x$ $(i = 1, \ldots, n)$, where n is the number of segments of length Δx. The sum consists of rectangles with a height of $f(a + i\Delta x)$ and a base of Δx, and if $q \neq 0$, we add the rectangle with a base of q and a height of $f(b)$:

$$\mathsf{S} = \sum_{i=1}^{n} f(a + i\Delta x)\Delta x + f(b)q.$$

As Henle and Kleinberg (1979) have pointed out, S is a function of Δx defined in $\mathbb{R}$, and thus it can be extended to a function $^*\mathsf{S}$ over hyperreal numbers.

> For any infinitesimal dx, the value of $\mathsf{S}_a^b f(x)(\cdot)$, $\mathsf{S}_a^b f(x)(\mathrm{d}x)$, is a perfectly well-defined hyperreal. (Henle & Kleinberg, 1979, 55)

Thus, we can introduce the following definition.

Definition 7.6.1 Let $f : [a, b] \to \mathbb{R}$, where $[a, b]$ is an interval of real numbers, such that $f(x) \geqslant 0$.

(1) f is *integrable* if and only if $\mathsf{S}_a^b f(x)\mathrm{d}x$ is finite and always has the same standard part for every positive infinitesimal $\mathrm{d}x = (b - a)/N$, where N is an unlimited positive hyperinteger.
(2) If $f : [a, b] \to \mathbb{R}$ is integrable, the *integral* of f over $[a, b]$ is the number

$$\mathrm{sh}\left(\mathsf{S}_a^b f(x)\mathrm{d}x\right),$$

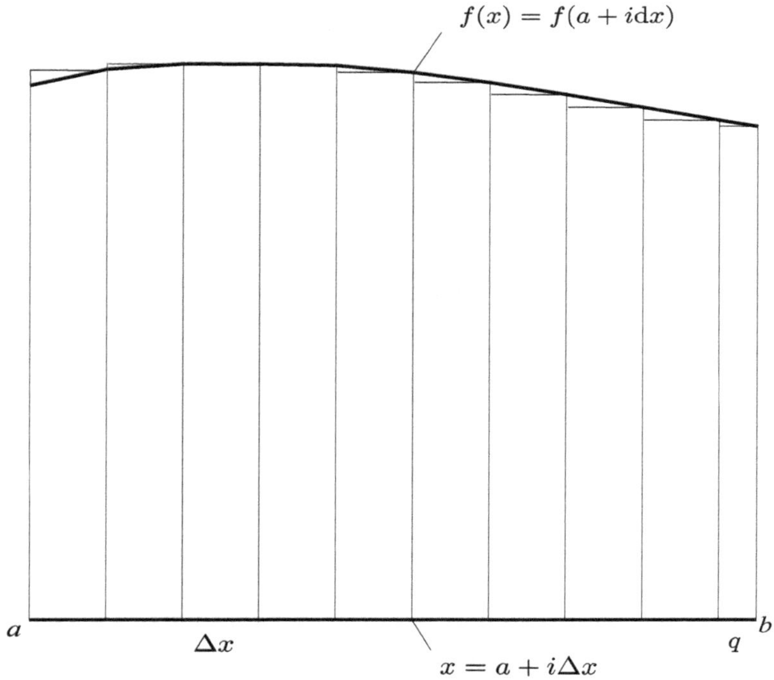

Fig. 7.5 Approximation using the sum of rectangles

which is denoted as

$$\int_a^b f(x)\,\mathrm{d}x.$$

(3) An *antiderivative* of a continuous function h is a function H such that

$$\int_a^b g(t)\mathrm{d}t = H(b) - H(a)$$

for all a, b (Fig. 7.6).

Theorem 7.6.2 *If f is a continuous function, then for any function H, H is an antiderivative of f if and only if f is the derivative of H.*

Proof Suppose that H is an antiderivative of f. Then, for any real number a and any $\Delta x > 0$, we have

$$\int_a^{a+\Delta x} f(t)\,\mathrm{d}t = H(a + \Delta x) - H(a).$$

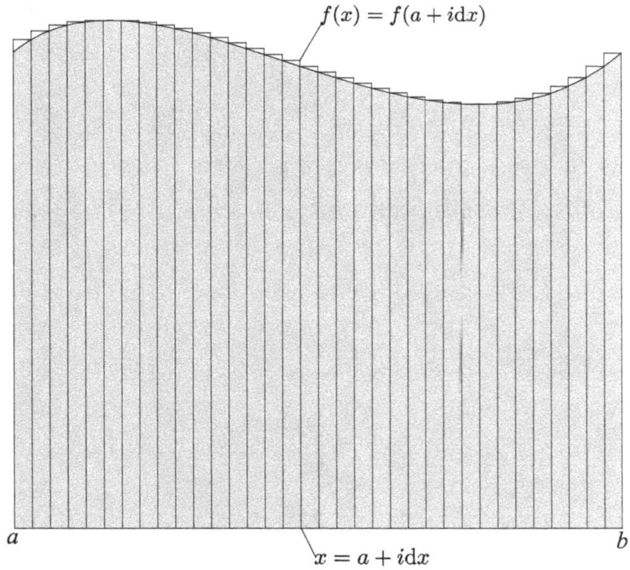

Fig. 7.6 Rectangles of infinitesimal width

Let $M = \max f(x)$ and $m = \min f(x)$ for $x \in [a, a + \Delta x]$. Since f is continuous, by Darboux's property, there exist points b and c such that $m = f(b)$ and $M = f(c)$. Therefore,

$$m\Delta x = \int_a^{a+\Delta x} f(t)\,\mathrm{d}t = H(a + \Delta x) - H(a) \leq M\Delta x.$$

Thus,

$$f(b) = m \leq \frac{H(a + \Delta x) - H(a)}{\Delta x} \leq M = f(c).$$

Since this inequality holds for real $\Delta x, b, c$, it also holds for hyperreal numbers, particularly when $\Delta x \to 0$. Then, since Δx is arbitrarily small and includes cases where $b = c = a$, we get $f(a) = f(b) = f(c)$, leading to

$$f(a) = \mathrm{sh}\left(\frac{H(a + \Delta x) - H(a)}{\Delta x}\right).$$

Since $f(a) \in \mathbb{R}$, we conclude that $H'(a) = f(a)$, completing the proof. ⊣

By means of the just introduced concepts, we can now define the following important function.

Definition 7.6.3 If $x > 0$, we define the *natural logarithm* of x as the integral

$$\ln x = \int_1^x \frac{1}{t} \mathrm{d}t.$$

Theorem 7.6.4 *$\ln x$ is an antiderivative of $1/x$ for $x > 0$, and $\ln 1 = 0$.*

Proof $\ln x$ would be an antiderivative of $1/x$ if, for all a, b,

$$\int_a^b \frac{1}{x} \mathrm{d}x = \ln b - \ln a.$$

By definition, we have

$$\ln b = \int_1^b \frac{1}{x} \mathrm{d}x,$$
$$\ln a = \int_1^a \frac{1}{x} \mathrm{d}x,$$

and

$$- \ln a = \int_a^1 \frac{1}{x} \mathrm{d}x.$$

Thus,

$$\int_a^b \frac{1}{x} \mathrm{d}x = \int_a^1 \frac{1}{x} \mathrm{d}x + \int_1^b \frac{1}{x} \mathrm{d}x$$
$$= - \ln a + \ln b.$$

In particular,

$$\ln 1 = \int_1^1 \frac{1}{t} \mathrm{d}t = 0. \dashv$$

Another important function in calculus is the exponential function, which can be introduced here as the inverse of the logarithmic function. In fact, the logarithmic function has an inverse. But before proving this, it is useful to establish the following proposition.

Theorem 7.6.5 *If f and g are inverse functions of each other,*

$$y = f(x) \quad \textit{and} \quad x = g(y),$$

and the derivatives of both functions exist and are nonzero, then

$$f'(x) = \frac{1}{g'(y)}.$$

Proof We have

$$f'(x) \cdot g'(y) = \mathrm{sh}\left(\frac{\Delta y}{\Delta x}\right) \cdot \mathrm{sh}\left(\frac{\Delta x}{\Delta y}\right).$$

Using the standard part property,

$$\mathrm{sh}\left(\frac{\Delta y}{\Delta x} \cdot \frac{\Delta x}{\Delta y}\right) = 1.$$

Thus,

$$f'(x) = \frac{1}{g'(x)}. \dashv$$

Theorem 7.6.6 *The function* $\ln x$ *has an inverse function.*

Proof Since $\ln x$ is monotonically non-decreasing and never constant (because $\mathrm{d}\ln x/\mathrm{d}x = 1/x > 0$ for all x), an inverse function for $\ln x$ exists. ⊣

The inverse function of $\ln x$ is called the *exponential function* and is denoted by $\exp(x)$ or e^x. An interesting property of the function $\exp(x)$ is that it is its own derivative.

Theorem 7.6.7 *The derivative of the exponential function is the function itself:*

$$\frac{\mathrm{d}}{\mathrm{d}x}\exp(x) = \exp(x).$$

Proof Let $y = \exp(x)$. Then,

$$\frac{\mathrm{d}}{\mathrm{d}x}\exp(x) = \frac{1}{\mathrm{d}\ln y}\mathrm{d}y.$$

Since $\mathrm{d}\ln y/\mathrm{d}y = 1/y$, we obtain $1/(1/y) = y$. Since $y = \exp(x)$, we conclude that

$$\frac{\mathrm{d}}{\mathrm{d}x}\exp(x) = \exp(x). \dashv$$

7.7 Calculus in Several Variables: Differentiation

We consider functions $f : D \to \mathbb{R}$, where D is now a subset of $\mathbb{R}^n$. These functions are important in economic theory, where D can represent (for example) a set of consumption menus and f a utility function. The graph of this function for $n = 2$ is a "sheet" in three-dimensional space, extending over the first quadrant and typically rising indefinitely from the origin. More generally, for $n \geqslant 2$, it forms a hypersurface (see Fig. 7.7).

Notice that at every point on the hypersurface, there are infinitely many tangent lines. Any pair of these lines determines a plane that is tangent at that point. The interesting property of this plane is that, within an infinitesimally small neighborhood, it is essentially identical to the hypersurface, except for some infinitesimal variation.[2] Thus, in that neighborhood, it serves as an excellent approximation to the hypersurface itself.

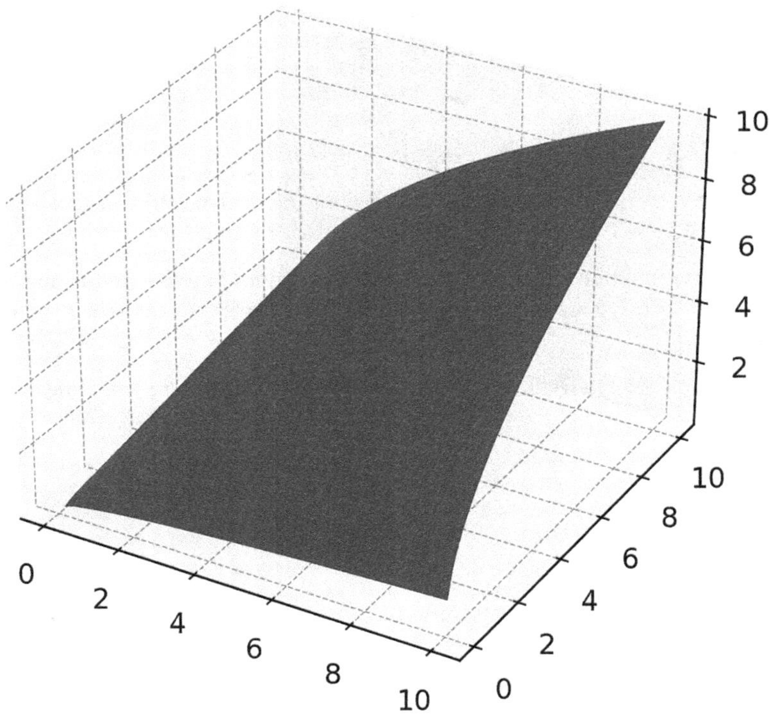

Fig. 7.7 Utility function of two variables

[2] In this way, we address the concern raised by Sydsaeter and Hammond (1974, 464) that "in mathematics textbooks for economists, it is common to require that dx and dy be "infinitesimals", or infinitely small, in the definition of the differential dz = $f_1'(x, y)$dx + $f_2'(x, y)$dy. It is claimed in such cases that Δz is equal to dz. Vague ideas of this kind have caused great confusion over the centuries, ever since Leibniz introduced them, and they have long been abandoned in mathematics".

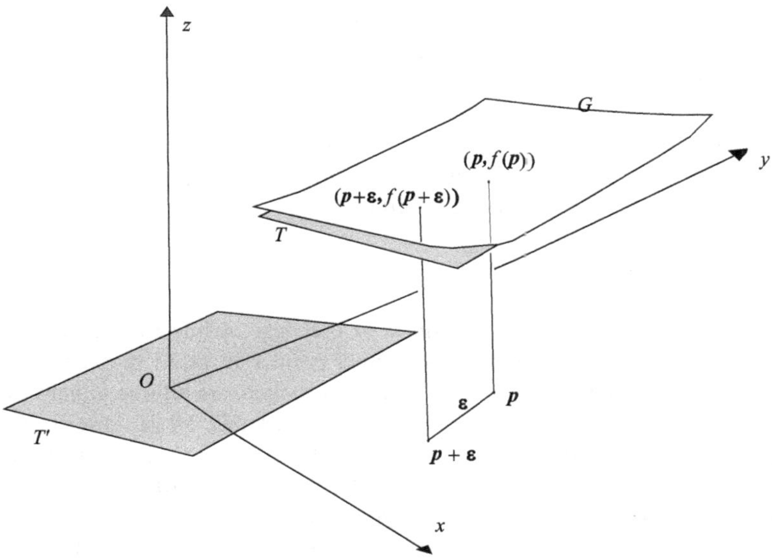

Fig. 7.8 T' is the plane T tangent to the graph G of the function f, shifted to the origin

This idea is easier to illustrate with Fig. 7.8. Let G be the graph of the function f and $\boldsymbol{\varepsilon}$ an infinitesimal vector; then T is the tangent plane at the point $(\boldsymbol{p}, f(\boldsymbol{p}))$. The point $(\boldsymbol{p} + \boldsymbol{\varepsilon}, f(\boldsymbol{p} + \boldsymbol{\varepsilon}))$ is at an infinitesimal distance from the point where the segment connecting it to $(\boldsymbol{p} + \boldsymbol{\varepsilon})$ intersects T. Thus, within an infinitesimal neighborhood, the graph of a function can be approximated almost identically by considering an infinitesimal difference. This is the idea that motivates the following development.[3]

It is not possible to define the derivative of these functions in the same way as we did for single-variable functions, since the tangent is no longer a line but a hyperplane. However, the formula

$$\Delta y = L(\Delta x) + \Delta x \cdot \eta(\Delta x)$$

suggests how we should approach differentiation in this context. We will denote vectors using bold letters: $\boldsymbol{p}, \boldsymbol{x}$, or $\boldsymbol{\delta}, \boldsymbol{\varepsilon}$ when they are infinitesimal vectors.

The goal of differentiating a multivariable function is to obtain a flat hypersurface that provides information about the neighborhood of the function's graph at any given point. It is important to determine the direction in which the hypersurface tilts or rises. This often represents the movement of some quantity—whether physical or economic—that follows a certain law of motion.

[3] I broadly follow the elegant presentation of Baxandall and Liebeck (1986, 113–131), modifying it to provide a nonstandard version of differentiation in multiple variables.

From a mathematical perspective, such a quantity is represented by a function f defined over a domain $D \subseteq \mathbb{R}^n$, which must have an interior or be open, since a function can only be differentiable (if at all) at interior points of its domain. Here, we will always assume that the domain is open.

Given a point $\boldsymbol{p}$ on the hypersurface G (the graph of f), we need to consider the distance from $f(\boldsymbol{p})$ to any other point $f(\boldsymbol{q})$ in the image of f. If $\boldsymbol{h}$ is the difference $\boldsymbol{q} - \boldsymbol{p}$, then we are interested in the difference:

$$f(\boldsymbol{p} + \boldsymbol{h}) - f(\boldsymbol{p}) = f(\boldsymbol{q}) - f(\boldsymbol{p}).$$

What we seek is to determine the slope of the tangent line to $(\boldsymbol{p}, f(\boldsymbol{p}))$ in the direction from $\boldsymbol{p}$ to $\boldsymbol{q}$. Note that for any other point $\boldsymbol{r}$ in D, there exists a tangent line at $(\boldsymbol{p}, f(\boldsymbol{p}))$ in the direction from $\boldsymbol{p}$ to $\boldsymbol{r}$. Thus, there are infinitely many tangent lines at $(\boldsymbol{p}, f(\boldsymbol{p}))$, scattered in all directions. Remarkably, all of these lines lie on the same hyperplane, and therefore, knowing n of them is sufficient to determine the hyperplane. Determining this hyperplane *is* determining the differential of f at $\boldsymbol{p}$.

The key to determining the tangent hyperplane lies precisely in the derivatives of the tangent lines at $(\boldsymbol{p}, f(\boldsymbol{p}))$ in the directions of the basis vectors of $\mathbb{R}^n$. These derivatives are known in the literature as the partial derivatives of f at $(\boldsymbol{p}, f(\boldsymbol{p}))$. These derivatives define a linear transformation $L_{f,\boldsymbol{p}}$, whose matrix is precisely the Jacobian matrix of f at $\boldsymbol{p}$. Thus, having this matrix means having the tangent plane at $(\boldsymbol{p}, f(\boldsymbol{p}))$ and, consequently, a representation of how f changes around $\boldsymbol{p}$.

To formulate all these ideas rigorously, we need to define the crucial concept of the differentiability of a function at a point, but before doing so, we must introduce the concept of the difference function.

The difference function is defined on the set of all differences $\boldsymbol{h}$ such that

$$f(\boldsymbol{p} + \boldsymbol{h}) - f(\boldsymbol{p}) = f(\boldsymbol{q}) - f(\boldsymbol{p}).$$

for some point $\boldsymbol{q}$ in D. If we call this set $D_{\boldsymbol{p}}$, we have:

$$D_{\boldsymbol{p}} = \{\boldsymbol{h} \mid \boldsymbol{h} = f(\boldsymbol{q}) - f(\boldsymbol{p}) \quad \text{for some} \quad \boldsymbol{q} \in D\}.$$

Naturally, we will take the extension ${}^*D_{\boldsymbol{p}}$ of this set in order to analyze the behavior of f in an infinitesimal neighborhood of $\boldsymbol{p}$. Note, however, that $D_{\boldsymbol{p}}$ is simply the translation of D to the origin, mapping $\boldsymbol{p}$ to $\boldsymbol{0}$.

The function that assigns to each difference $\boldsymbol{h} \in D_{\boldsymbol{p}}$ the value $f(\boldsymbol{p} + \boldsymbol{h}) - f(\boldsymbol{p})$ is, unsurprisingly, called the difference function.

Definition 7.7.1 The function $d_{f,\boldsymbol{p}} : D_{\boldsymbol{p}} \to \mathbb{R}$ that assigns to each $\boldsymbol{h} \in D_{\boldsymbol{p}}$ the value $f(\boldsymbol{p} + \boldsymbol{h}) - f(\boldsymbol{p})$ is called the *difference function*.

The difference function can be extended to a function:

$${}^*d_{f,\boldsymbol{p}} : {}^*D_{\boldsymbol{p}} \to {}^*\mathbb{R},$$

so that the value ${}^*d_{f,\mathbf{p}}(\boldsymbol{\varepsilon})$ is defined for infinitesimal vectors $\boldsymbol{\varepsilon}$. We can then define the *Newton quotient* for such vectors:

$$\frac{{}^*d_{f,\mathbf{p}}(\boldsymbol{\varepsilon})}{||\boldsymbol{\varepsilon}||} = \frac{{}^*f(\mathbf{p}+\boldsymbol{\varepsilon}) - {}^*f(\mathbf{p})}{||\boldsymbol{\varepsilon}||}.$$

If this quotient has the same standard part (or shadow) for all $\boldsymbol{\varepsilon}$, then the derivative of the straight line in the direction determined by the points $\mathbf{p}$ and $\boldsymbol{\varepsilon}$ exists. If this derivative exists for all infinitesimal $\boldsymbol{\varepsilon} \in^* D_{\mathbf{p}}$, then we say that the function f is *differentiable at* $\mathbf{p}$. Since the existence of such derivatives is equivalent to the existence of a tangent plane to $(\mathbf{p}, f(\mathbf{p}))$, we can define differentiability as follows:

Definition 7.7.2 Let f be a real-valued function on a domain $D \subseteq \mathbb{R}^m$. The function f is *differentiable* at the point $\mathbf{p} \in D$ if and only if $\mathbf{p}$ is an interior point of D and there exists a linear function $L_{f,\mathbf{p}} : \mathbb{H}^m \to \mathbb{H}$ and a function $\eta : \mathcal{H}(\mathbf{0}) \to \mathbb{I}$ from the set of infinitesimal vectors into the set of infinitesimal numbers, such that

$$ {}^*f(\mathbf{p}+\boldsymbol{\varepsilon}) - {}^*f(\mathbf{p}) = L_{f,\mathbf{p}}(\boldsymbol{\varepsilon}) + ||\boldsymbol{\varepsilon}||\eta(\boldsymbol{\varepsilon}) \tag{1}$$

for all $\boldsymbol{\varepsilon} \in \mathcal{H}(\mathbf{0})$. In this case, the linear function $L_{f,\mathbf{p}}$ is said to *closely approximate* the difference $f(\mathbf{p}+\boldsymbol{\varepsilon}) - f(\mathbf{p})$ near $\mathbf{0}$ (see the plane T' in Fig. 7.8).

What this definition states is that there exists a linear function $L_{f,\mathbf{p}}$ that matches the difference $f(\mathbf{p}+\boldsymbol{\varepsilon}) - f(\mathbf{p})$ with an infinitesimal error term $||\boldsymbol{\varepsilon}||\eta(\boldsymbol{\varepsilon})$ that depends on $\boldsymbol{\varepsilon}$. Clearly, if we divide both sides of Definition 7.7.2 (1) by $||\boldsymbol{\varepsilon}||$, we obtain the Newton quotient:

$$\frac{{}^*f(\mathbf{p}+\boldsymbol{\varepsilon}) - {}^*f(\mathbf{p})}{||\boldsymbol{\varepsilon}||}.$$

This quotient is equal to $L_{f,\mathbf{p}}(\boldsymbol{\varepsilon})$ with an infinitesimal error term $\eta(\boldsymbol{\varepsilon})$. Taking the standard part of these quantities, we find that $\mathrm{sh}(L_{f,\mathbf{p}}(\boldsymbol{\varepsilon}))$ is nothing but the derivative of f at $\mathbf{p}$.

Next, we will show that the linear approximation $L_{f,\mathbf{p}}(\boldsymbol{\varepsilon})$ is unique. However, before doing so, we will see that, just as in the case of $n = 1$ (Theorem 7.7.3), the differentiability of a function implies its continuity.

Theorem 7.7.3 *Any function $f : D \subseteq \mathbb{R}^n \to \mathbb{R}$ is continuous at the points where it is differentiable.*

Proof Let f be differentiable at $\mathbf{p} \in D$. Then there exists a linear function $L_{f,\mathbf{p}}$: $\mathbb{H}^m \to \mathbb{H}$ and a function $\eta : \mathcal{H}(\mathbf{0}) \to \mathbb{I}$ such that

$$ {}^*f(\mathbf{p}+\boldsymbol{\varepsilon}) - {}^*f(\mathbf{p}) = L_{f,\mathbf{p}}(\boldsymbol{\varepsilon}) + ||\boldsymbol{\varepsilon}||\eta(\boldsymbol{\varepsilon})$$

for all $\boldsymbol{\varepsilon} \in \mathcal{H}(\mathbf{0})$.

Since every linear transformation is continuous, $L_{f,\boldsymbol{p}}(\boldsymbol{\varepsilon}) \approx 0$, meaning that $L_{f,\boldsymbol{p}}(\boldsymbol{\varepsilon}) + ||\boldsymbol{\varepsilon}||\eta(\boldsymbol{\varepsilon})$ is infinitesimal for every infinitesimal vector $\boldsymbol{\varepsilon}$. This proves that f is continuous at $\boldsymbol{p}$. ⊣

The fact that the derivatives of the tangent lines at the point $\boldsymbol{p}$ are defined for any variation $\boldsymbol{\varepsilon}$ makes it intuitive that there cannot be more than one linear approximation to the difference $f(\boldsymbol{p}+\boldsymbol{\varepsilon}) - f(\boldsymbol{p})$.

Theorem 7.7.4 *Let $f : D \subseteq \mathbb{R}^m \to \mathbb{R}$ be differentiable at $\boldsymbol{p} \in D$. Then there is only one linear approximation to $f(\boldsymbol{p}+\boldsymbol{\varepsilon}) - f(\boldsymbol{p})$ near $\mathbf{0}$.*

Proof Let $L_{f,\boldsymbol{p}} : \mathbb{H}^m \to \mathbb{H}$ be a linear approximation to $f(\boldsymbol{p}+\boldsymbol{\varepsilon}) - f(\boldsymbol{p})$ near $\mathbf{0}$, and let $\eta : \mathcal{H}(\mathbf{0}) \to \mathbb{I}$ be a function such that

$$^*f(\boldsymbol{p}+\boldsymbol{\varepsilon}) -^* f(\boldsymbol{p}) = L_{f,\boldsymbol{p}}(\boldsymbol{\varepsilon}) + ||\boldsymbol{\varepsilon}||\eta(\boldsymbol{\varepsilon}).$$

for each infinitesimal $\boldsymbol{\varepsilon}$. Let $\boldsymbol{e}_1, \ldots, \boldsymbol{e}_m$ be the standard basis of $\mathbb{R}^m$. Since f is differentiable at $\boldsymbol{p}$, if ε_j is infinitesimal for each $j = 1, \ldots, m$, then:

$$^*f(\boldsymbol{p}+\varepsilon_j\boldsymbol{e}_j) -^* f(\boldsymbol{p}) = L_{f,\boldsymbol{p}}(\varepsilon_j\boldsymbol{e}_j) + ||\varepsilon_j\boldsymbol{e}_j||\eta(\varepsilon_j\boldsymbol{e}_j).$$

Thus,

$$^*f(\boldsymbol{p}+\varepsilon_j\boldsymbol{e}_j) -^* f(\boldsymbol{p}) = \varepsilon_j L_{f,\boldsymbol{p}}(\boldsymbol{e}_j) + |\varepsilon_j|\eta(\varepsilon_j\boldsymbol{e}_j),$$

so that, for $\varepsilon_j \neq 0$,

$$\frac{^*f(\boldsymbol{p}+\varepsilon_j\boldsymbol{e}_j) -^* f(\boldsymbol{p})}{\varepsilon_j} = L_{f,\boldsymbol{p}}(\boldsymbol{e}_j) \pm \eta(\varepsilon_j\boldsymbol{e}_j).$$

Since $\eta(\varepsilon_j\boldsymbol{e}_j)$ is infinitesimal, we have

$$\mathrm{sh}\left(\frac{^*f(\boldsymbol{p}+\varepsilon_j\boldsymbol{e}_j) -^* f(\boldsymbol{p})}{\varepsilon_j}\right) = L_{f,\boldsymbol{p}}(\boldsymbol{e}_j).$$

Now, if $L^*_{f,\boldsymbol{p}}$ were another linear approximation to f near $\boldsymbol{p}$, the previous argument would also show that

$$\mathrm{sh}\left(\frac{^*f(\boldsymbol{p}+\varepsilon_j\boldsymbol{e}_j) -^* f(\boldsymbol{p})}{\varepsilon_j}\right) = L^*_{f,\boldsymbol{p}}(\boldsymbol{e}_j).$$

Therefore, $L_{f,\boldsymbol{p}}(\boldsymbol{e}_j) = L^*_{f,\boldsymbol{p}}(\boldsymbol{e}_j)$ for each $j = 1, \ldots, m$. But a linear function is uniquely determined by its effect on a basis, so it follows that $L_{f,\boldsymbol{p}} = L^*_{f,\boldsymbol{p}}$. ⊣

If $L_{f,\boldsymbol{p}}$ is the differential of f at $\boldsymbol{p}$, then:

(1) All partial derivatives of f at $\boldsymbol{p}$ are well-defined, since they correspond to the values $L_{f,\boldsymbol{p}}(\boldsymbol{e}_j)$ for $j = 1, \ldots, n$.
(2) Since f is differentiable, there exists a function $\eta : \mathcal{H}(\boldsymbol{0}) \to \mathbb{I}$ such that:

$$\begin{aligned} f(\boldsymbol{p}+\boldsymbol{\varepsilon}) - f(\boldsymbol{p}) &= L_{f,\boldsymbol{p}}(\boldsymbol{\varepsilon}) + ||\boldsymbol{\varepsilon}||\eta(\boldsymbol{\varepsilon}) \\ &= L_{f,\boldsymbol{p}}(\varepsilon_1\boldsymbol{e}_1 + \cdots + \varepsilon_m\boldsymbol{e}_m) + ||\boldsymbol{\varepsilon}||\eta(\boldsymbol{\varepsilon}) \\ &= \varepsilon_1 L_{f,\boldsymbol{p}}(\boldsymbol{e}_1) + \cdots + \varepsilon_m L_{f,\boldsymbol{p}}(\boldsymbol{e}_m) + ||\boldsymbol{\varepsilon}||\eta(\boldsymbol{\varepsilon}) \\ &= \varepsilon_1 \frac{\partial f}{\partial x_1}(\boldsymbol{p}) + \cdots + \varepsilon_m \frac{\partial f}{\partial x_m}(\boldsymbol{p}) + ||\boldsymbol{\varepsilon}||\eta(\boldsymbol{\varepsilon}) \end{aligned}$$

for all $\boldsymbol{\varepsilon} \in \mathcal{H}(\boldsymbol{0})$.

If conditions (1) and (2) hold, then the function is differentiable at $\boldsymbol{p}$, since the linear approximation $L_{f,\boldsymbol{p}}$ to f at $\boldsymbol{p}$ is given by the Jacobian matrix $J_{f,\boldsymbol{p}}$. ⊣

Now we examine with more detail the hyperplane tangent to a given point on a hypersurface.

Definition 7.7.5 Let $f : D \subseteq \mathbb{R}^n \to \mathbb{R}$ be differentiable. The *graph* of f is the surface:

$$G = \{(\boldsymbol{x}, f(\boldsymbol{x})) \mid \boldsymbol{x} \in D\}.$$

If $(\boldsymbol{p}, f(\boldsymbol{p})) \in G$, the *tangent hyperplane* to the surface at the point $(\boldsymbol{p}, f(\boldsymbol{p}))$ is given by:

$$z = f(\boldsymbol{p}) + L_{f,\boldsymbol{p}}(\boldsymbol{x} - \boldsymbol{p}).$$

The surface T, consisting of all points satisfying $z = f(\boldsymbol{p}) + L_{f,\boldsymbol{p}}(\boldsymbol{x} - \boldsymbol{p})$ for $\boldsymbol{x} \in \mathcal{H}(\boldsymbol{p})$, provides a good approximation to G, since the distance between the point $f(\boldsymbol{p}) + L_{f,\boldsymbol{p}}(\boldsymbol{x} - \boldsymbol{p})$ on T and the point $f(\boldsymbol{x})$ on G is infinitesimal. In fact, if $\boldsymbol{x} \in \mathcal{H}(\boldsymbol{p})$, then:

$$f(\boldsymbol{p} + \boldsymbol{x} - \boldsymbol{p}) - f(\boldsymbol{p}) = L_{f,\boldsymbol{p}}(\boldsymbol{x} - \boldsymbol{p}) + ||\boldsymbol{x} - \boldsymbol{p}||\eta(\boldsymbol{x} - \boldsymbol{p}).$$

Thus,

$$f(\boldsymbol{x}) = f(\boldsymbol{p}) + L_{f,\boldsymbol{p}}(\boldsymbol{x} - \boldsymbol{p}) + ||\boldsymbol{x} - \boldsymbol{p}||\eta(\boldsymbol{x} - \boldsymbol{p}),$$

where $||\boldsymbol{x} - \boldsymbol{p}||\eta(\boldsymbol{x} - \boldsymbol{p})$ is infinitesimal.

The equation of the tangent plane to the graph of f at the point $(a, b, f(a, b))$ is:

$$z = f(a, b) + (x - a)\frac{\partial f}{\partial x}(a, b) + (y - b)\frac{\partial f}{\partial y}(a, b).$$

The concept of the total differential is particularly important in economic theory. The total differential is used to approximate the total change in a function given infinitesimal changes in its arguments. For example, if the utility function $u(x, y, z)$ depends on three variables, its total differential represents the change in u given infinitesimal changes in x, y, z.

Definition 7.7.6 If $y = f(\boldsymbol{x})$ is a differentiable function at $\boldsymbol{x}$, and $\mathrm{d}x_j$ is an infinitesimal change in the variable x_j, then defining the infinitesimal vector $\boldsymbol{\delta} = (\mathrm{d}x_1, \ldots, \mathrm{d}x_m)$, the *total differential* of f is:

$$\begin{aligned} \mathrm{d}y &= L_{f,x}(\boldsymbol{\delta}) \\ &= \mathrm{d}x_1 \frac{\partial f}{\partial x_1}(\boldsymbol{x}) + \cdots + \mathrm{d}x_m \frac{\partial f}{\partial x_m}(\boldsymbol{x}). \end{aligned}$$

Typically, $\mathrm{d}y$ will be an infinitesimal number, which makes it difficult to define the concept in standard analysis (see footnote 2 on page 186).

We previously stated that there are infinitely many tangent lines at a point on a hypersurface. This means that it is possible to determine the slope of any line regardless of direction. This leads to the next definition.

Definition 7.7.7 (Directional Derivative) Let $f : D \subseteq \mathbb{R}^m \to \mathbb{R}$ be differentiable, $\boldsymbol{x} \in \mathbb{R}^n$, and $\varepsilon \approx 0$. The *directional derivative* of f at $\boldsymbol{p}$ in the direction of $\boldsymbol{x}$ is the number:

$$D_{\boldsymbol{x}} f(\boldsymbol{p}) = \mathrm{sh}\left(\frac{f(\boldsymbol{p} + \varepsilon \boldsymbol{x}) - f(\boldsymbol{p})}{\varepsilon}\right).$$

Theorem 7.7.8 *Let $f : D \subseteq \mathbb{R}^m \to \mathbb{R}$ be differentiable, $\boldsymbol{x} \in \mathbb{R}^n$, and $\varepsilon \approx 0$. Then, for any real number $t \neq 0$:*

(1) *The partial derivative $\partial f/\partial x_i$ is the derivative in the direction of the standard basis vector $\boldsymbol{e}_i$, i.e.,*

$$D_{\boldsymbol{e}_i} f(\boldsymbol{p}) = D_i f(\boldsymbol{p}).$$

(2) *The derivative of f in the direction of the scaled vector $t\boldsymbol{x}$ is equal to the derivative in the direction of $\boldsymbol{x}$ multiplied by t, i.e.,*

$$D_{t\boldsymbol{x}} f(\boldsymbol{p}) = t D_{\boldsymbol{x}} f(\boldsymbol{p}).$$

Proof For (1), we simply use the following chain of equalities:

$$\begin{aligned} D_{\boldsymbol{e}_i} f(\boldsymbol{p}) &= \mathrm{sh}\left(\frac{f(\boldsymbol{p} + \varepsilon \boldsymbol{e}_i) - f(\boldsymbol{p})}{\varepsilon}\right) \\ &= D_i f(\boldsymbol{p}). \end{aligned}$$

The proof of (2) follows directly:

$$\begin{aligned} D_{tx} f(\boldsymbol{p}) &= \text{sh}\left(\frac{f(\boldsymbol{p}+\varepsilon(t\boldsymbol{x})) - f(\boldsymbol{p})}{\varepsilon}\right) \\ &= \text{sh}\left(\frac{t\,[f(\boldsymbol{p}+(\varepsilon t)\boldsymbol{x}) - f(\boldsymbol{p})]}{\varepsilon t}\right) \\ &= \text{sh}(t)\cdot \text{sh}\left(\frac{[f(\boldsymbol{p}+\delta\boldsymbol{x}) - f(\boldsymbol{p})]}{\delta}\right) \qquad \delta = \varepsilon t \\ &= t D_{\boldsymbol{x}} f(\boldsymbol{p}). \end{aligned}$$

Thus, we conclude our brief excursion into calculus, hoping that a more intuitive understanding has been achieved through the use of nonstandard notions and techniques. ⊣

Exercises

7.1 Show that the norm of a vector is infinitesimal iff all its coordinates are infinitesimal.

7.2 Prove that Theorem 7.5.6 still holds if n is replaced by any fraction.

7.3 Using Theorems 7.5.9 and 7.5.10, and a list of trigonometric identities, deduce the derivatives of all trigonometric functions.

References

Baxandall, P., & Liebeck, H. (1986). *Vector calculus*. Clarendon Press/Oxford University Press.
Henle, J. M., & Kleinberg, E. M. (1979). *Infinitesimal calculus*. The MIT Press.
Keisler, H. J. (2000). *Elementary calculus. An infinitesimal approach*. Creative Commons.
Robinson, A. (1974). *Non-standard analysis*. North Holland and American Elsevier.
Sydsaeter, K., & Hammond, P. (1996). *Matemáticas para el análisis económico*. Prentice Hall.

Chapter 8
Topology

Abstract Chapter 8 begins with a detailed exposition of the usual topological notions and theorems related to them. It thoroughly explains the nonstandard version of these topological concepts and proves the theorems required for the economic theory presented in later chapters. It adds some results concerning the topology of the extensions of real vector spaces, which were discussed in Chap. 7, such as the Heine–Borel Theorem. It then introduces nonstandard topologies and proves the relevant results concerning them for use in the subsequent chapters.

Keywords Nonstandard topology · Topology of hyperrreal vector spaces

8.1 Introduction

Let us recall that a *topology on a set* T is a family $\mathcal{T}$ of subsets of T that includes T and $\varnothing$ as elements and is closed under finite intersections and arbitrary unions. The pair $\mathfrak{T} = \langle T, \mathcal{T} \rangle$ is called a *topological space*, but it is typical to refer to it as "the topological space T" when the topology is understood. The elements of $\mathcal{T}$ are called *open sets* of T.

A subset S of T is *closed* if and only if its complement $T \setminus S$ is open. V is a *neighborhood* of the point $p \in T$ if there exists an open set S such that $p \in S \subseteq V$. A point p is an *interior point* of S if there exists a neighborhood V of p such that $V \subseteq S$. The *interior* of a set S, denoted by S°, is the set of all its interior points. The closure of a set S, S^-, is the smallest closed set containing S (that is, the intersection of all closed sets that contain S). A point p is a *boundary point* of S if every neighborhood of p has points in common with both S and $T \setminus S$. The *boundary* of a set S, denoted by ∂S, is the set of the boundary points of S. It is obvious from the definition that $\partial S = \partial(T \setminus S)$.

The following theorem enumerates some elementary facts about closures, interiors, and boundaries of sets in any topological space.

Theorem 8.1.1 *Let S be a subset of the topological space T. Then:*

A. García de la Sienra, *The Logical Foundations of Nonstandard Analysis*, Synthese Library 530, https://doi.org/10.1007/978-3-032-05677-1_8

(1) $(S^\circ)^\circ = S^\circ$.
(2) $(S^-)^- = S^-$.
(3) If S is open, then $S^\circ = S$.
(4) If S is closed, then $S^- = S$.
(5) $\partial S \subseteq S^-$.
(6) $S^\circ \cap \partial S = \varnothing$.
(7) $S^- = S^\circ \cup \partial S$.
(8) $S^\circ = S \setminus \partial S$.
(9) $\partial S = S^- \cap (T \setminus S)^-$.
(10) If S is open, then $S \cap \partial S = \varnothing$.
(11) If S is closed, then $\partial S \subseteq S$.
(12) $\partial S^- = \partial S$.
(13) $S^- \setminus \partial S = S^\circ$.
(14) $(S^-)^\circ = S^\circ$.

Proof

(1) If p is an interior point of S°, then there exists a neighborhood V of p such that $p \in V \subseteq S$. Hence, $S^\circ \subseteq S$. If $T = S^\circ$, it is easy to see that $(S^\circ)^\circ = T^\circ \subseteq T = S^\circ$. Moreover, $p \in (S^\circ)^\circ$ if and only if there exists a neighborhood V of p such that $V \subseteq S^\circ$, implying p is an interior point of S°.
(2) By definition, S^- is the smallest closed set containing S. Thus, $S \subseteq S^-$ and it is immediate that $S^- \subseteq (S^-)^-$.
(3) If S is open, then $S = S^\circ$ since S is a neighborhood for each of its points, making all its points interior.
(4) If S is closed, S itself is among the family of all closed sets containing S, and S^- is the smallest of them. Thus, $S^- \subseteq S \subseteq S^-$.
(5) Let $p \in \partial S$ and assume $p \notin S^-$. Then $p \in T \setminus S^-$, which is open. Since $S \subseteq S^-$, it follows that $T \setminus S^- \subseteq T \setminus S$. Thus, $T \setminus S^-$ is an open neighborhood of p contained in $T \setminus S$, contradicting $p \in \partial S$. Hence, $p \in S^-$.
(6) If $p \in S^\circ$, then not every neighborhood of p intersects the complement of S, hence $p \notin \partial S$. If $p \in \partial S$, then no neighborhood of p can be contained in S.
(7) It is immediate that $\partial S \subseteq S^-$ and $S^\circ \subseteq S^-$, so $S^\circ \cup \partial S \subseteq S^-$. Since $T \setminus S^- = (T \setminus S)^\circ$, if $p \in T \setminus S^-$, then $p \notin S$ and $p \notin \partial S$. Hence, $p \notin S^-$. This contradiction shows $S^\circ \cup \partial S = S^-$.
(8) From (6) and (7), $S^\circ = S \setminus \partial S$.
(9) It is clear that $\partial S \subseteq S^- \cap (T \setminus S)^-$. If $p \in S^- \cap (T \setminus S)^-$, then p cannot be in S° and hence $p \in \partial S$.
(10) If S is open, $S = S^\circ$ by (3), thus $S \cap \partial S = S^\circ \cap \partial S = \varnothing$ by (6).
(11) If S is closed, $S = S^-$ by (4), hence $\partial S \subseteq S^- = S$ by (5).
(12) We have the following chain of equalities:

$$
\begin{aligned}
\partial S^- &= (S^-)^- \cap (T \setminus S^-)^- \\
&= S^- \cap [(T \setminus S)^\circ]^-
\end{aligned}
$$

$$= S^- \cap (T \setminus S)^- = \partial S.$$

(13) By (7), $S^- = S^\circ \cup \partial S$, and since $S^\circ \cap \partial S = \varnothing$, we get:

$$S^- \setminus \partial S = (S^\circ \cup \partial S) \setminus \partial S = S^\circ.$$

(14) If $p \in (S^-)^\circ$, then there exists a neighborhood V of p such that $V \subseteq S^-$. Thus,

$$p \in (S^-)^\circ \leftrightarrow V \subseteq S^- \leftrightarrow p \in S^- \text{ and } p \notin \partial S^- \leftrightarrow p \in S^- \text{ and } p \notin \partial S$$

(by (12)), hence,

$$p \in S^- \setminus \partial S \leftrightarrow p \in S^\circ.$$

This completes the proof. ⊣

8.2 Nonstandard Topology

In this section we will study the structure ${}^*\mathfrak{T} = \langle {}^*T, {}^*\mathcal{T} \rangle$ in ${}^*\mathfrak{v}(\mathbb{X})$. If $\mathfrak{T} = \langle T, \mathcal{T} \rangle$ (or simply T) is a topological space in $\mathfrak{v}(\mathbb{X})$, the following facts hold: ${}^*\mathcal{T}$ is a standard set, and the elements of ${}^*\mathcal{T}$ are internal sets in ${}^*\mathfrak{v}(\mathbb{X})$, with some of them being standard, specifically those that are images of the elements of $\mathcal{T}$ under *. Robinson (1974, 89) calls ${}^*\mathcal{T}$ the *set of open sets* in ${}^*\mathfrak{T}$. The standard elements of ${}^*\mathcal{T}$ are of the form ${}^*B$ with $B \in \mathcal{T}$. Recall that ${}^*B \neq B$ whenever B is infinite.

Both $\varnothing$ and ${}^*T$ are elements of ${}^*\mathcal{T}$, and ${}^*\mathcal{T}$ is closed under finite intersections, though not under arbitrary unions. However, the union of an *internal* subset of elements of ${}^*\mathcal{T}$ is an element of ${}^*\mathcal{T}$. Since the intersection of any two sets in ${}^*\mathcal{T}$ is in ${}^*\mathcal{T}$, ${}^*\mathcal{T}$ forms a basis for a topology $\mathfrak{Q}$, called the *Q-topology* ; that is, $\mathfrak{Q}$ is such that every element S of $\mathfrak{Q}$ is a union of elements of ${}^*\mathcal{T}$. Note that every topology $\mathcal{T}$ in $\mathfrak{v}(\mathbb{X})$ has its associated Q-topology.

The standard point ${}^*p$ in ${}^*T$ is identified with the point $p \in T$. If p is a standard point in ${}^*T$ and $\mathcal{U}_p$ is the collection of all elements of $\mathcal{T}$ to which p belongs, we define the *monad* of p, $\mu(p)$, by

$$\mu(p) = \bigcap_{U \in \mathcal{U}_p}^{*} U;$$

that is, the intersection of all standard open neighborhoods of p. This leads to a series of interesting and suggestive theorems that hold for any topological space T.

From now on, we will assume that the enlargement $\mathfrak{b}(\mathbb{Y})$ is α-saturated and that the subsets of T are α-small.

Theorem 8.2.1 *Let p be a standard point. Then there exists an open set V in ${}^*\mathcal{T}$ such that V contains p and is contained in $\mu(p)$.*

Proof Let Rxy denote the relation "x is an open neighborhood of p and $y \in^* \mathcal{T}$ is an open neighborhood of p included in ${}^*x$". In order to show that R is concurrent, let $U_{\alpha_1}, \ldots, U_{\alpha_n}$ be any finite family of elements of $\mathcal{U}_p$, which happens to be the domain of R. For each α_i $(i = 1, \ldots, n)$, it is clear that $U_{\alpha_i} \subseteq^* U_{\alpha_i}$. As a matter of fact, $\bigcap_{i=1}^{n} U_{\alpha_i} \in \mathcal{U}_p$ and

$$p \in \bigcap_{i=1}^{n} U_{\alpha_i} \subseteq \bigcap_{i=1}^{n} {}^*U_{\alpha_i} = {}^*\bigcap_{i=1}^{n} U_{\alpha_i}.$$

Hence, there is a $V \in^* \mathcal{T}$ such that, for every $U_\alpha \in \mathcal{U}_p$,

$$\langle U_\alpha, V\rangle \in R;$$

that is to say $p \in V \subseteq^* U_\alpha$ for every $U_\alpha \in \mathcal{U}$. Therefore,

$$p \in V \subseteq \bigcap_{\alpha} {}^*U_\alpha = \mu(p). \dashv$$

Theorem 8.2.2 *Let p and q be standard points such that $q \in \mu(p)$. Then $\mu(q) \subseteq \mu(p)$.*

Proof Let $\mathcal{U}_p$ be the family of open neighborhoods of p and $\mathcal{V}_q$ the collection of open neighborhoods of q. Assume $q \in \mu(p)$. Then $q \in \bigcap^*_{U\in\mathcal{U}_p} U$ and thus every neighborhood of p is also a neighborhood of q; that is, $\mathcal{V}_q \subseteq \mathcal{U}_p$. Therefore,

$$\mu(q) = \bigcap_{U\in\mathcal{U}_p} {}^*U \subseteq \bigcap_{V\in\mathcal{V}_q} {}^*V = \mu(p). \dashv$$

Theorem 8.2.3 *A set S of points in T is open if and only if, for every point $p \in S$, $\mu(p) \subseteq^* S$.*

Proof If S is open, then S is an open neighborhood of each of its elements, so $S \in \mathcal{U}_p$ for each $p \in S$. It follows immediately that

$$\mu(p) = \bigcap_{U\in\mathcal{U}_p} {}^*U \subseteq {}^*S. \dashv$$

Theorem 8.2.4 *A set S in T is closed if and only if, for any standard point p not in S, the monad $\mu(p)$ has no points in common with ${}^*S$.*

Proof If S is closed and p is a standard point not in S, then $T \setminus S$ is open and $p \in T \setminus S$. By Theorem 8.2.3

$$\mu(p) \subseteq {}^*(T \setminus S) = {}^*T \setminus {}^*S,$$

and so $\mu(p) \cap {}^*S = \varnothing$.

Conversely, suppose that, for any standard points p not in S, that is in $T \setminus S$, $\mu(p) \cap {}^*S = \varnothing$. Then $\mu(p) \subseteq {}^*T \setminus {}^*S$ and so, by Theorem 8.2.3 , $T \setminus S$ is open. It follows that S is closed. ⊣

Theorem 8.2.5 *A point $p \in T$ belongs to the closure S^- of a set S in T if and only if $\mu(p)$ has a point in common with ${}^*S$.*

Proof Let p be any standard point and suppose $\mu(p) \cap {}^*S = \varnothing$. Then $p \notin {}^*S$ and, by Theorem 8.2.4, S is closed, so $S = S^-$. It follows that $p \notin S^-$. Therefore, $p \in S^-$ implies

$$\mu(p) \cap {}^*S \neq \varnothing. \dashv$$

Suppose now that $\mu(p)$ has a point in common with ${}^*S$ and that $p \notin S^-$. By Theorem 8.2.4, $\mu(p)$ has no points in common with ${}^*(S^-)$. But ${}^*S \subseteq {}^*(S^-)$, so $\mu(p)$ has no points in common with ${}^*S$. This contradiction shows that p must belong to S^-. ⊣

Theorem 8.2.6 *A point p belongs to the boundary of a set S in T if and only if $\mu(p)$ has points in common with both ${}^*S$ and ${}^*(T \setminus S) = {}^*T \setminus {}^*S$.*

Proof Suppose p belongs to the boundary of S. By Theorem 8.1.1 (9), $\partial S \subseteq S^-$ and $\partial S \subseteq (T \setminus S)^-$. Hence, by Theorem 8.2.5, both $\mu(p) \cap {}^*S \neq \varnothing$ and $\mu(p) \cap {}^*(T \setminus S) \neq \varnothing$.

If $\mu(p)$ has points in common with both ${}^*S$ and ${}^*(T \setminus S) = {}^*T \setminus {}^*S$, then, by the same Theorem 8.2.5, p belongs to both S^- and $(T \setminus S)^-$. Hence, p must be a boundary point of S. ⊣

It is possible to define, for every topological space T, the standard part of any subset $S \subseteq {}^*T$. This consists of all the standard points whose monads have elements in common with S; symbolically:

$$\text{st}(S) = \{p \in T \mid \mu(p) \cap S \neq \varnothing\}.$$

Regarding the topology of $\mathbb{R}^n$, we said that vector q is called *nearstandard* if q is in the halo of a standard point p. This notion can be generalized to any topology as follows. A point p in ${}^*T$ is called *nearstandard* if there exists a standard point q such that $p \in \mu(q)$. A set S is *nearstandard* if all its elements are nearstandard.

A *Hausdorff space* is characterized by the fact that, for every pair of points p, q, there exist respective neighborhoods V_p, V_q whose intersection $V_p \cap V_q$ is empty.

In such spaces every nearstandard point belongs to the monad of a single standard point. The standard part of the nearstandard point p is denoted $\text{st}(p)$.

Definition 8.2.7 A *sequence* in a standard topological space T is a function $p : \mathbb{N} \to T$. If T is a Hausdorff space, a point p is said to be the *limit* of the sequence (p_n) if, for every open neighborhood U of p, there exists a natural number N such that $p_n \in U$ for all $n > N$.

Note that the function p can be extended to a function from $^*\mathbb{N} \to^* T$, so that the sequence is expanded to one whose terms are of the form p_n with n possibly being an unlimited hyperinteger.

Theorem 8.2.8 *Let (p_n) be an infinite sequence in a Hausdorff space T. The point p is the limit of (p_n) in T if and only if, in $^*T$, $p_n \in \mu(p)$ for every unlimited positive integer n.*

Proof Let (p_n) be an infinite sequence in a Hausdorff space T and let p be the limit of (p_n) in T. Then, for every open neighborhood U of p, there exists $N \in \mathbb{N}$ such that $p_n \in U$ for all $n > N$. Symbolically:

$$(\forall U \in \mathcal{T})(\exists N \in \mathbb{N})(n > N \Rightarrow p_n \in U)$$

By transfer, we get:

$$(\forall U \in {}^*\mathcal{T})(\exists N \in {}^*\mathbb{N})(n > N \Rightarrow p_n \in U).$$

By Theorem 8.2.1, there exists a $V \in {}^*\mathcal{T}$ such that $V \subseteq \mu(p)$. But all points p_n with $n > N$ are in V.

Now suppose $p_n \in \mu(p)$ for every unlimited n. This means that such p_n are in the intersection of all sets of the form $^*U$, where U is an open neighborhood of p in T. Thus, we have:

$$(\exists N \in {}^*\mathbb{N})(\forall n \in {}^*\mathbb{N})(n > N \Rightarrow p_n \in {}^*U)$$

for every open neighborhood U of p in T. By transfer, it follows that:

$$(\exists N \in \mathbb{N})(\forall n \in \mathbb{N})(n > N \Rightarrow p_n \in U)$$

where U is any open neighborhood of p in T. This means that p is the limit of sequence (p_n). $\dashv$

In any topological space T, if $S \subseteq T$, a point p is an *accumulation point* or a *limit point*, also called a *cluster point* of S, if every neighborhood of p contains a point of S distinct from p. A point p is an *adherent point* of S if and only if there is a sequence of points of S whose limit is p.

Notice that every cluster point of S is an adherent point of S, but not vice versa. For instance, if p is an isolated point in S, there exist neighborhoods of p that contain no points of S other than p. However, the constant sequence $f(n) = p$ is a sequence of points of S that trivially converges to p. It is easy to see that the *closure* S^- of the subset S in T is equal to the set of all adherent points of S, as every neighborhood of an adherent point of S contains elements of S.

Theorem 8.2.9 *If a set S in a topological space T is closed, then S contains all its cluster points.*

Proof Suppose S is closed, let p be an accumulation point of S, and let (p_n) be a sequence of points in S whose limit is p. If $p \notin S$, then p would be an interior point of $T \setminus S$. Hence, there would exist a neighborhood V of p such that $p \in V \subseteq T \setminus S$. Then there would exist $N \in \mathbb{N}$ such that the points p_n with $n > N$ would be in V, that is, in $T \setminus S$. But this is impossible since $(p_n) \subseteq S$. Thus, $p \in S$. ⊣

A family $\mathscr{A}$ is a *cover* of a set B if and only if B is a subset of the union $\bigcup \mathscr{A}$; that is, if and only if every element of B belongs to some element of $\mathscr{A}$. The family $\mathscr{A}$ is an *open cover* of B if and only if every element of $\mathscr{A}$ is an open set. A *subcover* of $\mathscr{A}$ is a subfamily that is also a cover. A set B in T is said to be *compact* if every open cover of B has a finite subcover.

Theorem 8.2.10 *A topological space T is compact if and only if all points of ${}^*T$ are nearstandard.*

Proof Suppose T is a compact topological space and that there is a point $p \in {}^*T$ that is not nearstandard. This means that p is not in the monad of any standard point, so every point q in T has a neighborhood U_q such that $p \notin {}^*U_q$. Now, the family $\{U_q \mid q \in T\}$ is an open cover of T, and since T is compact, it has a finite subcover $U_q^1, \ldots, U_q^n$. Since the statement

$$U_q^1, \ldots, U_q^n = T$$

is true in $\mathfrak{v}(\mathbb{X})$, by transfer, the statement

$${}^*U_q^1, \ldots, {}^*U_q^n = {}^*T$$

is true in ${}^*\mathfrak{v}(\mathbb{X})$. But this is impossible since $p \in^* T$. This contradiction shows that p must be nearstandard.

Now suppose that all points of ${}^*T$ are nearstandard and that T is not compact. Then there exists an open cover $\mathscr{C}$ of T that has no finite subcover. Let Rxy be the relation in $\mathfrak{v}(\mathbb{X})$ defined by the condition

$$Rxy \leftrightarrow x \in \mathscr{C} \text{ and } y \in T \text{ and } y \notin x.$$

We will show that R is concurrent because, for every finite set $\{U_1, \ldots, U_n\} \subseteq \mathscr{C}$, there exists $p \in T$ such that $\langle U_i, p\rangle \in R$ for every $i = 1, \ldots, n$; that is,

$$p \in T \text{ and } p \notin U_i.$$

It follows by Theorem 4.6.8 that there exists p in the domain of $^*R$ such that $p \notin {}^*U$ for every $U \in \mathscr{C}$.

On the other hand, if q is any element of T, then $q \in V$ for some $V \in \mathscr{C}$. Since $p \notin V$, $p \notin \mu(q)$; that is, there exists a point that is not nearstandard in $^*T$. This contradiction proves that T must be compact. ⊣

Corollary 8.2.11 *A set S in T is compact if and only if $^*S$ is nearstandard.*

Lemma 8.2.12 *If S is a compact subset of a Hausdorff topological space T, then S is closed.*

Proof Suppose $S \subseteq T$ is compact and let p be an element of the boundary ∂S of S. Then, by Theorem 8.2.6 $\mu(p) \cap {}^*S \neq \varnothing$ and so there exists a point $q \in \mu(p) \cap {}^*S$. Since $^*S$ is nearstandard, there exists a standard point $r \in S$ such that $q \in \mu(r)$. Hence, $q \in \mu(p) \cap \mu(r)$. However, since T is Hausdorff, we must have $p = r$, as p is standard. It follows that $p \in S$ and, consequently, $\partial S \subseteq S$; that is, S is closed. ⊣

Let T be a Hausdorff topological space. We have proven the following proposition:

$$\forall x((x \subseteq T) \Rightarrow (x \text{ compact} \Rightarrow x \text{ closed}))$$

By transfer, we have

$$\forall x((x \subseteq {}^*T) \Rightarrow (x \text{ compact} \Rightarrow x \text{ closed}))$$

In other words,

Corollary 8.2.13 *Every internal compact subset of the enlargement of a Hausdorff topological space is closed.*

Theorem 8.2.14 *Let $\langle T, \mathscr{T}\rangle$ be a topological space, and let $B \subseteq {}^*T$ be an internal set in $\mathfrak{v}(\mathbb{Y})$, where this universe is an α-saturated enlargement. If the topology $\mathscr{T}$ is α-small, then* st(B) *is closed.*

Proof Let p be a closure point of st(B), so every neighborhood of p intersects st(B). We need to show that $p \in \text{st}(B)$; that is, $\mu(p) \cap B \neq \varnothing$.

Since $U \cap \text{st}(B) \neq \varnothing$ for every $U \in \mathscr{T}$ with $p \in U$, there exists $q \in U \cap \text{st}(B)$. Since q is standard, there exists $r \in \mu(q) \cap B \subseteq {}^*U \cap B$. Let $\mathscr{B}$ be the set:

$$\mathscr{B} = \{{}^*U \cap B \mid p \in U \text{ and } {}^*U \cap B \neq \varnothing\}.$$

We see that $\mathscr{B}$ has the finite intersection property (FIP), since for any finite family $\{({}^*U_1 \cap B), \ldots, ({}^*U_n \cap B)\}$, we observe that the intersection $U_1 \cap \cdots \cap U_n$ is also an open neighborhood of p and, hence,

$$({}^*U_1 \cap B) \cap \cdots ({}^*U_n \cap B) = ({}^*U_1 \cap \cdots \cap^* U_n) \cap B = {}^*(U_1 \cap \cdots \cap U_n) \cap B \neq \varnothing.$$

Since $\mathcal{B} \subseteq \mathcal{T}$ is α-small, we have $\bigcap \mathcal{B} \neq \varnothing$. It follows that

$$\left(\bigcap_{U \in \mathcal{U}_p} {}^*U \right) \cap B = \mu(p) \cap B \neq \varnothing. \dashv$$

Recall that a space T is *compact* if every open cover of T has a finite subcover. It suffices that every family of closed subsets of T has the finite intersection property (FIP) for T to be compact. Indeed, let $\{\mathcal{V}_\alpha\}$ be an open cover of T and suppose that every family of closed subsets of T has the FIP. If we define $F_\alpha = T \setminus V_\alpha$, we have the following chain of implications:

$$T = \bigcup_\alpha V_\alpha \Rightarrow T \setminus \bigcup V_\alpha = \varnothing \Rightarrow \bigcap_\alpha F_\alpha = \bigcap_\alpha (T \setminus V_\alpha) = \varnothing$$

$$\Rightarrow \mathcal{F} = \{F_\alpha\} \text{ does not have the FIP}$$

$$\Rightarrow \text{there exist } F_{\alpha_1}, \ldots, F_{\alpha_m} \in \mathcal{F} \text{ such that } \bigcap_{i=1}^n F_{\alpha_i} = \varnothing$$

$$\Rightarrow T = T \setminus \bigcap_{i=1}^n F_{\alpha_i}$$

$$\Rightarrow \bigcup_{i=1}^n V_{\alpha_i} = \bigcup_{i=1}^n (T \setminus F_{\alpha_i}) = T.$$

Thus, $\{V_{\alpha_1}, \ldots, V_{\alpha_n}\}$ is a finite subcover of T, and consequently, T is compact.

Any non-empty subset A of T is a topological space with the *relative topology*

$$\mathcal{T}_A = \{U \mid U = V \cap A \text{ for some } V \in \mathcal{T}\}.$$

Therefore, A is $\mathcal{T}_A$-compact. We will show that it is also $\mathcal{T}$-compact. Indeed, if $\{C_\alpha\}$ is an open cover of A, then $\{C_\alpha \cap A\}$ is an open cover in the relative topology since $C_\alpha \cap A$ is by definition open in the topology $\mathcal{T}_A$. Then, if A is such that every family of $\mathcal{T}_A$-closed subsets of A has the FIP, $\{C_\alpha \cap A\}$ has a finite subcover $\{C_{\alpha_1} \cap A, \ldots, C_{\alpha_n} \cap A\}$. But then $\{C_{\alpha_1}, \ldots, C_{\alpha_n}\}$ is also a finite subcover of $\{C_\alpha\}$ since $C_{\alpha_i} \cap A \subseteq C_{\alpha_i}$.

Furthermore, if A is closed in T, every closed subset F in A is also closed in T. Indeed, if $F \subseteq A$ is closed in A, F is of the form $F = A \setminus (A \cap U)$ with $U \in \mathcal{T}$. But we have the following chain of equalities:

$$F = A \setminus (A \cap U) = (A \setminus A) \cup (A \setminus U) = A \setminus U = A \cap (T \setminus U).$$

Since A is closed in T and $T \setminus U$ is also closed in T, we conclude that F is closed in T. $\dashv$

Based on these considerations, we can prove the following important theorem, which is required in the proofs of the existence of fixed points for correspondences.

Theorem 8.2.15 *Let $\langle T, \mathcal{T} \rangle$ be a Hausdorff topological space and let $B \subseteq {}^*T$ be an internal set in $\mathfrak{v}(\mathbb{Y})$ consisting of nearstandard points, where this universe is an α-saturated enlargement. If the topology $\mathcal{T}$ is α-small, then* st(B) *is compact.*

Proof Let $A = \mathrm{st}(B)$ and note that A is a closed set by Theorem 8.2.14. It suffices to show that if $\mathcal{F} = \{F_\alpha\}$ is a collection of closed subsets of A such that every subcollection has a non-empty intersection, then the intersection of the collection $\mathcal{F}$ is also non-empty.

Let $\mathcal{F} = \{F_\alpha\}$ be any collection of closed subsets of A such that every subcollection $\{F_{\alpha_1}, \ldots, F_{\alpha_n}\}$ of $\mathcal{F}$ has a non-empty intersection. If a is an element of $\bigcap_{i=1}^n F_{\alpha_i}$, then there exists $b \in B$ such that $b \in \mu(a)$ or—equivalently—$a = \mathrm{st}(b)$. Consequently, if $U \in \mathcal{T}$ is such that $U^- \cap F_{\alpha_i} = \varnothing$ for every i ($i = 1, \ldots, n$), then $b \notin^* U$. Otherwise, a would be in the closure $({}^*U)^-$ and hence in U^- (since a is standard), contradicting the assumption that $U^- \cap F_{\alpha_i}$ is empty.

Let Ryz be the relation defined by the following condition:

$$Ryz \leftrightarrow y \in \mathcal{F} \text{ (i.e., } \mathcal{F} \text{ is the domain of } R\text{) and,}$$
$$\text{for every open set } U \in \mathcal{T} \text{ with } U^- \cap y = \varnothing,$$
$$\text{there exists } z \in B \text{ such that } z \notin {}^*U.$$

We have shown that if $F_{\alpha_1}, \ldots, F_{\alpha_n}$ is any finite family of elements of $\mathcal{F}$, then there exists $b \in B$ such that

$$\langle F_{\alpha_1}, b\rangle, \ldots, \langle F_{\alpha_n}, b\rangle \in R.$$

In other words, we have shown that R is concurrent. Since $\mathfrak{v}(\mathbb{Y})$ is α-saturated, there exists $\hat{b} \in B$ such that $\langle F_\alpha, \hat{b}\rangle \in R$ for every $F_\alpha \in \mathcal{F}$.

Let p be a point of T such that $\hat{b} \in \mu(p)$. If $p \notin F_\alpha$ for some α, there would exist an open neighborhood U of p with $U^- \cap F_\alpha = \varnothing$, which implies that $\hat{b} \notin {}^*U$. But this is impossible since $\hat{b} \in \mu(p)$. $\dashv$

Another property that is relevant in economic theory is convexity. Quasiconvexity is a similar property that we will find important in nonstandard economics. A set S is *quasiconvex* if for all $p, q \in \mathrm{sh}(S)$ and $\lambda \in [0, 1]$, there exists $r \in S$ such that $r \in \mu[\lambda p + (1 - \lambda)q]$.

Theorem 8.2.16 *Let $\langle T, \mathcal{T} \rangle$ be a topological space. If S is a quasiconvex subset of ${}^*T$, then* st(S) *is convex.*

Proof Suppose S is quasiconvex. Let p, q be any elements of st(S), and let $\lambda \in [0, 1]$. Then there exists $r \in S$ such that $r \in \mu[\lambda p + (1-\lambda)q]$. It follows that $\lambda p + (1-\lambda)q \in \text{st}(S)$. ⊣

8.3 The Topology of $\mathbb{R}^n$

We introduced the fundamental features of the topology of $\mathbb{R}^n$ in Sect. 7.1 of Chap. 7. Other important notions are defined next.

It is easy to see that the *closure* S^- of the subset S of $\mathbb{R}^n$ is equal to the set of all adherent points of S, since every neighborhood of an adherent point of S contains elements of S.

Theorem 8.3.1 *A set S in $\mathbb{R}^n$ is closed if and only if S contains all its cluster points.*

Proof Suppose S is closed and let p be an accumulation point of S. If $p \notin S$, then p would be an interior point of $\mathbb{R}^n \setminus S$ and therefore, there would exist $\varepsilon > 0$ such that $p \in B_\varepsilon(p) \subseteq \mathbb{R}^n \setminus S$. Since $B_\varepsilon(p)$ is a neighborhood of p, by hypothesis, there exists a point $q \in S$ such that $q \in B_\varepsilon(p)$, which is impossible. It follows that $p \in S$.

Now suppose that S contains all its accumulation points and let p be an arbitrary point of ∂S (the boundary of S). Then p is an accumulation point of S, since every neighborhood of p contains points of S different from p. Therefore, $p \in S$ and consequently $\partial S \subseteq S$. ⊣

We have said that subset S of T is compact if and only if every open cover of S has a finite subcover. In the special case of $\mathbb{R}^n$ with the usual topology, it is possible to characterize compactness in terms of boundedness and closure, as stated in the following theorem. Recall that a set $S \subseteq \mathbb{R}^n$ is *bounded* if and only if there exists an open ball $B_\varepsilon(p)$ that contains S.

Theorem 8.3.2 (Heine-Borel) *A set $S \subseteq \mathbb{R}^n$ is compact if and only if S is bounded and closed.*

Proof We have shown (Corollary to Theorem 8.2.10) that a set $S \subseteq T$ in any topological space T is compact if and only if ${}^*S$ is nearstandard.

We have also shown (Theorem 7.2.2) that a set $S \subseteq \mathbb{R}^n$ is closed and bounded if and only if all points of ${}^*S$ are nearstandard. It follows that a set $S \subseteq \mathbb{R}^n$ is compact if and only if it is closed and bounded. ⊣

Having concluded this brief excursion through the usual topology of $\mathbb{R}^n$, our task now is to see how this relates to the topologies relevant in nonstandard economic theory.

8.4 The Usual Nonstandard Topologies

As we have just seen, the typical topology of $\mathbb{R}^n$, $\mathcal{T}$, has as a basis the family of all open balls defined in the usual way. In nonstandard analysis (*cf.* Brown & Robinson, 1975),[1] it is common to encounter two topologies defined on the natural extension ${}^*\mathbb{R}^n$ of $\mathbb{R}^n$: the S-topology and the Q-topology. Unlike the definition of a monad in general topology, which is defined exclusively for standard points, in the metric space defined by the extension of the Euclidean metric it is possible to define the concept for any point of the space. The *halo* of a point p, hal(p), is defined as the set of all points q such that the distance ${}^*\rho(q, p)$ is infinitesimal; in that case, we write $p \approx q$. A point p is *limited* or *finite* if there exists a standard point q such that ${}^*\rho(p, q)$ is infinitesimal. Thus, a nearstandard point is limited.

The S-topology on ${}^*\mathbb{R}^n$ has as a basis the set of open balls defined as follows.

Definition 8.4.1 Let ${}^*\rho$ be the nonstandard extension of the Euclidean metric ρ and r a positive real number. The *S-ball* centered at the point $p \in^* \mathbb{R}^n$ is the set

$$B_r(p) = \{q \in^* \mathbb{R}^n \mid^* \rho(p, q) < r\}.$$

We know that a necessary and sufficient condition for a family $\mathcal{B}$ of subsets to be a basis for the set $T = \bigcup\{B \mid B \in \mathcal{B}\}$ is that for every two members U and V of $\mathcal{B}$ and every point $p \in U \cap V$, there exists $W \in \mathcal{B}$ such that $p \in W \subseteq U \cap V$.[2] It is feasible to show that the family of all S-balls satisfies this condition.

Lemma 8.4.2 *Let $B_r(p)$ and $B_{r'}(p')$ be S-balls. If $q \in B_r(p) \cap B_{r'}(p')$, then there exists an S-ball $B_s(q)$ such that $B_s(q) \subseteq B_r(p) \cap B_{r'}(p')$.*

Proof Since $q \in B = B_r(p) \cap B_{r'}(p')$, q is an interior point of B and so there exists $s > 0$ in $\mathbb{R}$ such that $B_s(q) \subseteq B$. ⊣

All general topological notions are defined in the usual way for this topology. A point p is an *interior point* of S if there exists a real number r such that $B_r(p) \subseteq S$. A set $S \subseteq^* \mathbb{R}^n$ is *open* if all its points are interior to S.

Theorem 8.4.3 *A set $S \subseteq^* \mathbb{R}^n$ is closed-S if and only if S is closed under limits-F.*

Proof Suppose S is closed-S and f is a convergent sequence of points of S whose limit is p. Since every neighborhood of p contains a point of S different from p, p is an accumulation point of S. The argument in the first part of the proof of Theorem 8.2.9 shows that $p \in S$.

Now suppose S is closed under limits-F and p is the limit of a convergent-F sequence of points of S. Then p is an adherent point of S and therefore $p \in S^- = S$, since S is closed. ⊣

---

[1] See also Robinson (1974, 99).

[2] See Theorem 11 in Kelly (1955, 47).

Definition 8.4.4 Let ${}^*\rho$ be the nonstandard extension of the Euclidean metric ρ and ε a positive hyperreal number. The *Q-ball* centered at the point $p \in^* \mathbb{R}^n$ is the set

$$B_\varepsilon(p) = \{q \in^* \mathbb{R}^n \mid^* \rho(p, q) < \varepsilon\}.$$

Theorem 8.4.5 *Let $\mathcal{T}$ be the usual topology of $\mathbb{R}^n$. Then ${}^*\mathcal{T}$ is a basis for $\mathcal{Q}$.*

Proof By the transfer principle and the definition of Q-balls and open sets. ⊣

Theorem 8.4.6 *The monad of any standard point ${}^*p$ is equal to its halo.*

Proof By the transfer principle and the definition of monads and halos, the inclusion $\mu(p) \subseteq \text{hal}(p)$ and vice versa are established, showing that they are equal. ⊣

The following assertion follows immediately.

Corollary 8.4.7 *Let $S \subseteq \mathbb{R}^n$. A point $p \in S$ is an interior point of S if and only if the halo of p is a subset of ${}^*S$.*

Theorem 8.4.8 *A subset S of $\mathbb{R}^n$ is bounded if and only if all points of ${}^*S$ are limited.*

Proof By transfer of the statement about boundedness and the behavior of hyperreal distances, showing that boundedness implies all elements are limited and vice versa. ⊣

Theorem 8.4.9 *The shadow of any element of an S-closed set $S \subseteq {}^*\mathbb{R}^n$ belongs to S.*

Proof If S is S-closed and q is any element of S, then q is of the form $p+\varepsilon$, where ε is an infinitesimal vector. If $p \notin S$, then p would belong to the complement ${}^*\mathbb{R}^n \setminus S$, which is open. This would imply $q \notin S$, which is a contradiction. Thus, $p \in S$. ⊣

Theorem 8.4.10 (Robinson) *Let Π be the product of the family of topological spaces $\{\mathbf{T}\}_{\nu \in I}$, namely the sets of all functions f on I such that $f(\nu) \in A_\nu$ for all $\nu \in I$ and A_ν is the set of points in $\mathbf{T}_\nu$. Let $p = f$ and $q = g$ be two points in ${}^*\Pi$ such that p is standard. Then, for all standard $\nu \in \mathbb{N}$,*

$$q \in \mu(p) \textit{ iff } g(\nu) \in f(\nu).$$

Proof It is necessary to show that $g \in \text{hal}(f)$ if and only if, for arbitrary $k \in \mathbb{N}$, the value $g(k)$ lies in every set of the form ${}^*U$, where $U \subseteq \mathbb{R}^m$ is a neighborhood of $f(k)$.

Suppose that $g \in \text{hal}(f)$, and let U be an arbitrary open neighborhood of $f(k)$ in $\mathbb{R}^m$; we want to show that $g(k) \in {}^*U$. By hypothesis, $g \in {}^*\pi_k^{-1}(U)$, since $\pi_k^{-1}(U)$ is an open neighborhood of f, as U is open and π_k is continuous. Therefore, the following statement holds in the standard universe:

$$(\forall h \in (\mathbb{R}^m)^{\mathbb{N}})\ (h \in \pi_k^{-1}(U) \leftrightarrow h(k) \in U).$$

Then, by transfer, the following statement is true in the nonstandard universe:

$$(\forall h \in {}^*[(\mathbb{R}^m)^{\mathbb{N}}])\ (h \in^* \pi_k^{-1}(U) \leftrightarrow h(k) \in {}^*U).$$

From this and the hypothesis, it follows that $g(k) \in {}^*U$; and since U was an arbitrary neighborhood, we conclude that $g(k) \in \text{hal}[f(k)]$.

Now assume that $g(k) \in \text{hal}[f(k)]$ for all $k \in \mathbb{N}$. This means that $g(k)$ belongs to every set of the form ${}^*U$, where U is an open neighborhood of $f(k)$ in $\mathbb{R}^m$.

Let V be an arbitrary open neighborhood of f. We must show that $g \in {}^*V$. Let B be any basic open set such that $f \in B \subseteq V$. Then there exist open subsets A_{k_i} (for $i = 1, \ldots, n$) of $\mathbb{R}^m$ such that:

$$B = \pi_{k_1}^{-1}(A_{k_1}) \cap \cdots \cap \pi_{k_n}^{-1}(A_{k_n}).$$

If k is a natural number such that $f(k) \in A_{k_i}$ for each i, then each A_{k_i} is an open neighborhood of $f(k)$, and by hypothesis, $g(k) \in {}^*A_{k_i}$ for all i. By virtue of the transferred statement (above), we conclude that $g \in {}^*\pi_{k_i}^{-1}(A_{k_i})$. Therefore:

$$\begin{aligned} g &\in {}^*\pi_{k_1}^{-1}(A_{k_1}) \cap \cdots \cap {}^*\pi_{k_n}^{-1}(A_{k_n}) \\ &= {}^*[\pi_{k_1}^{-1}(A_{k_1}) \cap \cdots \cap \pi_{k_n}^{-1}(A_{k_n})] \\ &= {}^*B \\ &\subseteq {}^*V. \end{aligned}$$

This establishes the desired conclusion. ⊣

In this way, we conclude our review of the topological results required for the economic theory that shall be presented in the following chapters.

References

Brown, D. J., & Robinson, A. (1975). Nonstandard exchange economies. *Econometrica, 43*(1), 41–55.

Kelly, J. L. (1955). *General topology*. Springer.

Robinson, A. (1974). *Non-standard analysis*. North Holland and American Elsevier.

Chapter 9
Nonstandard Mathematical Methods for Economic Theory

Abstract Chapter 9 begins with a brief review of the basic notions of convex space theory. It defines various nonstandard notions of convexity and proves a series of lemmas about them that are relevant for establishing the Separating Hyperplane Theorem. After introducing the standard proof of this theorem in a detailed and complete manner, it proceeds to demonstrate a nonstandard version of it. The chapter then explains in detail the meaning and application of the Loeb–Lyapunov Theorem in economic theory. It concludes with a proof of the nonstandard version of the Fan–Glicksberg Fixed Point Theorem and derives Kakutani's theorem as a corollary.

Keywords Separating hyperplane · Loeb-Lyapunov theorem · Fixpoint theorems

9.1 Convexity

A notion of particular relevance in economic theory is that of a convex set. Convexity, for instance, in the context of a set of technological production possibilities, implies the absence of external diseconomies. Generally, the proof of the existence of an equilibrium requires assuming that the relevant sets are convex.

There are several notions of convexity involved in nonstandard analysis, which are captured in the following definition.

Definition 9.1.1 The following are diverse notions of convexity for sets.

(1) A set S is a *convex* subset of $\mathbb{R}^n$ if, for all $\boldsymbol{x}, \boldsymbol{y} \in S$ and $\lambda \in [0, 1]$, we have $\lambda\boldsymbol{x} + (1 - \lambda)\boldsymbol{y} \in S$.
(2) A set S is a *S-convex* subset of ${}^*\mathbb{R}^n$ if, for all $\boldsymbol{x}, \boldsymbol{y} \in S$ and $\lambda \in [0, 1]$, we have $\lambda\boldsymbol{x} + (1 - \lambda)\boldsymbol{y} \in S$.
(3) A set S is a *Q-convex* subset of ${}^*\mathbb{R}^n$ if, for all $\boldsymbol{x}, \boldsymbol{y} \in S$ and $\lambda \in^* [0, 1]$, we have $\lambda\boldsymbol{x} + (1 - \lambda)\boldsymbol{y} \in S$.
(4) A set S is a *quasi-convex* subset of ${}^*\mathbb{R}^n$ if, for all $\boldsymbol{x}, \boldsymbol{y} \in S$ and $\lambda \in [0, 1]$, there exists $\boldsymbol{z} \in S$ such that $\boldsymbol{z} \approx \lambda\boldsymbol{x} + (1 - \lambda)\boldsymbol{y}$.

A. García de la Sienra, *The Logical Foundations of Nonstandard Analysis*, Synthese Library 530, https://doi.org/10.1007/978-3-032-05677-1_9

The following notation is standard in convex analysis in $\mathbb{R}^n$ (see Rockafellar, 1970, 16 ff.):

$\lambda S = \{\lambda \boldsymbol{x} \mid \boldsymbol{x} \in S\}$
$S + \boldsymbol{a} = \{\boldsymbol{x} + \boldsymbol{a} \mid \boldsymbol{x} \in S\}$
$S_1 + S_2 = \{\boldsymbol{x}_1 + \boldsymbol{x}_2 \mid \boldsymbol{x}_1 \in S_1 \text{ and } \boldsymbol{x}_2 \in S_2\}$

The reader can verify that if S, S_1, and S_2 are convex sets in $\mathbb{R}^n$, $\boldsymbol{a} \in \mathbb{R}^n$, and λ is any real number, then the sets λS, $S + \boldsymbol{a}$, and $S_1 + S_2$ are convex. The set λS represents an expansion or contraction of S that keeps the origin invariant, while the set $S + \boldsymbol{a}$ is a translation of S.

In this section, we will use the letter B to denote the closed ball of radius 1 centered at the origin $\mathbf{0}$. Note that the closed ball B_ϵ ($\epsilon > 0$) centered at $\mathbf{0}$ is equal to the set ϵB. If we add the vector $\boldsymbol{x}$ to ϵB, we obtain a closed ball of radius ϵ centered at $\boldsymbol{x}$. Given a set S, the set $S + \epsilon B$ is, in fact, the union of the family of all closed balls of radius ϵ centered at some point of S, that is,

$$S + \epsilon B = \bigcup \{\boldsymbol{x} + \epsilon B \mid \boldsymbol{x} \in S\}.$$

If $\lambda \in [0, 1]$, then $\lambda S_1 + (1 - \lambda) S_2$ is the set

$$\{\lambda \boldsymbol{x}_1 + (1 - \lambda)\boldsymbol{x}_2 \mid \boldsymbol{x}_1 \in S_1 \text{ and } \boldsymbol{x}_2 \in S_2\}$$

of all convex combinations $\lambda \boldsymbol{x}_1 + (1 - \lambda)\boldsymbol{x}_2$, where $\boldsymbol{x}_1 \in S_1$ and $\boldsymbol{x}_2 \in S_2$. All these notions extend naturally to convex sets in ${}^*\mathbb{R}^n$.

Recall that if S is nearstandard, then all elements of S are bounded. The following lemmas, as well as the separation theorem, all refer to nearstandard sets.

Lemma 9.1.2 *If $S \subseteq {}^*\mathbb{R}^n$ is S-convex, then its shadow* $\text{sh}(S) \subseteq \mathbb{R}^n$ *is convex.*

Proof Let $\boldsymbol{x}$ and $\boldsymbol{y}$ be elements of $\text{sh}(S)$ and let $\lambda \in [0, 1]$. Then there exist $\boldsymbol{x}', \boldsymbol{y}' \in S$ such that $\boldsymbol{x}' \approx \boldsymbol{x}$ and $\boldsymbol{y}' \approx \boldsymbol{y}$. By the convexity-S of S, we have

$$\lambda \boldsymbol{x}' + (1 - \lambda)\boldsymbol{y}' \in S.$$

Since $\lambda \boldsymbol{x} + (1 - \lambda)\boldsymbol{y} \approx \lambda \boldsymbol{x}' + (1 - \lambda)\boldsymbol{y}'$, it follows that

$$\lambda \boldsymbol{x} + (1 - \lambda)\boldsymbol{y} \in \text{sh}(S). \dashv$$

Lemma 9.1.3 *If S is S-convex in ${}^*\mathbb{R}^n$, $\boldsymbol{x} \in S^\circ$, and $\boldsymbol{y} \in S^-$, then:*

$$(1 - \lambda)\boldsymbol{x} + \lambda \boldsymbol{y} \in S^\circ$$

for any real λ such that $0 \leq \lambda < 1$.

Proof We need to show that:

$$(1-\lambda)\boldsymbol{x}+\lambda\boldsymbol{y}+\epsilon B\subseteq S$$

for some $\epsilon>0$, i.e., there exists a closed ball of radius ϵ centered at $(1-\lambda)\boldsymbol{x}+\lambda\boldsymbol{y}$ that is contained in S.[1]

Since $\boldsymbol{y}\in S^{-}$, every closed ball of radius $\epsilon>0$ centered at $\boldsymbol{y}$ contains elements of S. If $\boldsymbol{z}\in B_{\epsilon}(\boldsymbol{y})\cap S$, then $\boldsymbol{y}\in B_{\epsilon}(\boldsymbol{z})=\boldsymbol{z}+\epsilon B$ as well, since $|\boldsymbol{z}-\boldsymbol{y}|<\epsilon$. Thus:

$$\boldsymbol{y}\in\bigcup\{\boldsymbol{z}+\epsilon B\mid \boldsymbol{z}\in S\}=S+\epsilon B.$$

So,

$$\lambda\boldsymbol{y}\in\lambda(S+\epsilon B)=\{\lambda\boldsymbol{w}\mid \boldsymbol{w}\in S+\epsilon B\}.$$

Therefore, for all $\epsilon>0$,

$$(1-\lambda)\boldsymbol{x}+\lambda\boldsymbol{y}+\epsilon B\subseteq(1-\lambda)\boldsymbol{x}+\lambda S+\lambda\epsilon B+\epsilon B.$$

Rewriting,

$$(1-\lambda)\boldsymbol{x}+\lambda S+(\lambda+1)\epsilon B=(1-\lambda)[\boldsymbol{x}+\epsilon(\lambda+1)(1-\lambda)^{-1}B]+\lambda S.$$

Since $\boldsymbol{x}\in S^{\circ}$, for sufficiently small ϵ,

$$\boldsymbol{x}+\epsilon(\lambda+1)(1-\lambda)^{-1}B\subseteq S.$$

Thus,

$$(1-\lambda)[\boldsymbol{x}+\epsilon(\lambda+1)(1-\lambda)^{-1}B]\subseteq(1-\lambda)S.$$

Since $(1-\lambda)S+\lambda S=S$, it follows that

$$(1-\lambda)\boldsymbol{x}+\lambda\boldsymbol{y}+\epsilon B\subseteq S.\dashv$$

Lemma 9.1.4 *If S is nearstandard and S-convex in ${}^{*}\mathbb{R}^{n}$, then S° is also S-convex.*

Proof Assume S is nearstandard and S-convex. Let $\boldsymbol{x},\boldsymbol{y}$ be any points in S°, and let $\lambda\in[0,1]$. We must show that:

$$(1-\lambda)\boldsymbol{x}+\lambda\boldsymbol{y}\in S^{\circ}.$$

[1] This proof is adapted from Theorem 6.1 in Rockafellar (1970, 45); see also Lemma 3 in Brown (1975: 47).

Since $S^\circ \subseteq S \subseteq S^-$, we have $\boldsymbol{x} \in S^\circ$ and $\boldsymbol{y} \in S^-$. By Lemma 9.1.3, it follows that:

$$(1-\lambda)\boldsymbol{x} + \lambda\boldsymbol{y} \in S^\circ$$

for any real λ such that $0 \leq \lambda < 1$. However, if $\lambda = 1$,

$$(1-\lambda)\boldsymbol{x} + \lambda\boldsymbol{y} = \boldsymbol{y} \in S^\circ.$$

Thus, S° is S-convex. ⊣

Lemma 9.1.5 *If S is nearstandard and S-convex in* $^*\mathbb{R}^n$, *then* S^- *is also S-convex.*

Proof Let $\boldsymbol{x}, \boldsymbol{y}$ be arbitrary elements of S^- and let $\lambda \in [0, 1]$. We need to show that:

$$\lambda\boldsymbol{x} + (1-\lambda)\boldsymbol{y} \in S^-.$$

Since $\boldsymbol{x}, \boldsymbol{y} \in S^-$, we have $\text{hal}(\boldsymbol{x}) \cap S \neq \varnothing$ and $\text{hal}(\boldsymbol{y}) \cap S \neq \varnothing$. If $\boldsymbol{x}' \in \text{hal}(\boldsymbol{x}) \cap S$ and $\boldsymbol{y}' \in \text{hal}(\boldsymbol{y}) \cap S$, then by the S-convex property of S:

$$\lambda\boldsymbol{x}' + (1-\lambda)\boldsymbol{y}' \in S.$$

Since

$$\lambda\boldsymbol{x} + (1-\lambda)\boldsymbol{y} \approx \lambda\boldsymbol{x}' + (1-\lambda)\boldsymbol{y}',$$

it follows that $\lambda\boldsymbol{x} + (1-\lambda)\boldsymbol{y} \in S^-$. ⊣

9.2 Separating Hyperplane Theorem

One of the most widely applied propositions in economic theory is the separating hyperplane theorem. In this section, we will present a standard proof with some completeness before moving on to its nonstandard version.

Brown (1975: 46) warns that since the convexity and topology notions used in nonstandard economics are not internal, we cannot prove separator hyperplane theorems in $^*\mathbb{R}^n$ by a simple transfer from $\mathbb{R}^n$. Instead, Brown establishes a separation theorem for nearstandard S-convex sets, requiring a series of preliminary lemmas, including a standard proof of the separating hyperplane theorem.

Given the great importance of this theorem in economic theory, it is worth providing a complete treatment. We begin by recalling the concept of inner product in a linear space.

Definition 9.2.1 Let V be a real linear space. The *inner product* (or *dot product*) of vectors $\boldsymbol{x}$, $\boldsymbol{y}$ is a binary operation:

$$V \times V \to \mathbb{R}$$

that assigns to each pair of vectors $\boldsymbol{x}, \boldsymbol{y} \in V$ a number $(\boldsymbol{x}, \boldsymbol{y})$, satisfying:

(1) *Symmetry*: $(\boldsymbol{x}, \boldsymbol{y}) = (\boldsymbol{y}, \boldsymbol{x})$.
(2) *Linearity*: $(\alpha \boldsymbol{x}, \boldsymbol{y}) = \alpha(\boldsymbol{x}, \boldsymbol{y})$.
(3) *Additivity*: $(\boldsymbol{x} + \boldsymbol{y}, \boldsymbol{z}) = (\boldsymbol{x}, \boldsymbol{z}) + (\boldsymbol{y}, \boldsymbol{z})$.
(4) *Positivity*: $(\boldsymbol{x}, \boldsymbol{x}) \geq 0$ and $(\boldsymbol{x}, \boldsymbol{x}) = 0$ if and only if $\boldsymbol{x} = 0$.

A useful and well-known result on inner products is the following inequality.

Theorem 9.2.2 $(\boldsymbol{x}, \boldsymbol{y})^2 \leqslant |\boldsymbol{x}|^2 |\boldsymbol{y}|^2$.

Proof Let α be an arbitrary real number. By condition (3) of Definition 9.2.1, we have:

$$\begin{aligned}
(\boldsymbol{x} - \alpha \boldsymbol{y}, \boldsymbol{x} - \alpha \boldsymbol{y}) &= (\boldsymbol{x}, \boldsymbol{x} - \alpha \boldsymbol{y}) + (-\alpha \boldsymbol{y}, \boldsymbol{x} - \alpha \boldsymbol{y}) \\
&= [(\boldsymbol{x}, \boldsymbol{x}) + (\boldsymbol{x}, -\alpha \boldsymbol{y})] + (-\alpha \boldsymbol{y}, \boldsymbol{x} - \alpha \boldsymbol{y}) \\
&= [(\boldsymbol{x}, \boldsymbol{x}) - \alpha(\boldsymbol{x}, \boldsymbol{y})] - \alpha(\boldsymbol{y}, \boldsymbol{x} - \alpha \boldsymbol{y}) \\
&= [(\boldsymbol{x}, \boldsymbol{x}) - \alpha(\boldsymbol{x}, \boldsymbol{y})] - \alpha[(\boldsymbol{y}, \boldsymbol{x}) + (\boldsymbol{y}, -\alpha \boldsymbol{y})] \\
&= [(\boldsymbol{x}, \boldsymbol{x}) - \alpha(\boldsymbol{x}, \boldsymbol{y})] - \alpha[(\boldsymbol{x}, \boldsymbol{y}) - \alpha(\boldsymbol{y}, \boldsymbol{y})] \\
&= (\boldsymbol{x}, \boldsymbol{x}) - 2\alpha(\boldsymbol{x}, \boldsymbol{y}) + \alpha^2(\boldsymbol{y}, \boldsymbol{y}) \\
&\geqslant 0.
\end{aligned}$$

Thus, we obtain a quadratic trinomial in α:

$$(\boldsymbol{y}, \boldsymbol{y})\alpha^2 - 2(\boldsymbol{x}, \boldsymbol{y})\alpha + (\boldsymbol{x}, \boldsymbol{x}).$$

Since it can never be negative for any value of α, it cannot have two distinct real roots, and therefore its discriminant is non-positive:

$$4(\boldsymbol{x}, \boldsymbol{y})^2 - 4(\boldsymbol{x}, \boldsymbol{x})(\boldsymbol{y}, \boldsymbol{y}) \leqslant 0,$$

which simplifies to:

$$(\boldsymbol{x}, \boldsymbol{y})^2 \leqslant |\boldsymbol{x}|^2 |\boldsymbol{y}|^2. \dashv$$

The following proposition serves as the foundation for proving the separating hyperplane theorem. The next lemma is due to Berge (1963: 162), but my proof is more detailed and explicit.

Lemma 9.2.3 *If C is a convex, closed set that does not contain the origin, then there exists a vector $\boldsymbol{x}_0$ such that:*

$$(\boldsymbol{x}_0, \boldsymbol{y}) \geqslant |\boldsymbol{x}_0|^2 \quad \textit{for all } \boldsymbol{y} \in C.$$

Proof Let $B_\lambda(\mathbf{0})$ be a closed ball centered at the origin that intersects C. Since the set $C \cap B_\lambda(\mathbf{0})$ is compact, the continuous function $f(\boldsymbol{x}) = |\boldsymbol{x}|$ attains a minimum at some point $\boldsymbol{x}_0$ in this set. Since $\boldsymbol{x}_0 \in C$, we have $|\boldsymbol{x}_0| > 0$.

For every point $\boldsymbol{x}$ in $C \cap B_\lambda(\mathbf{0})$, we have $\lambda \geqslant |\boldsymbol{x}| \geqslant |\boldsymbol{x}_0|$, and since $|\boldsymbol{x}| > \lambda$ for all other points in C, we conclude that $|\boldsymbol{x}| \geqslant |\boldsymbol{x}_0|$ for all $\boldsymbol{x} \in C$. Now, we show by *reductio ad absurdum* that:

$$(\boldsymbol{x}_0, \boldsymbol{y}) \geqslant |\boldsymbol{x}_0|^2$$

for all $\boldsymbol{y} \in C$. Suppose there exists $\varepsilon > 0$ such that:

$$(\boldsymbol{x}_0, \boldsymbol{y}) = |\boldsymbol{x}_0|^2 - \varepsilon.$$

Choosing t sufficiently small such that:

$$\frac{t}{1-t}(|\boldsymbol{y}|^2 - |\boldsymbol{x}_0|^2) < 2\varepsilon,$$

which implies:

$$t^2(|\boldsymbol{y}|^2 - |\boldsymbol{x}_0|^2) < 2t(1-t)\varepsilon,$$

we obtain

$$|(1-t)\boldsymbol{x}_0 + t\boldsymbol{y}|^2 < |\boldsymbol{x}_0|^2.$$

However, by assumption,

$$|(1-t)\boldsymbol{x}_0 + t\boldsymbol{y}|^2 \geqslant |\boldsymbol{x}_0|^2 > 0,$$

which is a contradiction. Hence, the original assumption must be false, proving that:

$$(\boldsymbol{x}_0, \boldsymbol{y}) \geqslant |\boldsymbol{x}_0|^2 \quad \forall \boldsymbol{y} \in C. \dashv$$

The following lemma is due to Karlin (1959).[2] However, I use Berge's Lemma in my proof, which differs from Karlin's original argument.

Lemma 9.2.4 *Let X be a convex set and let $\boldsymbol{y}$ be a point outside the closure X^-. Then, there exists a vector $\boldsymbol{a}$ such that:*

$$\inf_{x \in X} (\boldsymbol{a}, \boldsymbol{x}) > (\boldsymbol{a}, \boldsymbol{y}).$$

Proof Define the set:

$$C \equiv \left\{\boldsymbol{x} - \boldsymbol{y} \mid \boldsymbol{x} \in X^-\right\}.$$

Let $\boldsymbol{x}^0$ be a point in E such that:

$$|\boldsymbol{x}^0 - \boldsymbol{y}| = \inf_{x \in X} |\boldsymbol{x} - \boldsymbol{y}|.$$

To show that $\boldsymbol{x}^0 \in X^-$, consider the closed ball $B_\lambda(\boldsymbol{y})$ centered at $\boldsymbol{y}$, and define:

$$D = B_\lambda(\boldsymbol{y}) \cap X^-.$$

Since D is compact and the function $f(\boldsymbol{x}) = |\boldsymbol{x} - \boldsymbol{y}|$ is continuous on D, there exists a point $\boldsymbol{x}^0 \in D$ such that:

$$f(\boldsymbol{x}^0) = \min_{x \in D} |\boldsymbol{x} - \boldsymbol{y}|.$$

This clearly implies:

$$\boldsymbol{x}^0 - \boldsymbol{y} = \inf_{x \in X} |\boldsymbol{x} - \boldsymbol{y}|.$$

Since C is closed, by Berge's Lemma:

$$\boldsymbol{x} - \boldsymbol{y} \in C \Rightarrow (\boldsymbol{x}^0 - \boldsymbol{y}, \boldsymbol{x} - \boldsymbol{y}) \geqslant |\boldsymbol{x}^0 - \boldsymbol{y}|^2 > 0,$$

because $\boldsymbol{y} \notin X^-$ and thus $\boldsymbol{0} \notin C$.

Let $\boldsymbol{a} = \boldsymbol{x}^0 - \boldsymbol{y}$. Then, for any $\boldsymbol{x} \in X^-$:

$$\begin{aligned}(\boldsymbol{a}, \boldsymbol{x}) > (\boldsymbol{a}, \boldsymbol{y}) &\leftrightarrow (\boldsymbol{x}^0 - \boldsymbol{y}, \boldsymbol{x}) > (\boldsymbol{x}^0 - \boldsymbol{y}, \boldsymbol{y}) \\ &\leftrightarrow (\boldsymbol{x}^0 - \boldsymbol{y}, \boldsymbol{x}) - (\boldsymbol{x}^0 - \boldsymbol{y}, \boldsymbol{y}) > 0 \\ &\leftrightarrow (\boldsymbol{x}^0 - \boldsymbol{y}, \boldsymbol{x} - \boldsymbol{y}) > 0.\end{aligned}$$

[2] Lemma B.1.1, p. 397.

If $\boldsymbol{x}$ is an arbitrary element of X^-, we get:

$$\begin{aligned}(\boldsymbol{a}, \boldsymbol{x}) - (\boldsymbol{a}, \boldsymbol{y}) &= (\boldsymbol{a}, \boldsymbol{x} - \boldsymbol{y}) \\ &\geqslant (\boldsymbol{a}, \boldsymbol{a}) \\ &= (\boldsymbol{a}, \boldsymbol{x}^0) - (\boldsymbol{a}, \boldsymbol{y}).\end{aligned}$$

Thus, $(\boldsymbol{a}, \boldsymbol{x}) \geqslant (\boldsymbol{a}, \boldsymbol{x}^0)$ for all $\boldsymbol{x} \in X^-$, which establishes that:

$$(\boldsymbol{a}, \boldsymbol{x}^0) = \inf_{\boldsymbol{x} \in X} (\boldsymbol{a}, \boldsymbol{x}) > (\boldsymbol{a}, \boldsymbol{y}). \dashv$$

Recall that the rank r of a matrix is the maximum number of linearly independent rows. It can also be shown that r is the maximum number of linearly independent columns.

Definition 9.2.5 Let $V = \mathbb{R}^n$ be the Euclidean space of n dimensions. A *k-dimensional plane* is the set of all points in V whose coordinates satisfy the system of equations:

$$\begin{aligned} a_{11}x_1 + a_{12}x_2 + \cdots + a_{1n}x_n &= \alpha_1, \\ a_{21}x_1 + a_{22}x_2 + \cdots + a_{2n}x_n &= \alpha_2, \\ &\vdots \\ a_{m1}x_1 + a_{m2}x_2 + \cdots + a_{mn}x_n &= \alpha_m, \end{aligned}$$

where the rank of the coefficient matrix is r and $k = n - r$. One-dimensional planes are called *lines*, and $(n - 1)$-dimensional planes are called *hyperplanes*.

If the rank of the matrix is $r = 1$, then $k = n - 1$, meaning all rows are linearly dependent on a single row. In this case, the equation of the hyperplane simplifies to:

$$a_1x_1 + a_2x_2 + \cdots + a_nx_n = \alpha.$$

The set H of vectors that satisfy the system of equations is a convex set, given by:

$$H = \{\boldsymbol{x} \mid (\boldsymbol{a}^i, \boldsymbol{x}) = \alpha_i\}.$$

A *hyperplane*:

$$H = \{\boldsymbol{x} \mid (\boldsymbol{a}, \boldsymbol{x}) = \alpha\}$$

determines two *closed half-spaces*:

$$H_1 = \{\boldsymbol{x} \mid (\boldsymbol{a}, \boldsymbol{x}) \geqslant \alpha\}, \quad H_2 = \{\boldsymbol{x} \mid (\boldsymbol{a}, \boldsymbol{x}) \leqslant \alpha\}.$$

Both of these are convex and closed sets.

Lemma 9.2.6 *If $X \subseteq E$ is a convex set and $\mathbf{y}$ is a boundary point of X, then there exists a supporting plane passing through $\mathbf{y}$, i.e., there exists a vector $\boldsymbol{a} \neq \mathbf{0}$ such that:*

$$\inf_{x \in X} (\boldsymbol{a}, \boldsymbol{x}) = (\boldsymbol{a}, \boldsymbol{y}).$$

Proof Let $(\boldsymbol{y}^\nu)$ be a sequence of points in $V \setminus X^-$ converging to $\boldsymbol{y}$. By Lemma 9.2.4, for each ν, there exists $\boldsymbol{a}^\nu$ such that:

$$\inf_{x \in X} (\boldsymbol{x}, \boldsymbol{a}^\nu) > (\boldsymbol{y}^\nu, \boldsymbol{a}^\nu) \quad (\nu = 1, 2, \ldots).$$

Each $\boldsymbol{a}^\nu$ can be normalized so that:

$$(\boldsymbol{a}^\nu, \boldsymbol{a}^\nu) = 1,$$

for

$$\begin{aligned}(\boldsymbol{x}, \boldsymbol{a}^\nu) > (\boldsymbol{y}^\nu, \boldsymbol{a}^\nu) &\Leftrightarrow \frac{1}{|\boldsymbol{a}^\nu|}(\boldsymbol{x}, \boldsymbol{a}^\nu) > \frac{1}{|\boldsymbol{a}^\nu|}(\boldsymbol{y}^\nu, \boldsymbol{a}^\nu) \\ &\Leftrightarrow \left(\boldsymbol{x}, \frac{1}{|\boldsymbol{a}^\nu|}\boldsymbol{a}^\nu\right) > \left(\boldsymbol{y}^\nu, \frac{1}{|\boldsymbol{a}^\nu|}\boldsymbol{a}^\nu\right),\end{aligned}$$

but

$$\begin{aligned}\left(\frac{1}{|\boldsymbol{a}^\nu|}\boldsymbol{a}^\nu, \frac{1}{|\boldsymbol{a}^\nu|}\boldsymbol{a}^\nu\right) &= \frac{1}{|\boldsymbol{a}^\nu|}\left(\boldsymbol{a}^\nu, \frac{1}{|\boldsymbol{a}^\nu|}\boldsymbol{a}^\nu\right) \\ &= \left(\frac{1}{|\boldsymbol{a}^\nu|}\right)^2 (\boldsymbol{a}^\nu, \boldsymbol{a}^\nu) \\ &= \frac{1}{(\boldsymbol{a}^\nu, \boldsymbol{a}^\nu)}(\boldsymbol{a}^\nu, \boldsymbol{a}^\nu) \\ &= 1.\end{aligned}$$

By taking limits, we obtain a vector $\boldsymbol{a} \neq \mathbf{0}$ satisfying:

$$\inf_{x \in X} (\boldsymbol{a}, \boldsymbol{x}) = (\boldsymbol{a}, \boldsymbol{y}). \dashv$$

Lemma 9.2.7 *A closed convex set is the intersection of all the half-spaces that contain it, with each each half-space being determined by a supporting hyperplane.*

Proof Let X be a closed and convex set in $\mathbb{R}^n$, and let $\mathscr{H}$ be the family of all half-spaces H that contain X and are determined by a supporting hyperplane of X. Clearly, $\mathscr{H}$ is closed and $X \subseteq \bigcap \mathscr{H}$. If $y \in \bigcap \mathscr{H}$ but $y \notin X$, then there exists a vector $\boldsymbol{a}$ such that

$$(\boldsymbol{a}, \boldsymbol{x}) > (\boldsymbol{a}, \boldsymbol{y})$$

for all $\boldsymbol{x} \in X$. Moreover, X is contained in one side H_1 of the half-space determined by the hyperplane

$$H = \{z \in \mathbb{R}^n \mid (\boldsymbol{a}, z) = (\boldsymbol{a}, \boldsymbol{x}^0)\}$$

for some $\boldsymbol{x}^0$ on the boundary of X. It follows that $\boldsymbol{y}$ belongs to the other half-space H_2 but not to H. However, this is impossible because $\bigcap \mathscr{H} \subseteq H_1$. ⊣

Theorem 9.2.8 (Separating Hyperplane Standard) *If X and Y are two convex sets with no interior point in common, then there exists a hyperplane H that separates X from Y; that is, there exists a nonzero vector $\mathbf{a}$ and a scalar α such that*

$$(\boldsymbol{a}, \boldsymbol{x}) \geqslant \alpha \quad \textit{for all } \boldsymbol{x} \in X, \quad \textit{and} \quad (\boldsymbol{a}, \boldsymbol{y}) \leqslant \alpha \quad \textit{for all } \boldsymbol{y} \in Y.$$

Proof Consider the set

$$X - Y = \{\boldsymbol{x} - \boldsymbol{y} \mid \boldsymbol{x} \in X, \boldsymbol{y} \in Y\}.$$

Observe that this set is convex and does not contain the zero vector in its interior. By the previous lemmas, there exists a vector $\boldsymbol{a}$ such that

$$(\boldsymbol{a}, \boldsymbol{x}) - (\boldsymbol{a}, \boldsymbol{y}) = (\boldsymbol{a}, \boldsymbol{x} - \boldsymbol{y}) \geqslant (\boldsymbol{a}, \mathbf{0}) = 0 \quad (\boldsymbol{x} \in X, \boldsymbol{y} \in Y),$$

or equivalently,

$$(\boldsymbol{a}, \boldsymbol{x}) \geqslant (\boldsymbol{a}, \boldsymbol{y}) \quad \text{for all } \boldsymbol{x} \text{ and } \boldsymbol{y}.$$

That is, the numbers in the set $\{(\boldsymbol{a}, \boldsymbol{x})\}$ are upper bounds for the set $\{(\boldsymbol{a}, \boldsymbol{y})\}$. The infimum of the first set is greater than or equal to the supremum of the second, so we can choose α as any number between these two values. ⊣

In the following proofs, all topological notions should be understood with respect to the S-topology of ${}^*\mathbb{R}^n$, whose basis consists of S-balls (cf. Definition 8.4.1).

Lemma 9.2.9 *If $S \subseteq^* \mathbb{R}^n$ and $\mathbf{0} \notin ({}^\circ S)$, then* $\text{hal}(\mathbf{0}) \cap \text{hal}(\boldsymbol{x}) = \varnothing$ *for every* $\boldsymbol{x} \in ({}^\circ S)$.

Proof Since $\mathbf{0} \notin (^\circ S)$, hal($\mathbf{0}$) contains points from $^*\mathbb{R}^n \setminus S$. But if hal($\mathbf{0}$) had points in common with hal($\boldsymbol{x}$), then hal($\boldsymbol{x}$) would be identical to hal($\mathbf{0}$), which is impossible since hal($\boldsymbol{x}$) contains no points from $^*\mathbb{R}^n \setminus S$. ⊣

Lemma 9.2.10 *If $S \subseteq^* \mathbb{R}^n$ and $\mathbf{0} \notin (^\circ S)$, then $\mathbf{0} \notin$ sh($(^\circ S)$).*

Proof If $\mathbf{0}$ belonged to sh($(^\circ S)$), then $\mathbf{0}$ would be in the interior of S, but this contradicts Lemma 9.2.9. ⊣

Lemma 9.2.11 *If S is an S-nearstandard convex set in $^*\mathbb{R}^n$ and $\mathbf{0} \notin (^\circ S)$, then there exists a $\boldsymbol{p} \in \mathbb{R}^n$ such that $\boldsymbol{p} \neq \mathbf{0}$ and, for every $\boldsymbol{x} \notin$ sh($(^\circ S)$), $\boldsymbol{p} \cdot \boldsymbol{x} \geqslant 0$.*

Proof By Lemma 9.1.4, $(^\circ S)$ is S-convex. By Lemma 9.1.2, sh($(^\circ S)$) is convex. Let X, Y be subsets of $\mathbb{R}^n$ such that $X =$ sh($(^\circ S)$) and $Y = \{\mathbf{0}\}$. By the standard Separating Hyperplane Theorem (Theorem 9.2.8), there exist a vector $\boldsymbol{p} \neq \mathbf{0}$ and a scalar $\alpha \in \mathbb{R}$ such that $\boldsymbol{p} \cdot \boldsymbol{x} \geqslant \alpha$ and $\boldsymbol{p} \cdot \mathbf{0} = 0 \leqslant \alpha$; that is, $\boldsymbol{p} \cdot \boldsymbol{x} \geqslant 0$. ⊣

The following proposition is called S-Separation Lemma by Brown and Robinson (1975, 47).

Theorem 9.2.12 (Separating Hyperplane Nonstandard) *If S is an S-nearstandard convex subset of $^*\mathbb{R}^n$ and $\mathbf{0} \notin (^\circ X)$, then there exists a standard vector $\boldsymbol{p} \in^* \mathbb{R}^n$ such that, for every $\boldsymbol{y} \in (^\circ S)$, $\boldsymbol{p} \cdot \boldsymbol{y}$ is greater than* 0 *or exceeds* 0 *by at most an infinitesimal amount.*

Proof Suppose that S is an S-nearstandard convex subset of $^*\mathbb{R}^n$, and that $\mathbf{0} \notin (^\circ X)$. By Lemma 9.1.4, since S is S-nearstandard and S-convex, $(^\circ S)$ is also S-convex. By Lemma 9.2.11, there exists a $\boldsymbol{p} \in \mathbb{R}^n$ such that $\boldsymbol{p} \neq \mathbf{0}$ and $\boldsymbol{p} \cdot \boldsymbol{x} \geqslant 0$.

Since A is nearstandard, every element $\boldsymbol{y} \in (^\circ A)$ can be written as

$$\boldsymbol{y} = \boldsymbol{x} + \boldsymbol{\varepsilon},$$

where $\boldsymbol{x} \in$ st($(^\circ A)$) and $\boldsymbol{\varepsilon}$ is a vector with infinitesimal coordinates. Therefore,

$$\boldsymbol{p} \cdot \boldsymbol{y} = \boldsymbol{p} \cdot \boldsymbol{x} + \boldsymbol{p} \cdot \boldsymbol{\varepsilon} = \boldsymbol{p} \cdot \boldsymbol{x} + \delta \geqslant 0 + \delta = \delta,$$

where δ is infinitesimal. ⊣

The above theorem can be reformulated using the expression

$$\boldsymbol{p} \cdot \boldsymbol{y} \gtrsim 0 \quad \text{for all } \boldsymbol{y} \in (^\circ X).$$

Indeed, Brown and Robinson introduce a special notation in addition to the one we have been using throughout this book. If $\boldsymbol{y} \in$ hal($\boldsymbol{x}$), we have written $\boldsymbol{x} \approx \boldsymbol{y}$. The symbols $\boldsymbol{x} > \boldsymbol{y}$, $\boldsymbol{x} \geq \boldsymbol{y}$, and $\boldsymbol{x} \geqq \boldsymbol{y}$ retain their conventional meanings. The notation $\boldsymbol{x} \gtrsim \boldsymbol{y}$ means that $\boldsymbol{x}$ is greater than $\boldsymbol{y}$, or that $\boldsymbol{y}$ exceeds $\boldsymbol{x}$ by at most an infinitesimal amount in each coordinate. The notation $\boldsymbol{x} \ggcurly \boldsymbol{y}$ means that $\boldsymbol{x}$ is greater than $\boldsymbol{y}$ by a non-infinitesimal amount in each coordinate.

Brown and Robinson (1975) define a (pure exchange) nonstandard economy $\mathfrak{E}$ as a structure $\langle \{\boldsymbol{x}_k\}_{k=1}^{\kappa}, \{\succ_k\}_{k=1}^{\kappa} \rangle$, where K is an hyperfinite set $\{1, 2, \ldots, \kappa\}$ of agents, $\boldsymbol{x}_k$ is the initial endowment of agent $k \in K$ and $\succ_k$ is its preference relation. Theorem 9.2.12 is used by Brown and Robinson (1975) to prove Edgeworth's conjecture: In a nonstandard economy $\mathfrak{E}$, an allocation X is in the core of $\mathfrak{E}$ if and only if there exists a price vector $\boldsymbol{p}$ that, together with X, constitutes a competitive equilibrium. The key to the proof lies in showing that there exists a coalition of agents U such that $\mathbf{0}$ does not belong to the convex hull of the set of optimal allocations with respect to the agents' preference relations. See Brown and Robinson (1975, 47–50) for details.

9.3 Fixed Points

Fixed point theorems are essential tools in economic methodology. In a result published in 1950 in the *Proceedings of the National Academy of Sciences* (Nash, 1950), which was communicated by Solomon Lefschetz on November 16, 1949, the author observed that equilibria in finite games could be conceptualized as fixed points of a certain type of correspondences. This approach contributed to the logical structuring of games in general. In that communication, it was noted that Kakutani's Theorem (Kakutani, 1941) was suitable for establishing the existence of equilibrium in mixed strategies for the mentioned class of games, provided certain conditions were met.

The nonstandard economies introduced by Robinson and Brown (1975), Brown (1976), and Rashid (1978) make use of a nonstandard version of Kakutani's Theorem. However, formulating the logical foundations of nonstandard economics requires a nonstandard version of the Fan-Glicksberg Theorem, which is more general than Kakutani's. This will be discussed in the next chapter.

The Fan-Glicksberg Theorem (Fan, 1952; Glicksberg, 1952) is a generalization of Kakutani's fixed point theorem. Unlike Kakutani's theorem, which deals with correspondences between subsets of $\mathbb{R}^n$ (or ${}^*\mathbb{R}^n$ in the nonstandard case), the Fan-Glicksberg theorem applies more generally to correspondences between subsets of Hausdorff topological vector spaces that are linear and convex. This raises the question: what would be the nonstandard version of this theorem? An exact formulation of the standard version is as follows.

Theorem 9.3.1 *Let $\varphi : S \to S$ be an upper semicontinuous (usc) correspondence. If S is a nonempty, convex, and compact subset of a Hausdorff topological vector space, and $\varphi(x)$ is nonempty, convex, and closed for all $x \in S$, then there exists $x^* \in S$ such that $x^* \in \varphi(x^*)$.*

The nonstandard version of the theorem is stated as follows.

Theorem 9.3.2 *Let S be a nonempty, quasi-convex, S-nearstandard, and closed internal subset of ${}^*T$, where T is a Hausdorff topological vector space. Let φ :*

$S \twoheadrightarrow S$ be an internal upper semicontinuous (usc) correspondence, meaning that its graph Φ is an internal subset of $^*(T \times T)$. If $\varphi(x)$ is nonempty, closed, and quasi-convex for all $x \in S$, then there exists $x^* \in S$ such that $x^* \in \varphi(x^*)$.

The aim of this section is to prove a nonstandard version of the Fan-Glicksberg Theorem and derive the nonstandard version of Kakutani's Theorem as a corollary. The proof requires several preliminary lemmas. We will consider the standard parts (also called "shadows") of the sets S, $\varphi(x)$, and Φ, denoted respectively as $\mathrm{sh}(S)$, $\mathrm{sh}[\varphi(x)]$, and $\mathrm{sh}(\Phi)$.

The idea is to show, based on the properties attributed to S and φ in the theorem's statement, the following:

(1) $\mathrm{sh}(S)$ is compact.
(2) $\mathrm{sh}(S)$ is convex.
(3) $\varphi \restriction \mathrm{sh}(S)$ is upper semicontinuous (usc).
(4) $\mathrm{sh}[\varphi(x)]$ is closed for all $x \in S$.
(5) $\mathrm{sh}[\varphi(x)]$ is convex for all $x \in S$.

We will demonstrate (1)–(4) in detail and then apply the standard version of the Fan-Glicksberg Theorem to the shadows. Finally, we will show that the nonstandard theorem follows as a transfer from the standard result.

The following lemmas assume that T is a Hausdorff topological vector space, that $\varphi : S \twoheadrightarrow S$ is a correspondence from a nearstandard and quasi-convex internal subset S of $^*T$ into itself, that φ is internal and upper semicontinuous, and that $\varphi(x)$ is quasi-convex for all $x \in S$.

Lemma 9.3.3 $\mathrm{sh}(S)$ *is compact.*

Proof By hypothesis, S is an internal set of nearstandard points in $^*T$. Since $\langle T, \mathcal{T}\rangle$ is regular (as it is a topological vector space), we are working in an α-saturated enlargement with $\mathrm{card}(\mathcal{T}) < \alpha$. By Luxemburg's Theorem (Theorem 8.2.15), it follows that $\mathrm{sh}(S)$ is compact. $\dashv$

Lemma 9.3.4 $\mathrm{sh}(S)$ *is convex.*

Proof Let $x, y \in \mathrm{sh}(S)$ and $\lambda \in [0, 1]$. Since x and y are standard and the space T is convex, we have

$$\lambda x + (1 - \lambda)y \in T.$$

That is, the combination $\lambda x + (1 - \lambda)y$ is standard. Since S is quasi-convex, there exists $z \in S$ such that

$$z \in \mathrm{hal}(\lambda x + (1 - \lambda)y).$$

But this means that

$$\lambda x + (1 - \lambda)y \in \mathrm{sh}(S). \dashv$$

Lemma 9.3.5 *$\varphi \restriction \mathrm{sh}(S)$ is upper semicontinuous (usc).*

Proof Let $\psi = \varphi \restriction \mathrm{sh}(S)$, the restriction of φ to $\mathrm{sh}(S)$. First, note that Ψ is closed in the product topology $\mathcal{P}$, whose basis consists of all sets of the form $U \times V$, where $U, V \in \mathcal{T}$.

Since $\nu(\mathbb{R})$ is α-saturated, $\mathrm{card}(\mathcal{P}) < \alpha$, and Φ is an internal set, it follows that $\mathrm{sh}(\Phi) = \Psi$ is closed by Luxemburg's Theorem (8.2.14).

To show that ψ is usc, suppose Ψ is closed. Consider sequences (x_n) and (y_n) in $\mathrm{sh}(S)$ such that $y_n \in \varphi(x_n)$ and $x_n \to x_0$, $y_n \to y_0$. Since (x_n, y_n) converges in Ψ, it follows that $(x_0, y_0) \in \Psi$, meaning $y_0 \in \psi(x_0)$. Thus, ψ is usc. ⊣

Having established these preparatory lemmas, we are now ready to proceed with the proof of the main result.

Theorem 9.3.6 (Fan-Glicksberg, Nonstandard) *Let S be a nonempty, quasi-convex, nearstandard, and closed internal subset of $^*T$, where T is a Hausdorff topological vector space that is linear and convex. Let $\varphi : S \to S$ be an internal upper semicontinuous (usc) correspondence, meaning that its graph Φ is an internal subset of $^*(T \times T)$. If $\varphi(x)$ is nonempty, closed, and quasi-convex for all $x \in S$, then there exists $x^* \in S$ such that $x^* \in \varphi(x^*)$.*

Proof The shadow $\mathrm{sh}(S)$ of S is a compact and convex subset of X. The shadow of Φ, denoted as $\Psi = \mathrm{sh}(\Phi)$, is the graph of the restriction $\psi : \mathrm{sh}(S) \to \mathrm{sh}(S)$ of the correspondence φ to the shadow of S. Since Ψ is closed, the correspondence ψ is usc. Applying the standard Fan-Glicksberg Theorem, we obtain the existence of $(x^*, x^*) \in \mathrm{sh}(\Phi)$.

Indeed, $x^* \in S$ because, assuming $x^* \in^* T \setminus S$, since S is closed, $^*T \setminus S$ is open, meaning that x^* would be an interior point of $^*T \setminus S$, and the halo of x^*, $\mathrm{hal}(x^*)$, would be contained in $^*T \setminus S$. But this is impossible because, as x^* is the shadow of some element of S, there must be an element in S that belongs to the halo of x^*. Therefore, x^* is a fixed point of φ. ⊣

9.4 Kakutani's Theorem

As we shall see, a special case of the Fan-Glicksberg Theorem is Kakutani's (1941), which can be formulated as follows in its nonstandard version.

Corollary 9.4.1 *Let S be a nonempty, quasi-convex, nearstandard, and closed internal subset of $^*\mathbb{R}^n$ with the usual topology, and let $\varphi : S \to S$ be an internal upper semicontinuous (usc) correspondence, meaning that its graph Φ is an internal subset of $^*(\mathbb{R}^n \times \mathbb{R}^n)$. If $\varphi(x)$ is nonempty, closed, and quasi-convex for all $x \in S$, then there exists $x^* \in S$ such that $x^* \in \varphi(x^*)$.*

Proof It is sufficient to observe that the usual topological space $\mathbb{R}^n$ is a Hausdorff topological vector space that is linear and convex. ⊣

9.5 The Lyapunov-Loeb Theorem

Brown (1976) proves the existence of a fixed point for the choice correspondence φ using Kakutani's Theorem. To do so, he requires that φ be nonempty, S-convex, and S-closed. For the proof of convexity, he relies on the Lyapunov-Loeb Theorem. Here, we will explore how this important result is used to establish the convexity of φ.

Definition 9.5.1 $\mathfrak{M}$ is an *infinitesimal scalar measure space* if and only if there exist κ, $K, \mathcal{I}(K)$, $\{\varepsilon_k\}_{k=1}^{\kappa}$, and μ such that:

(0) $\mathfrak{M} = \langle K, \mathcal{I}(K), \mu\rangle$.
(1) κ is an unlimited number in $^*\mathbb{N}$.
(2) K is a set in $^*\nu(\mathbb{R})$ with $\text{card}(K) = \kappa$.
(3) $\mathcal{I}(K)$ is the family of all internal subsets of K.
(4) $\{\varepsilon_k\}_{k=1}^{\kappa}$ is an internal set of infinitesimal numbers such that $\sum_{k=1}^{\kappa} \varepsilon_k$ is a limited number.
(5) $\mu(\varnothing) = 0$.
(6) For every $S \in \mathcal{I}(K)$, if $S \neq \varnothing$, then

$$\mu(S) = \sum_{k\in S} \varepsilon_k.$$

The economic motivation behind Definition 9.5.1 is as follows: κ is an unlimited positive integer that represents the number of agents in a large competitive economy; that is, κ is the cardinality of the set K of agents. $\mathcal{I}(K)$ represents the set of all possible coalitions of agents. The numbers ε_k represent possible quantities of some good in the agents' initial endowments. The measure $\mu(S)$ represents the total amount of the good in question. Brown assumes that this number is limited, while Rashid (1987) considers that the quantities of goods in endowments are appreciable, making the total aggregate demand unlimited (an unlimited positive integer).

We observe that $\mathfrak{M}$ is indeed a measure space because $\mathcal{I}(K)$ is a Boolean algebra and, if $S, R \in \mathcal{I}(K)$ with $S \cap R = \varnothing$, then

$$\mu(S \cup R) = \sum_{k\in S\cup R} \varepsilon_k = \sum_{k\in S} \varepsilon_k + \sum_{k\in R} \varepsilon_k = \mu(S) + \mu(R).$$

An *infinitesimal vector measure* is an n-dimensional vector $\boldsymbol{\mu} = \langle \mu_1, \ldots, \mu_n\rangle$, where $\mu_1, \ldots, \mu_n$ are infinitesimal scalar measures.

Theorem 9.5.2 (Lyapunov-Loeb) *If $\boldsymbol{\mu}$ is an infinitesimal vector measure, then for all $S, R \in \mathcal{I}(K)$ and $\lambda \in^* \mathbb{R}$ with $0 < \lambda < 1$, there exists a set $Q \in \mathcal{I}(K)$ such that*

$$\boldsymbol{\mu}(Q) \approx \lambda\boldsymbol{\mu}(S) + (1 - \lambda)\boldsymbol{\mu}(R).$$

For a proof, see Loeb (1973). A useful consequence of the Lyapunov-Loeb Theorem is the following:

Theorem 9.5.3 *If $\boldsymbol{\mu}$ is an infinitesimal vector measure, then for all $S \in \mathcal{I}(K)$ and $\lambda \in^* \mathbb{R}$ with $0 < \lambda < 1$, there exists a set $R \in \mathcal{I}(K)$ such that $R \subseteq S$ and*

$$\boldsymbol{\mu}(R) \approx \lambda\boldsymbol{\mu}(S).$$

Proof Let S be any element of $\mathcal{I}(K)$ and $\lambda \in^* \mathbb{R}$ with $0 < \lambda < 1$. By Theorem 9.5.2, there exists $R \in \mathcal{I}(K)$ such that

$$\boldsymbol{\mu}(R) \approx \lambda\boldsymbol{\mu}(S) + (1-\lambda)\boldsymbol{\mu}(\varnothing) = \lambda\boldsymbol{\mu}(S). \dashv$$

In a social system, where $A = \bigtimes_k A_k$ is the set of all possible action profiles of agents, a profile $\boldsymbol{a} \in A$ determines a set

$$\varphi(\boldsymbol{a}) = \bigtimes_{k \in K} \varphi_k(\boldsymbol{a})$$

of response profiles or best responses of agents to $\boldsymbol{a}$. To apply Kakutani's Theorem, it is required (among other things) that $\varphi(\boldsymbol{a})$ be convex for each $\boldsymbol{a}$. Usually, to obtain the convexity of $\varphi(\boldsymbol{a})$, it is assumed that each of the best response sets $\varphi_k(\boldsymbol{a})$ is convex. However, this is not necessary if the Lyapunov-Loeb Theorem is used.

Each of the sets $\varphi(\boldsymbol{a})$ has an associated family $\eta_{\varphi_{\boldsymbol{a}}}$ of choice function $\mathcal{C} : K \to^* \mathbb{R}^n$, which is defined as follows:

$$\eta_{\varphi_{\boldsymbol{a}}} \equiv \{\mathcal{C} \mid \mathcal{C} : K \to^* \mathbb{R}^n, \mathcal{C} \text{ is internal, and}$$
$$\text{for every } k \in K, \mathcal{C}(k) \in \varphi_k(\boldsymbol{a}) \text{ and } \mathcal{C}(k) \text{ is limited}\}.$$

Each of these $\mathcal{C}$ represents a profile of choices—one for each k—in $\varphi_k(\boldsymbol{a})$, where $\varphi_k(\boldsymbol{a})$ is a set of hyperreal vectors constituting the best responses of k to $\boldsymbol{a}$. Each choice function $\mathcal{C}$ in $\eta_{\varphi_{\boldsymbol{a}}}$ determines an average aggregate choice

$$(1/\kappa) \sum_{k \in K} \mathcal{C}(k),$$

which is the aggregate average choice of all agents when $\mathcal{C} \in \eta_{\varphi_{\boldsymbol{a}}}$ prevails.

The set

$$(1/\kappa) \sum_{k \in K} \mathcal{C}_{\boldsymbol{a}}(k)$$

denotes all aggregate average choices determined by choice functions in $\eta_{\varphi_{\boldsymbol{a}}}$, i.e.,

$$(1/\kappa) \sum_{k \in K} \mathcal{C}_{\boldsymbol{a}}(k) = \{\boldsymbol{x} \in^* \mathbb{R}^n \mid \boldsymbol{x} = (1/\kappa) \sum_{k \in K} \mathcal{C}(k) \text{ for some } \mathcal{C} \in \eta_{\varphi_{\boldsymbol{a}}}\}.$$

We observe that $\mathscr{C} \in (1/\kappa)\sum_{k\in K}\varphi_{a}$ determines an infinitesimal vector measure

$$\boldsymbol{\mu}_{\mathscr{C}} = \langle \mu^{1}_{\mathscr{C}}, \ldots, \mu^{n}_{\mathscr{C}} \rangle$$

where, for $S \in \mathcal{I}(K)$,

$$\mu^{i}_{\mathscr{C}}(S) = \sum_{k\in S}(1/\kappa)\mathscr{C}^{i}(k)$$

is the i-th coordinate of the vector

$$\sum_{k\in S}(1/\kappa)\mathscr{C}^{i}(k).$$

Thus,

$$\boldsymbol{\mu}_{\mathscr{C}}(K) = \langle \mu^{1}_{\mathscr{C}}(K), \ldots, \mu^{n}_{\mathscr{C}}(K) \rangle = \boldsymbol{x}. \dashv$$

The proof of the next theorem requires an application of the Lyapunov-Loeb Theorem (Theorem 9.5.2). I follow Brown's (1976) demonstration.

Theorem 9.5.4

$$(1/\kappa)\sum_{k\in K}\mathscr{C}_{\boldsymbol{a}}(k) \textit{ is S-convex.}$$

Proof Let $\boldsymbol{x}, \boldsymbol{y}$ be elements of $(1/\kappa)\sum_{k\in K}\mathscr{C}_{\boldsymbol{a}}(k)$ and let $\lambda \in [0, 1]$. We need to show that there exists $\boldsymbol{z}$ in $(1/\kappa)\sum_{k\in K}\mathscr{C}_{\boldsymbol{a}}(k)$ such that

$$\lambda\boldsymbol{x} + (1-\lambda)\boldsymbol{y} \approx \boldsymbol{z}.$$

If $\mathfrak{c}$ and $\mathfrak{d}$ are elements of $\eta_{\varphi_{\boldsymbol{a}}}$ such that $\boldsymbol{x} = \boldsymbol{\mu}_{\mathfrak{c}}(K)$ and $\boldsymbol{y} = \boldsymbol{\mu}_{\mathfrak{d}}(K)$, we need to show that there exists an $\mathfrak{e} \in \eta_{\varphi_{\boldsymbol{a}}}$ such that

$$\boldsymbol{z} = \boldsymbol{\mu}_{\mathfrak{e}}(K) \approx \lambda\boldsymbol{\mu}_{\mathfrak{c}}(K) + (1-\lambda)\boldsymbol{\mu}_{\mathfrak{d}}(K).$$

Let $\boldsymbol{\mu}$ be the infinitesimal vector measure (of dimension $2n$), given by $\langle \boldsymbol{\mu}_{\mathfrak{c}}, \boldsymbol{\mu}_{\mathfrak{d}} \rangle$. We observe that if $\lambda = 0$, then $\boldsymbol{\mu}(\varnothing) \approx \lambda\boldsymbol{\mu}(K)$, if $\lambda = 1$, then $\boldsymbol{\mu}(K) \approx \lambda\boldsymbol{\mu}(K)$, and if $0 < \lambda < 1$, there exists $S \in \mathscr{I}(K)$ such that $\boldsymbol{\mu}(S) \approx \lambda\boldsymbol{\mu}(K)$. Thus, for every $\lambda \in [0, 1]$, there exists a set S such that

$$\boldsymbol{\mu}(S) \approx \lambda\boldsymbol{\mu}(K) = \langle \lambda\boldsymbol{\mu}_{\mathfrak{c}}(K), \lambda\boldsymbol{\mu}_{\mathfrak{d}}(K) \rangle = \langle \lambda\boldsymbol{x}, \lambda\boldsymbol{y} \rangle.$$

Clearly,

$$\begin{aligned}\boldsymbol{\mu}(K \setminus S) &= \boldsymbol{\mu}(K) - \boldsymbol{\mu}(S)\\ &\approx \boldsymbol{\mu}(K) - \lambda\boldsymbol{\mu}(K)\\ &= (1-\lambda)\boldsymbol{\mu}(K)\\ &= \langle (1-\lambda)\boldsymbol{x}, (1-\lambda)\boldsymbol{y}\rangle.\end{aligned}$$

Now, define the following infinitesimal vector measure:

$$\boldsymbol{\mu}' = \begin{cases} \boldsymbol{\mu}_{\mathfrak{c}} & \text{if } k \in S,\\ \boldsymbol{\mu}_{\mathfrak{d}} & \text{if } k \in K \setminus S.\end{cases}$$

Then, we obtain:

$$\begin{aligned} \boldsymbol{z} &= \boldsymbol{\mu}'(K)\\ &= \boldsymbol{\mu}'[S \cup (K \setminus S)]\\ &= \boldsymbol{\mu}'(S) + \boldsymbol{\mu}'(K \setminus S)\\ &= \boldsymbol{\mu}_{\mathfrak{c}}(S) + \boldsymbol{\mu}_{\mathfrak{d}}(K \setminus S)\\ &\approx \lambda\boldsymbol{x} + (1-\lambda)\boldsymbol{y}.\end{aligned}$$

Now, define the function $\mathfrak{e} : K \to^{*} \mathbb{R}$ as:

$$\mathfrak{e}(k) = \begin{cases} \mathfrak{c}(k) & \text{if } k \in S,\\ \mathfrak{d}(k) & \text{if } k \in K \setminus S.\end{cases}$$

Then, we have:

$$\begin{aligned} \boldsymbol{z} &= \boldsymbol{\mu}_{\mathfrak{c}}(S) + \boldsymbol{\mu}_{\mathfrak{d}}(K \setminus S)\\ &= (1/\kappa)\sum_{k\in S}\mathfrak{c}(k) + (1/\kappa)\sum_{k\in K\setminus S}\mathfrak{d}(k)\\ &= (1/\kappa)\left[\sum_{k\in S}\mathfrak{c}(k) + \sum_{k\in K\setminus S}\mathfrak{d}(k)\right]\\ &= (1/\kappa)\sum_{k\in K}\mathfrak{e}(k).\end{aligned}$$

Indeed, $\mathfrak{e} \in \eta_{\varphi_{\boldsymbol{a}}}$ because $\mathfrak{e}(k) = \mathfrak{c}(k) \in \varphi_{\boldsymbol{a}}(k)$ if $k \in S$, $\mathfrak{e}(k) = \mathfrak{d}(k) \in \varphi_{\boldsymbol{a}}(k)$ if $k \in K \setminus S$, and hence, in all cases, $\mathfrak{e}(k) \in \varphi_{\boldsymbol{a}}(k)$. Since both $\mathfrak{c}$ and $\mathfrak{d}$ are bounded and internal, $\mathfrak{e}$ is also bounded and internal, as it is the combination of two bounded internal functions. $\dashv$

Theorem 9.5.4 is used to prove the existence of equilibria as follows. An *allocation* is a function $X : K \to^{*} \mathbb{R}^n$ that assigns a consumption bundle $X(k)$ to each agent. The *initial endowment* of agent k is an allocation $I(k)$ describing their wealth at the beginning of the analysis.

The set of all possible price systems is the simplex:

$$P = \left\{ \boldsymbol{p} \in^{*} \mathbb{R}^n \mid \boldsymbol{p} \geqslant \boldsymbol{0}, \sum_{i=1}^{n} p_i = 1 \right\}.$$

If $I(k)$ is the initial endowment of agent k, the *budget set* $B_{\boldsymbol{p}}(k)$ of agent k under price system $\boldsymbol{p} \in P$ is defined by:

$$B_{\boldsymbol{p}}(k) = \{\boldsymbol{x} \in^{*} \mathbb{R}^n \mid \boldsymbol{p} \cdot \boldsymbol{x} \leqslant \boldsymbol{p} \cdot I(k)\}.$$

A bound $\nu\boldsymbol{e}$ is imposed on admissible consumption bundles, where $\boldsymbol{e} = \langle 1, \ldots, 1\rangle$ and $\nu \in^{*} \mathbb{N} \setminus \mathbb{N}$, chosen such that

$$\nu = \max_{K} ||I(k)||.$$

Define the set of bounded consumption bundles:

$$M = \{\boldsymbol{x} \in^{*} \mathbb{R}^n \mid \boldsymbol{x} \leqslant \nu\boldsymbol{e}\}.$$

Define now the set $D_{\boldsymbol{p}}(k)$ of menus satisfying $\boldsymbol{x} \leqslant \boldsymbol{e}$ and optimal under agent preferences. Then, define the correspondence $\psi : P \to M$ by:

$$\psi(\boldsymbol{p}) = \left\{ \boldsymbol{x} \in^{*} \mathbb{R}^n \mid \boldsymbol{x} = (1/\kappa) \sum_{k \in K} X(k), \ X(k) \in D_{\boldsymbol{p}}(k) \right\}.$$

Since $D_{\boldsymbol{p}}(k)$ is compact and if preferences satisfy a weak order condition, $\psi(\boldsymbol{p})$ is nonempty, internal, upper semi-continuous, and convex due to the Lyapunov-Loeb Theorem. For further details, see Rashid (1978, 1987). The next chapter will elaborate on the foundations of nonstandard economic theory.

Exercises

9.1 Show that if S, S_1, and S_2 are convex sets in $\mathbb{R}^n$, $\boldsymbol{a} \in \mathbb{R}^n$, and λ is any real number, then the sets λS, $S + \boldsymbol{a}$, and $S_1 + S_2$ are convex.

9.2 Prove that $\mathcal{I}(K)$ is a Boolean algebra.

References

Berge, C. (1963). *Topological spaces*. Edinburgh y Londres: Oliver & Boyd.

Brown, D. J. (1976). Existence of a competitive equilibrium in a nonstandard exchange economy. *Econometrica, 44*(3), 537–546.

Brown,D. J., & Robinson, A. (1975). Nonstandard exchange economies. *Econometrica, 43*(1), 41–55.

Fan, K. (1952). Fixed-point and minimax theorems in locally convex topological linear spaces. *Proceedings of the National Academy of Sciences, 38*, 121–126.

Glicksberg, I. L. (1952). A further generalization of the kakutani fixed point theorem, with application to nash equilibrium points. *Proceedings of the American Mathematical Society, 3*(1), 170–174.

Kakutani, S. (1941). A generalization of Brouwer's fixed point theorem. *Duke Mathematical Journal, 8*(3), 457–459.

Karlin, S. (1959). *Mathematical methods and theory in games, programming, and economics*. Addison- Wesley Publishing Company.

Loeb, P. A. (1973). A combinatorial analog of Lyapunov's theorem for infinitesimally generated atomic vector measures. *Proceedings of the American Mathematical Society, 39*(3), 585–586.

Nash, J.F. (1950). Equilibrium points in N-person games. *Proceedings of the National Academy of Sciences, 36*(1), 48–49.

Rashid, S. (1978). Existence of equilibrium in infinite economies with production. *Econometrica, 46*(5), 1155–1164.

Rashid, S. (1987). *Economies with many agents. An approach using nonstandard analysis*. The Johns Hopkins University Press.

Robinson, A., & Brown, D. J. (1975). Non-standard exchange economies. *Econometrica, 43*(1), 41–55.

Rockafellar, R. T. (1970). *Convex analysis*. Princeton University Press.

Chapter 10
Nonstandard Economics

Abstract The final Chap. 10 develops in detail Richter's Theorem (Richter, Preferences, Utility, and Demand, pp. 29–58. Harcourt Brace Jovanovich, New York, 1971) concerning the existence of nonstandard utility representations. It then introduces a nonstandard formulation of Gerard Debreu's concept of a social system, proves the existence of equilibria for social systems, and concludes with an application of the concept to economic theory, specifically to the theory of pure exchange. It is shown in detail that equilibrium in pure exchange structures is a special case of equilibrium of a social system.

Keywords Richter's theorem · Equilibria of social systems · Pure exchange economics

10.1 Richter's Theorem

A *regular preference structure* is a pair of the form $\mathfrak{A} = \langle A, R \rangle$ that satisfies the following two sentences:

$$\forall x \forall y (Rxy \vee Ryx)$$

$$\forall x \forall y \forall z (Rxy \wedge Ryz) \rightarrow Rxz).$$

That is, $\mathfrak{A}$ is a weak order. A strict regular preference structure is a pair of the form $\mathfrak{A} = \langle A, P \rangle$ that satisfies:

$$\forall x \forall y (x \neq y \rightarrow (Pxy \vee Pyx))$$

$$\forall x \forall y (Pxy \rightarrow \neg Pyx).$$

$$\forall x \forall y \forall z ((Pxy \wedge Pyz) \rightarrow Pxz).$$

That is, $\mathfrak{A}$ is a strict simple order. Since the quotient set of every weak order, modulo the indifference relation, can be embedded into a strict simple order, I will

A. García de la Sienra, *The Logical Foundations of Nonstandard Analysis*, Synthese Library 530, https://doi.org/10.1007/978-3-032-05677-1_10

henceforth consider only structures of this type, which I will call strict preference structures.

Dana Scott and Patrick Suppes (1958: 115) defined a *theory of measurement* as a class $\mathcal{M}$ of relational systems of type $\mathfrak{a} = \{1\}$, closed under isomorphism, for which there exists a numerical structure $\mathfrak{N}$, also of type $\mathfrak{a}$, whose relation is binary, such that all structures in $\mathcal{M}$ can be embedded in $\mathfrak{N}$. In the case of preference structures, this measurement theory can be called a *utility theory*, as suggested by Skala (1975: 10). If all structures in the class $\mathcal{M}$ can be embedded in the numerical system $\mathfrak{N}$, we say that $\mathcal{M}$ is a *utility theory with respect to* $\mathfrak{N}$.

A natural question that arises, and one that is of interest in economic theory, is whether the class $\mathcal{M}$ of all strict preference structures, with cardinality at most that of the continuum, constitutes a utility theory with respect to the standard model of analysis $\mathfrak{R} = \langle \mathbb{R}, < \rangle$. Since every finite or countable strict simple order can be embedded in the structure $\mathfrak{Q} = \langle \mathbb{Q}, < \rangle$, where $\mathbb{Q}$ is the set of rational numbers, the class $\mathcal{K}$ of all strict simple orders of cardinality $\leqslant \aleph_0$ forms a utility theory with respect to $\mathfrak{Q}$. Unfortunately, it is known that lexicographic orders cannot be embedded in $\mathfrak{R}$, and thus the answer to whether the class $\mathcal{M}$ of all strict preference structures of cardinality $\leqslant 2^{\aleph_0}$ is a utility theory with respect to the standard model of analysis is negative.

Richter's Theorem (1971) states precisely that the class $\mathcal{M}$ of all strict preference structures, with cardinality at most that of the continuum, forms a utility theory with respect to the nonstandard model of analysis $^*\mathfrak{R} = \langle ^*\mathbb{R}, ^*< \rangle$ (whose cardinality, incidentally, is $2^{\aleph_0}$). In other words, for every regular strict preference structure $\mathfrak{A} = \langle A, P \rangle$, with cardinality at most that of the continuum, there exists a function $u : A \to {}^*\mathbb{R}$ such that Pab if and only if $u(a) < u(b)$. The proof of this theorem requires establishing a couple of preliminary lemmas.

Recall that if $(\mathfrak{A}, a)_{a \in A}$ is an expansion of the model $\mathfrak{A}$ by adding constants a, the diagram of $\mathfrak{A}$ is the set Δ_A of all atomic statements and their negations that are true in $\mathfrak{A}$.

Lemma 10.1.1 *If the language $\mathcal{L}$ of $\mathfrak{A}$ lacks function or constant symbols, then $\mathfrak{A}$ can be isomorphically embedded in the ultraproduct of its finite submodels.*

Proof Let Δ_A be the diagram of $(\mathfrak{A}, a)_{a \in A}$, and let I be the family of all finite subsets of Δ_A. If $\mathfrak{A}_i$ is a finite submodel of $\mathfrak{A}$, then $(\mathfrak{A}_i, a)_{a \in A_i}$ is also a finite submodel of $(\mathfrak{A}, a)_{a \in A}$ and is a model of any finite subset $i \in I$ contained in Δ_{a_i}. Since $\mathcal{L}$ lacks function or constant symbols, A_i generates a finite submodel of $\mathfrak{A}$. By the Compactness Theorem, there exists an ultrafilter $\mathcal{U}$ over I such that the ultraproduct $\prod_{i \in I} \mathfrak{A}_i / \mathcal{U}$ is a model of Δ_a. The reduct $\mathfrak{B}$ of $\prod_{i \in I} \mathfrak{A}_i / \mathcal{U}$ to $\mathcal{L}$ is a model in which $\mathfrak{A}$ is isomorphically embedded, by virtue of the composition $\psi = \vartheta \circ \varphi$. The situation is described in the following commutative diagram.

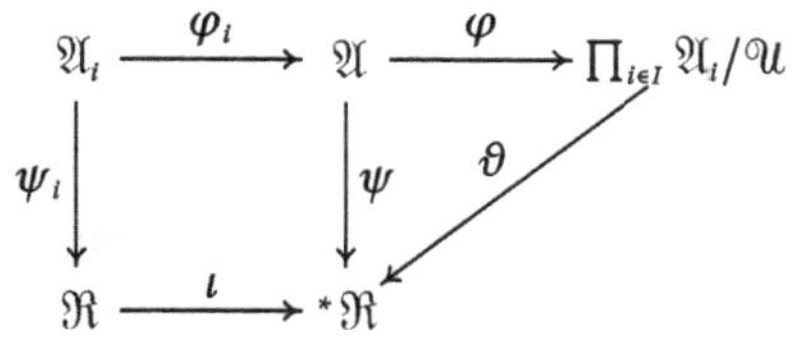

The demonstration of the lemma is complete.⊣

Lemma 10.1.2 *Let $\mathfrak{A} = \langle A, R_\xi \rangle$ be a model such that each finite submodel of $\mathfrak{A}$ admits a weak homomorphism (respectively strong homomorphism or isomorphism) into the model $\mathfrak{R} = \langle \mathbb{R}, \leqslant \rangle$. Then there exists a weak homomorphism (respectively strong homomorphism or isomorphism) of $\mathfrak{A}$ into a nonstandard model of $\mathfrak{R}$.*

Proof Let $\{\mathfrak{A}_i\}_{i \in I}$ be the family of all finite submodels of $\mathfrak{A}$, and for each $i \in I$, let $\varphi_i : A_i \to A$ be a homomorphism from $\mathfrak{A}_i$ to $\mathfrak{A}$. By Lemma 10.1.1, $\mathfrak{A}$ can be embedded in the ultraproduct $\prod_{i \in I} \mathfrak{A}_i / \mathcal{U}$. By the Homomorphism Theorem, since each $\mathfrak{A}_i$ is homomorphic to $\mathfrak{R}$, it follows that $\prod_{i \in I} \mathfrak{A}_i / \mathcal{U}$ is homomorphic to the ultrapower ${}^*\mathfrak{R} = \prod_{i \in I} \mathfrak{R} / \mathcal{U}$. Consequently, $\mathfrak{A}$ is also homomorphic to ${}^*\mathfrak{R}$.⊣

We proceed now to the main result.

Theorem 10.1.3 (Richter 1971) *The class $\mathcal{M}$ of all strict preference structures, with cardinality at most that of the continuum, forms a utility theory with respect to the nonstandard model of analysis ${}^*\mathfrak{R} = \langle {}^*\mathbb{R}, {}^*< \rangle$.*

Proof Two methods can be used: one relies on Maltsev's Lemma, while the other on compactness via ultraproducts. Both are interesting in their own right.

To use Maltsev's Lemma, we extend the language $L(\mathcal{M})$ of the similarity class $\mathcal{M}$, to which both the preference structure $\mathfrak{A}$ and the standard model of analysis $\mathfrak{R}$ belong. Note that both have the same type of similarity, with respective binary relations $\prec$ (in $\mathfrak{A}$) and $<$ (in $\mathfrak{R}$), which serve as interpretations of the binary predicate P.

For each element a in A, we introduce an individual constant symbol c_a and consider the diagram of $\mathfrak{A}$, that is, the set Γ_Δ of all formulas of the form $P(c_a, c_b)$ for all $a, b \in A$ such that $a \prec b$, and of the form $\neg P(c_a, c_b)$ for all $a, b \in A$ where $a \not\prec b$.

Note that the number of these constant symbols does not exceed the cardinality of the continuum. Let Γ_Θ be the set of all statements in the expanded language $L(\mathcal{M})$ that are true in the standard model $\mathfrak{R}$, and define $\Gamma = \Gamma_\Delta \cup \Gamma_\Theta$.

By the Compactness Theorem, since every finite subset of Γ is satisfiable (for example, in $\mathfrak{R}$), the entire set Γ is satisfiable; that is, Γ has a model $\mathfrak{B}$. The function u that assigns to each element a of A the interpretation of c_a in $\mathfrak{B}$ is thus a utility function.

In order to use ultraproducts, we take the asymmetric part of $\precsim$, denoted $\prec$, and consider equivalence classes of indifference as elements of A. Since every finite submodel of $\mathfrak{A}$ is isomorphically embeddable into $\mathfrak{R}$, it follows that $\mathfrak{A}$ itself can be embedded in an ultrapower of $\mathfrak{R}$.⊣

10.2 Social Systems

The hope that nonstandard analysis applied to economics would simplify mathematics has turned out to be an illusion. Any text on nonstandard economics has proven to be far more demanding and difficult to follow than standard texts. The problem I aim to address in this chapter is whether it is possible to simplify nonstandard mathematics applied to economics in a way that allows for the exploitation of the intuitive advantages that this approach was supposed to offer. One way to tackle this problem is to provide a nonstandard version of a very general and intuitive formulation that encompasses, among others, the most well-known models.

In 1952 Gerard Debreu introduced an abstract concept of a social system, emphasizing the idea of coordination of social actions. This concept is indeed general and intuitive and possesses natural properties that guarantee the existence of equilibria in social systems. In this chapter, I will provide a nonstandard version of Debreu's concept of a social system, which I hope will also be natural and intuitive. I will then prove the existence of equilibria by applying the Fan-Glicksberg theorem.

Debreu's concept of a social system is theoretically significant because it includes as special cases a wide range of games, particularly economic structures. The economies introduced by Robinson and Brown (1975), Brown, (1976, 1981), and Rashid (1978) are, in fact, special cases of nonstandard social systems. Thus, defining a nonstandard concept of a social system would represent progress in establishing the foundations of economic theory based on Robinsonian analysis.

In his definition of a social system, Debreu considers only a finite set of agents, each with an action space S_k, which is a subset of $\mathbb{R}^n$ endowed with the topology induced by the Euclidean metric. In contrast, the nonstandard concept of a social system requires a hyperfinite set of agents, each with an action space contained in ${}^*\mathbb{R}^n$. This implies that action profiles must be sequences of actions in a topological vector space of hyperfinite dimension.

In the Sect. 10.3 I introduce the topological linear space over which the agents' action space will be defined. In Sect. 10.4 I define the nonstandard concept of a social system, and in Sect. 10.5 one I use a nonstandard version of the Fan-Glicksberg theorem to prove the existence of an equilibrium in the social system. Another method that has been used to prove equilibria in nonstandard economies is the Lyapunov-Loeb theorem, but it remains to be seen whether it provides sufficient generality to establish equilibria in all types of social systems.

A methodological issue that I do not address in this chapter is the empirical applicability of the concept of a social system—that is, determining the scope of potential applications of the concept and the methods that could be used to develop them.[1]

---

[1] A general discussion of these methods can be found in García de la Sienra (2019).

10.3 The Topological Environment

Our starting point is the real vector space $\mathbb{R}^n$ with the topology induced by the Euclidean metric. Let $T = (\mathbb{R}^n)^\omega$ be the space of all sequences $s : \mathbb{N} \to \mathbb{R}^n$ that assign to each natural number i a vector in $\mathbb{R}^n$. We shall endow T with the product topology and the operations $\oplus$ and $\odot$ defined as follows, for scalars λ in $\mathbb{R}$ (we will represent any element $s \in T$ in the form $\langle s(i) : i \in \mathbb{N}\rangle$):

$$s \oplus s' = \langle s(i) + s'(i) : i \in \mathbb{N}\rangle,$$
$$\lambda \odot s = \langle \lambda s(i) : i \in \mathbb{N}\rangle,$$
$$\vec{0} = \langle s(i) : s(i) = 0 \,\text{and}\, i \in \mathbb{N}\rangle.$$

Since the context will always prevent confusion, we will use notation $s + s'$ instead of $s \oplus s'$, λs instead of $\lambda \odot s$, and 0 instead of $\vec{0}$. T, equipped with these operations, forms a linear space of dimension ω over the field of scalars $\mathbb{R}$. Moreover, with the product topology, it turns out to be a Hausdorff topological vector space.

Lemma 10.3.1 *T with the product topology $\mathcal{T}$ is a convex Hausdorff topological vector space.*

Proof The topological space T is Hausdorff because it is the product of Hausdorff topological spaces. Let $\lambda \in [0, 1]$ and let s, s' be any elements of T. Then

$$\begin{aligned}\lambda s + (1-\lambda)s' &= \langle \lambda s(i) : i \in \mathbb{N}\rangle + \\ &\quad + \langle (1-\lambda)s'(i) : i \in \mathbb{N}\rangle \\ &= \langle \lambda s(i) + (1-\lambda)s'(i) : i \in \mathbb{N}\rangle \\ &\in T,\end{aligned}$$

since $\lambda s(i) + (1-\lambda)s'(i) \in \mathbb{R}^n$ for all $i \in \mathbb{N}$. This shows that T is convex.

In order to show that the vector operations are continuous with respect to topology $\mathcal{T}$, recall that the base sets of $\mathcal{T}$ are of the form

$$B = \left\{s \in T \mid s(i_1) \in U_{i_1} \text{ and } \cdots \text{ and } s(i_m) \in U_{i_m}\right\},$$

where $U_{i_1}, \cdots, U_{i_m}$ are open subsets of $\mathbb{R}^n$. Let s^0, s'^0 be any elements of T and B^0 any base set with $s^0 + s'^0 \in B^0$. It will suffice to show that there are base sets B and B' such that $s \in B$ and $s' \in B'$ imply $s + s' \in B^0$. B^0 is of the form

$$B^0 = \left\{s'' \in T \mid s''(i_1) \in U^0_{i_1} \text{ and } \cdots \text{ and } s''(i_m) \in U^0_{i_m}\right\},$$

Since $s^0 + s'^0 \in B^0$ we have, for all $j = 1, \ldots, m$,

$$s^0(i_j) + s'^0(i_j) \in U^0_{i_j} \subseteq \mathbb{R}^n.$$

Given that $\mathbb{R}^n$ is a topological vector space, there exist neighborhoods V_{i_j} of $s^0(i_j)$ and V'_{i_j} of $s'^0(i_j)$ such that $s(i_j) \in V_{i_j}$ and $s'(i_j) \in V'_{i_j}$ imply that $s(i_j)+s'(i_j) \in U^0_{i_j}$. Let

$$B = \{s'' \in T | s''(i_1) \in V_{i_1} \text{ and } \cdots \text{ and } s''(i_m) V_{i_m}\}$$

and

$$B' = \{s'' \in T | s''(i_1) \in V'_{i_1} \text{ and } \cdots \text{ and } s''(i_m) \in V'_{i_m}\}.$$

Clearly, $s \in B$ and $s' \in B'$ implies that $s(i_j) + s'(i_j) \in U^0_{i_j}$ for $j = 1, \ldots, m$, and so $s + s' \in B^0$ as desired. A similar argument shows that the scalar multiplication operation $\lambda(\cdot)$ is also continuous in T.⊣

Nonstandard economic theory requires that the set of agents be represented by an initial segment of the set $^*\mathbb{N}$ of hypernatural numbers,

$$K = \{1, \ldots, \kappa\} = \{k \in {}^*\mathbb{N} | k \leqslant \kappa\},$$

where $\kappa \in {}^*\mathbb{N}_\infty$ is an unlimited hypernatural number. That is, unlike the original concept of a social system, in which there is a finite number of agents, we now want to consider a hyperfinite but unlimited number of agents. We will represent the set of agents by K.

For each agent $k \in K$, the set of possible actions for k is a subset S_k of $^*\mathbb{R}^n$. Hence, the set of strategy profiles or action profiles of the agents is given by:

$$S = \bigtimes_{i \in {}^*\mathbb{N}} S_i \qquad \text{with } S_i = \{0\} \text{ for } i > \kappa.$$

Hence, if we ignore the dummy sets $S_i = \{0\}$, a strategy profile or combination of strategies is a function $s : K \to {}^*\mathbb{R}^n$. Notice that S is a subset of $^*T$, the enlargement of T, which is the space of all sequences $s : {}^*\mathbb{N} \to {}^*\mathbb{R}^n$.

Sets $^*T$ and $^*\mathcal{T}$, which are enlargements of T and $\mathcal{T}$ respectively, do not constitute a topological space, since only *internal* families of open sets in $^*\mathcal{T}$ are closed under arbitrary unions. Yet, the internal open sets of $^*T$ constitute a base for a topology in $^*T$ called *Q-topology*.

In this topology the halo or monad of a standard element $s \in {}^*T$, denoted as $\mu(s)$, is defined as the intersection:

$$\mu(s) = \bigcap_{U \in \mathcal{U}_s} {}^*U,$$

where $\mathcal{U}_s$ is the collection of all open neighborhoods of s in T.

In this way, we conclude our review of the topological results required for the theory that shall be presented in the following sections.

10.4 The Concept of a Social System

Each agent $k \in K$ has a feasibility correspondence $\mathfrak{f}_k$, which assigns to each profile $s = \langle s_1, s_2, \ldots, s_i, \ldots \rangle \in S$ the set of feasible actions for agent k given s.

Additionally, each agent $k \in K$ has a preference relation $\succ_k$ that is regular, convex, and continuous. This preference relation gives rise to a utility function u_k that is continuous and quasiconcave.

The concept of a social system is precisely defined as follows:

Definition 10.4.1 $\mathfrak{S}$ is a *social system* iff K is an initial segment of ${}^*\mathbb{N}$ and there exist $\{S_k\}_{k \in K}$, $\{\mathfrak{f}_k\}_{k \in K}$ and $\{u_k\}_{k \in K}$ such that, for all $k \in K$,

(0) $\mathfrak{S} = \langle K, \{S_k\}_{k \in K}, \{\mathfrak{f}_k\}_{k \in K}, \{u_k\}_{k \in K} \rangle$.
(1) S_k is a nonempty internal subset of the non-negative orthant of ${}^*\mathbb{R}^n$ such that $0 \in S_k$.
(2) S_k is S-convex.
(3) S_k is S-closed in the S-topology of ${}^*\mathbb{R}^n$.
(4) S_k is nearstandard.
(5) $\mathfrak{f}_k : S \to S_k$ is an internal lower semicontinuous correspondence such that $\mathfrak{f}_k(s'_k, s_{-k}) = \mathfrak{f}_k(s_k, s_{-k})$ for every $s_k, s'_k \in S_k$ and $\mathfrak{f}_k(s)$ is a nonempty, internal convex and nearstandard subset of S_k for every $s \in S$.
(6) $u_k : \times_{k \in N} S_k \to {}^*\mathbb{R}$ is an internal, continuous, and quasiconcave function.

In order to clarify ideas, in the case where $\mathfrak{S}$ represents, for example, an exchange economy, one of the sets S_k corresponds to the set of price systems, while the other S_k are sets of consumption possibilities.

In this type of model, if S_k is the set of consumption possibilities for agent k, then $\mathfrak{f}_k(s)$ represents the budget set of agent k given profile s. Clearly, in this case, $\mathfrak{f}_k(s)$ is invariant under any transformation of s, except for the price system.

Since the sets S_k are nearstandard, it is possible to establish an unlimited upper bound on the set M of all feasible actions of the agents. This suggests the application of the Lyapunov-Loeb Theorem, eliminating the need to assume convexity of sets S_k.

For this, we would need to assume that the elements of the sets S_k are infinitesimal. In the case of nonstandard economies as described by Brown, this assumption holds because initial endowments are infinitesimal. However, as Rashid points out:

> It is an axiom of the neoclassical world view that individuals are of primary importance and the correct modeling of individuals is *sine qua non* for an acceptable model. From such a viewpoint, it is definitely preferable to permit all individuals to own one apple and face the consequences than to insist that society can have only one apple and let the individuals adapt to this fact. (Rashid, 1987, 45)

The consequence of allowing all individuals to own "one apple" (i.e. an appreciable amount of goods as initial endowment) is that the option sets S_k remain nearstandard but do not consist purely of infinitesimals, and the average sums

$$(1/\kappa) \sum_{k \in K} I(k)$$

of initial endowments $I(k)$ are not infinitesimal. This prevents the use of the Lyapunov-Loeb Theorem for proving equilibrium, which is why I have preferred to follow a more general approach.

The *best response* or *reaction correspondence* of agent $k \in K$ is a function $\varphi_k : S \to S_k$ such that, for every $s \in \bigtimes_{k \in K} S_k$:

$$\varphi_k(s) = \{s'_k \in \mathfrak{f}_k(s) \mid u_k(s'_k) = \max_{s''_k \in \mathfrak{f}_k(s)} u_k(s''_k)\}.$$

For $i > \kappa$, we define $\varphi_i(s) = \{0\}$.

In terms of this definition, the concept of equilibrium takes on a simple form.

Definition 10.4.2 An *equilibrium* of the social system $\mathfrak{S}$ is a strategy profile $s^* = (s_i^*) \in S$ such that, for each $k \in K$,

$$s_k^* \in \varphi_k(s^*).$$

It is easy to see that s^* is an equilibrium of $\mathfrak{S}$ if and only if s^* is a fixed point of the correspondence $\varphi : S \to S$, where

$$\varphi(s) = \bigtimes_{i \in {}^*\mathbb{N}} \varphi_i(s).$$

We shall see that the conditions imposed on the sets S_k guarantee the existence of an equilibrium.

10.5 Existence of Social Equilibria

In this section, we proceed to prove the existence of a social equilibrium. To achieve this, it will be necessary to derive, from the properties of the agents' action sets, that the set of strategy profiles S and the best response correspondences possess the following properties:

(1) S is an internal closed subset of ${}^*T$.
(2) S is nearstandard.
(3) S is S-convex.
(4) φ is upper semicontinuous (usc).
(5) $\varphi(s)$ is S-convex for every $s \in S$.

We proceed now to show that structure $\mathfrak{S}$ indeed satisfies these conditions.

Lemma 10.5.1 *S is an internal closed subset of ${}^*T$.*

Proof Let π_i be the ith projection $\pi_i : {}^*T \to {}^*T_i$, and $\phi(B)$ be the formula

$$\begin{aligned} &B \in {}^*\mathcal{T} \wedge \\ &(\exists! i \in {}^*\mathbb{N})(((i \in K \to \pi_i(B) = {}^*\mathbb{R}^n \setminus S_i) \wedge \\ &\quad (i \in {}^*\mathbb{N} \setminus K \to \pi_i(B) = {}^*\mathbb{R}^n \setminus \{0\})) \wedge \\ &\qquad (\forall j \in {}^*\mathbb{N})(j \neq i \to \pi_j(B) = {}^*\mathbb{R}^n)). \end{aligned}$$

This formula expresses that B is an (open) set in the base of the Q-topology, that there exists precisely one hypernatural number i such that the image of B under the projection π_i is the complement of S_i in ${}^*\mathbb{R}^n$, and that the image of B under any other projection π_j is the whole space ${}^*\mathbb{R}^n$.

The constants of this formula are ${}^*\mathbb{N}$, ${}^*\mathbb{R}^n$, 0, and the projections π, which are enlargements of the corresponding projections of T. Since all these constants denote internal sets, by the internal set formation principle, we infer that the set

$$\Phi = \{B \in {}^*\mathcal{T} \mid \phi(B)\}$$

is internal. Note that the elements of Φ are all sets of the form

$${}^*\mathbb{R}^n \times \cdots \times {}^*\mathbb{R}^n \times ({}^*\mathbb{R}^n \setminus S_i) \times {}^*\mathbb{R}^n \times \cdots$$

where all factors in the product are copies of ${}^*\mathbb{R}^n$, except for the ith factor, which is the open set ${}^*\mathbb{R}^n \setminus S_i$. In fact, such sets are sub-basic for the Q-topology.

We now define the family Ψ of the agents' strategy sets as the collection of sets that are the complements of elements of Φ. Let $\psi(B)$ be the formula:

$${}^*T \setminus B \in \Phi$$

and define:

$$\Psi = \{A \in {}^*\mathcal{T} \mid A = \psi(B) \text{ for some } B \in {}^*\mathcal{T}\}.$$

Again, since all constants appearing in the formula $\psi(A)$ denote internal sets, set Ψ is internal, and all its elements are closed in ${}^*T$.

The intersection $\bigcap \Psi$ is therefore an internal closed set. But $S = \times_i S_i$ is precisely this intersection, from which we conclude that the set S is internal and closed.⊣

Lemma 10.5.2 *S is nearstandard.*

Proof It is necessary to show that if s' is any element of S, then there exists a standard profile $s \in {}^*T$ such that $s' \in \mu(s)$. Let s' be any element of S. Since S_k is nearstandard for all $k \in K$, s'_k belongs to $\mu(s_k)$ for some standard point in ${}^*T$ which is also in S_k because S_k is S-closed Hence, let $s : {}^*\mathbb{N} \to {}^*\mathbb{R}^n$ be the sequence defined by

$$s_i = \begin{cases} s_k & \text{if } s'_k \in \mu(s_i) \text{ for } i \in K, \\ 0 & \text{otherwise.} \end{cases}$$

This sequence—which can be seen to be standard—satisfies condition

$$s'_i \in \mu(s_i) \quad \text{for all } i \in \mathbb{N}.$$

An application of Robinson's Theorem allows us to conclude that $s' \in \mu(s)$.⊣

Lemma 10.5.3 *S is S-convex.*

Proof Let s, s' be any elements of sh(S), and let $\lambda \in [0, 1]$. It is necessary to show that there exists a $s'' \in S$ such that $s'' \in \mu[\lambda s + (1 - \lambda)s']$.

For every $k \in K$, $\lambda \in [0, 1]$, and $s_k, s'_k \in \text{sh}(S_k)$, there exists $s''_k \in S_k$ such that $s''_k \in \mu[\lambda s_k + (1 - \lambda)s'_k]$, since S_k is S-convex.

Let $s : {}^*\mathbb{N} \to {}^*\mathbb{R}^n$ be the sequence defined by

$$s''_i = \begin{cases} s''_k & \text{if } i = k, \\ = 0 & \text{otherwise.} \end{cases}$$

Since s_k and s'_k are standard and T is convex, the combination $\lambda s_k + (1 - \lambda)s'_k$ is also standard and so the fact that, for every $k \in K$, we have

$$s''_k \in \mu[\lambda s_k + (1 - \lambda)s'_k]$$

implies

$$s'' \in \mu[\lambda s + (1 - \lambda)s'].⊣$$

If s^0 is a standard element of S, we say that the infinite sequence (s^l) of elements of S *converges* to s^0 if $s^l \in \mu(s^0)$ for all $l \in \mathbb{N}_\infty$. ⊣

Lemma 10.5.4 *φ is upper semicontinuous (usc).*

Proof First, we must show that φ_i is usc for every $i \in K$ (it is trivially so for all $i \notin K$). Let (x_k^l) be a sequence of points in S_k that converges to x_k^0, and let (y_k^l) be a sequence of points also in S_k that converges to y_k^0, assuming that $y_k^l \in \varphi_k(x^l)$ for all $l \in \mathbb{N}$. Notice that, since S_k is S-closed, limits x_k^0 and y_k^0 belong to S_k.

Since $y_k^l \in \varphi_k(x_k^l)$, y_k^l is optimal in $\mathfrak{f}_k(x^l)$ and, since u_k is continuous, we have

$$\lim_{\mu\to\infty} u_k(y_k^l) = u_k(y_k^0).$$

Now, let z_k^0 be any element of $\mathfrak{f}_k(x_k^0)$. Since $\mathfrak{f}_k$ is lsc, there exists a sequence (z_k^l) such that $z_k^l \to z_k^0$ with $z_k^l \in \mathfrak{f}(x_k^l)$. Since $u_k(y_k^l) \geq u_k(z_k^l)$ for all l, we get

$$u_k(y_k^0) = \lim_{\mu\to\infty} u_k(y_k^l) \geq \lim_{\mu\to\infty} u_k(z_k^l) = u_k(z_k^0).$$

Since z_k^0 was chosen arbitrarily in $\mathfrak{f}_k(x_k^0)$, this proves that $y_k^0 \in \varphi_k(x_k^0)$.

Suppose now that the sequence of profiles (s^l) converges to s^0, the sequence s'^l converges to s'^0, and that $s'^l \in \varphi(s^l)$ for all $l \in \mathbb{N}$. Without loss of generality, we can assume that s^0 and s'^0 are standard, since S is nearstandard.

The fact that $s'^l \in \varphi(s^l)$ implies that

$$s_k'^l \in \varphi_k(s_k^l)$$

for all $l \in {}^*\mathbb{N}$. Since φ_k is usc, $s_k'^0 \in \varphi_k(s_k^0)$. By Robinson's Theorem, we conclude that $s'^0 \in \varphi(s^0)$.⊣

Lemma 10.5.5 *$\varphi(s)$ is S-convex for all $s \in S$.*

Proof sh$[\varphi(s)]$ is the set of all standard profiles whose monads contain some element of $\varphi(s)$. Let s_1, s_2 be elements of sh$[\varphi(s)]$, and let $\lambda \in [0, 1]$. We must show that there exists an $s_3 \in \varphi(s)$ such that

$$s_3 \in \mu[\lambda s_1 + (1-\lambda)s_2].$$

Both s_1 and s_2 are standard profiles such that $s_1{}' \in \mu(s_1)$ and $s_2{}' \in \mu(s_2)$ for certain elements $s_1{}', s_2{}' \in \varphi(s)$. This implies that

$$s'_{1k} \in \mathfrak{f}_k(s_{1k}) \quad \text{with} \quad u_k(s'_{1k}) = u^0$$

and

$$s'_{2k} \in \mathfrak{f}_k(s_{2k}) \quad \text{with} \quad u_k(s'_{2k}) = u^0,$$

where

$$u^0 = \max_{x \in \mathfrak{f}_k(s_{ik})} u_k(x), \quad i = 1, 2.$$

Moreover, by Robinson's Theorem, for all $k \in \mathbb{N}$, we have $s'_{1k} \in \varphi_k(s_{1k})$ and $s'_{2k} \in \varphi_k(s_{2k})$.

Since $\mathfrak{f}_k(s_k)$ is convex, we get

$$\lambda s'_{1k} + (1 - \lambda)s'_{2k} \in \mathfrak{f}_k(s_k).$$

Furthermore, since u_k is quasi-concave,

$$u_k[\lambda s'_{1k} + (1 - \lambda)s'_{2k}] \geq u^0,$$

which implies that

$$\lambda s'_{1k} + (1 - \lambda)s'_{2k} \in \varphi_k(s_k).$$

Since for all $k \in \mathbb{N}$, we have $s'_{1k} \in \mu(s_{1k})$ and $s'_{2k} \in \mu(s_{2k})$, and since $s_{1k}, s_{2k}, s'_{1k}, s'_{2k}$ are vectors in ${}^*\mathbb{R}^n$, it follows that

$$|s'_{1k} - s'_{2k}| \approx 0, \quad |s_{1k} - s_{2k}| \approx 0.$$

Since $\lambda \geq 0$, we also have

$$|\lambda s'_{1k} - \lambda s'_{2k}| \approx 0, \quad |(1 - \lambda)s_{1k} - (1 - \lambda)s_{2k}| \approx 0.$$

Thus,

$$\begin{aligned}
&|[\lambda s'_{1k} + (1 - \lambda)s'_{2k}] - [\lambda s_{1k} + (1 - \lambda)h(k)]| \\
&= |\lambda s'_{1k} - \lambda s_{1k} + (1 - \lambda)s'_{2k} - (1 - \lambda)s_{2k}| \\
&\quad \leqslant |\lambda s'_{1k} - \lambda s_{1k}| + |(1 - \lambda)s'_{2k} - (1 - \lambda)s_{2k}| \\
&\quad \approx 0.
\end{aligned}$$

That is, for all $k \in \mathbb{N}$,

$$\lambda s'_{1k} + (1 - \lambda)s'_{2k} \in \mu[\lambda s_{1k} + (1 - \lambda)s_{2k}].$$

If we define s_3 as $\lambda s'_{1k} + (1 - \lambda)s'_{2k}$ then, by Robinson's Theorem, we obtain

$$s_3 \in \mu[\lambda s_1 + (1 - \lambda)s_2].$$

On the other hand, since $s_{3k} \in \varphi_k(s_k)$ for all k, it follows that $s_3 \in \varphi(s)$.$\dashv$

Since Lemmas 5–9 establish that a social system satisfies the conditions required for φ to possess a fixed point, the following theorem follows immediately.

Theorem 10.5.6 *Every social system has an equilibrium.*

Proof The defining conditions of a social system imply the existence of a fixed point $s^* = (s_i^*) \in S$ such that $s_i^* \in \varphi_i(s^*)$ for all $i \in {}^*\mathbb{N}$. This, in particular, implies that for all $k \in K$ we have $s_k^* \in \varphi_k(s^*)$.⊣

Once the nonstandard concept of a social system has been precisely formulated, its empirical applicability remains to be considered. This should be left for another occasion, but it is worth noting that the idea of an infinite number of agents—even a continuum of them—is not foreign to economic theory (see Aumann, 1966). A discussion of the methodology of economies with a continuum of agents can be found, for example, in Mas-Colell and and Vives (1993). I am not aware of any analogous treatment for economies with a hyperfinite number of agents.

10.6 Pure Exchange Economies

As a particular case of a social system, I shall discuss the concept of a pure exchange economy. The concept of a pure exchange economy aims to represent a group of consumer agents who can improve their utility through the exchange of goods. Except for a special agent called the *market participant*, the agents are represented by an initial segment K of ${}^*\mathbb{N}$, each of whom is defined by an initial endowment $\boldsymbol{\omega}_k$ and a preference relation $\succ_k$ over a space of consumption menus. These preference relations are supposed to be represented by utility functions u_k. Formally, the concept can be defined as follows.

Definition 10.6.1 $\mathfrak{E}$ is a *pure exchange economy* iff there exist K, X_k, $\succ_i$, $\boldsymbol{\omega}_k$ and P such that

(0) $\mathfrak{E} = \langle (X_k, \succ_k)_{k \in K}, (\boldsymbol{\omega}_k)_{k \in K}, P \rangle$
(1) K is an initial segment $\{1, \ldots, \kappa\}$ of ${}^*\mathbb{N}$.
(2) For every $k \in K$, X_k is an internal, quasiconvex and S-closed subset of the nonnegative orthant ${}^*\Omega$ of ${}^*\mathbb{R}^n$ that contains the origin.
(3) For every $k \in K$, $(X_k, \succ_k)$ is a preference structure such that $\succ_k$ is quasiconvex, continuous, and lacks a saturation point; i.e. there is no vector $\boldsymbol{x}_k \in X_k$ such that $\boldsymbol{x}_k \succ_i \boldsymbol{x}'_k$ for every $\boldsymbol{x}'_k \in X_k$.
(4) For every $k \in K$, $\boldsymbol{\omega}_k$ is a standard vector in X_k and, for some $\boldsymbol{x}_k \in X_k$, $\boldsymbol{0} < \boldsymbol{x}_k < \boldsymbol{\omega}_k$.
(5) P is the unit simplex of ${}^*\Omega$.

A *state* of the economy is an ordered tuple $(\boldsymbol{x}_k, \boldsymbol{p})$ consisting of elements of ${}^*\mathbb{R}^n$, which represents a possible allocation of consumption menus $\boldsymbol{x}_k$ to consumers and a price $\boldsymbol{p}$ set by the market. I will denote the allocation $(\boldsymbol{x}_k)$ using the notation $\vec{\boldsymbol{x}}$.

The *total initial endowment* is the sum

$$\boldsymbol{\omega} = \sum_{k \in K} \boldsymbol{\omega}_k.$$

The *excess demand* under the allocation $(\boldsymbol{x}_k)$ is given by

$$\mathbf{z} = \sum_{k \in K} \boldsymbol{x}_k - \boldsymbol{\omega}.$$

The set of possible aggregate demands is the sum

$$X = \sum_{k \in K} X_k.$$

The set of excess demands is

$$Z = X - \{\boldsymbol{\omega}\}.$$

A *feasible allocation* is a list $(\boldsymbol{x}_k)$ such that

$$\sum_{k \in K} \boldsymbol{x}_k \leqq \boldsymbol{\omega} \quad \text{and} \quad \boldsymbol{x}_k \in X_k.$$

We say that an action (consumption) $\boldsymbol{x}_k$ is *achievable* or *feasible* if there exists a feasible allocation $\vec{\boldsymbol{x}}$ such that $\boldsymbol{x}_k$ is the k-th coordinate of $\vec{\boldsymbol{x}}$.

The set of achievable actions for agent k is denoted by $\hat{X}_k$, and the set of all achievable actions is denoted by $\hat{X}$.

Lemma 10.6.2 *The set of possible aggregate actions is closed-S.*

Proof Recall that a set is closed-S if and only if it is closed under F-limits. Therefore, it suffices to show that if $f : {}^*\mathbb{N} \to {}^*\mathbb{R}^n$ is a convergent sequence of points in X, with $f(\nu) = \boldsymbol{x}^\nu$ and limit $\mathbf{y}$, then $\mathbf{y} \in X$.

Let $\boldsymbol{x}^\nu$ be a convergent sequence of points in X. Each point in the sequence is of the form

$$\boldsymbol{x}^\nu = \sum_{k \in K} \boldsymbol{x}_k^\nu,$$

where $\boldsymbol{x}_k^\nu \in X_k$ for each $k \in K$. If each corresponding sequence $\boldsymbol{x}_k^\nu$ converges to a respective limit $\boldsymbol{x}_k^0$, then, since X_k is closed-S, we have $\boldsymbol{x}_k^0 \in X_k$, and consequently,

$$\mathbf{y} = \lim_{\nu \to \infty} \boldsymbol{x}^\nu = \lim_{\nu \to \infty} \sum_{k \in K} \boldsymbol{x}_k^\nu = \sum_{k \in K} \lim_{\nu \to \infty} \boldsymbol{x}_k^\nu = \sum_{k \in K} \boldsymbol{x}_k^0 \in X.$$

This means that

$$\sum_{k\in K} \boldsymbol{x}_k^{\nu} \approx \mathbf{y} \quad \text{for } \nu \in {}^*\mathbb{N}_\infty,$$

or, for an arbitrary k,

$$\boldsymbol{x}_k^0 - \boldsymbol{x}_k^{\nu} + \left(\sum_{\substack{l\in K\\ l\neq k}} \boldsymbol{x}_l^0 - \sum_{\substack{l\in K\\ l\neq k}} \boldsymbol{x}_l^{\nu} \right) \approx \mathbf{0}.$$

Assuming, without loss of generality, that

$$\sum_{\substack{l\in K\\ l\neq k}} \boldsymbol{x}_l^0 - \sum_{\substack{l\in K\\ l\neq k}} \boldsymbol{x}_l^{\nu} \geqq \mathbf{0},$$

we conclude that $\boldsymbol{x}_k^0 - \boldsymbol{x}_k^{\nu}$ must be infinitesimal. Consequently, the sequence $\boldsymbol{x}_k^{\nu}$ is convergent with limit $\boldsymbol{x}_k^0$. ⊣

Lemma 10.6.3 *For every $k \in K$, $\hat{X}_k$ is S-closed.*

Proof Let f be a sequence of points in $\hat{X}_k$ that converges to a limit $\boldsymbol{x}_k^0$. Then, there exists a sequence $(\vec{\boldsymbol{x}}^{\nu})$ of feasible states, $(\vec{\boldsymbol{x}}^{\nu}) = (\boldsymbol{x}_{-k}^{\nu}, \boldsymbol{x}_k^{\nu}) \in \hat{X}$, which means that

$$\boldsymbol{x}_k^{\nu} + \sum_{\substack{l\in K\\ l\neq k}} \boldsymbol{x}_l^{\nu} \leqq \boldsymbol{\omega} \quad \forall \nu \in {}^*\mathbb{N};$$

that is, the sequence $\vec{\boldsymbol{x}}^{\nu} = \sum_{k\in K} \boldsymbol{x}_k^{\nu}$ is bounded. Since the set $X = \sum_{k\in K} X_k$ is closed, there exists a subsequence $\boldsymbol{x}^{\eta}$ of $\boldsymbol{x}^{\nu}$ that converges to a limit $\boldsymbol{x}^0$ in X.

Taking the limit, we obtain

$$\boldsymbol{x}_k^0 + \sum_{\substack{l\in K\\ l\neq k}} \boldsymbol{x}_l^0 = \lim_{\nu\to\infty} \left[\boldsymbol{x}_k^{\eta} + \sum_{\substack{l\in K\\ l\neq k}} \boldsymbol{x}_l^{\eta} \right] \leqq \boldsymbol{\omega},$$

and therefore, $(\boldsymbol{x}_{-k}^0, \boldsymbol{x}_k^0)$ is feasible, since

$$\lim_{\eta\to\infty} \boldsymbol{x}_k^{\eta} = \lim_{\nu\to\infty} \boldsymbol{x}_k^{\nu}.$$

It follows that $\boldsymbol{x}_k^0 \in \hat{X}_k$, and thus $\hat{X}_k$ is S-closed.⊣

Lemma 10.6.4 *For every $k \in K$, $\hat{X}_k$ is quasiconvex.*

Proof The statement follows from the convexity of $\hat{X}_k$. Indeed, if $\boldsymbol{x}_k$ and $\boldsymbol{x}'_k$ are arbitrary elements of $\hat{X}_k$ and $\lambda \in [0, 1]$, we have

$$\lambda \boldsymbol{x}_k + \lambda \sum_{\substack{l \in K \\ l \neq k}} \boldsymbol{x}_l \leq \lambda \boldsymbol{\omega} \quad \text{and} \quad (1-\lambda)\boldsymbol{x}'_k + (1-\lambda) \sum_{\substack{l \in K \\ l \neq k}} \boldsymbol{x}'_l \leqq (1-\lambda)\boldsymbol{\omega},$$

from which it follows that

$$\lambda \boldsymbol{x}_k + (1-\lambda)\boldsymbol{x}'_k + \sum_{\substack{l \in K \\ l \neq k}} [\lambda \boldsymbol{x}_l + (1-\lambda)\boldsymbol{x}'_l] \leq \boldsymbol{\omega};$$

that is,

$$\lambda \boldsymbol{x}_k + (1-\lambda)\boldsymbol{x}'_k \in \hat{X}_k. \dashv$$

Lemma 10.6.5 *For every $k \in K$, $\hat{X}_k$ is nearstandard.*

Proof This follows from the fact that $\hat{X}_k$ is bounded, that is,

$$\mathbf{0} \leq \boldsymbol{x}_k \leq \boldsymbol{\omega}_k \quad \text{for all } \boldsymbol{x}_k \in \hat{X}_k,$$

where $\boldsymbol{\omega}_k$ is a standard vector and therefore limited. $\dashv$

For each $k \in K$, we define the correspondence $B_k : P \to \hat{X}_k$ as

$$B_k(\boldsymbol{p}) = \{\boldsymbol{x}_k \in \hat{X}_k \mid \boldsymbol{p}\boldsymbol{x}_k \leqslant \boldsymbol{p}\boldsymbol{\omega}_k\}.$$

Thus, $B_k(\boldsymbol{p})$ represents the *opportunity set* given the price system $\boldsymbol{p}$.

Lemma 10.6.6 *The correspondence $B_k(\boldsymbol{p})$ is scs (upper semi-continuous) for all $k \in K$.*

Proof Let $\boldsymbol{p}^\nu$ be a sequence of elements in P that converges to $\boldsymbol{p}^0$, and let $\boldsymbol{x}_k^\nu$ be a sequence of elements in $\hat{X}_k$ that converges to $\boldsymbol{x}_k^0$, with $\boldsymbol{x}_k^\nu \in B_k(\boldsymbol{p}^\nu)$ for all ν. That $\boldsymbol{x}_k^0 \in B_k(\boldsymbol{p}^0)$ is shown in the following chain of inequalities:

$$\begin{aligned}
\boldsymbol{p}^0\boldsymbol{x}_k^0 &= \lim_{\nu\to\infty} \boldsymbol{p}^\nu \cdot \lim_{\nu\to\infty} \boldsymbol{x}_k^\nu \\
&= \lim_{\nu\to\infty} \boldsymbol{p}^\nu\boldsymbol{x}_k^\nu \\
&\leqslant \lim_{\nu\to\infty} \boldsymbol{p}^\nu\boldsymbol{\omega}_k \\
&= \lim_{\nu\to\infty} \boldsymbol{p}^\nu \cdot \lim_{\nu\to\infty} \boldsymbol{\omega}_k \\
&= \boldsymbol{p}^0\boldsymbol{\omega}_k. \dashv
\end{aligned}$$

If we designate the market participant as $\kappa+1$, we can define the correspondence $\mathfrak{f}_k : \hat{X} \to \hat{X}_k$ by the following condition:

$$\mathfrak{f}_k(\vec{\boldsymbol{x}}, \boldsymbol{p}) = \begin{cases} B_k(\boldsymbol{p}) & \text{if } k \in K, \\ P & \text{if } k = \kappa + 1. \end{cases}$$

We also define the correspondence

$$\varphi_k : \vec{X} \times P \to \hat{X}_k$$

by the condition:

$$\varphi_k(\vec{\boldsymbol{x}}, \boldsymbol{p}) = \left\{ \boldsymbol{x}_k \in \mathfrak{f}_k(\vec{\boldsymbol{x}}, \boldsymbol{p}) \mid u_k(\boldsymbol{x}_k) = \max_{\boldsymbol{x}'_k \in \mathfrak{f}_k(\vec{\boldsymbol{x}}, \boldsymbol{p})} u_k(\boldsymbol{x}'_k) \right\}$$

and

$$\varphi_\kappa : \vec{X} \times P \to \hat{X}_k$$

by the condition:

$$\varphi_\kappa(\vec{\boldsymbol{x}}, \boldsymbol{p}) = \left\{ \boldsymbol{p} \in \mathfrak{f}_\kappa(\vec{\boldsymbol{x}}, \boldsymbol{p}) \mid \boldsymbol{p}\mathbf{z} = \max_{\boldsymbol{p}' \in P} \boldsymbol{p}'\mathbf{z} \right\}.$$

We observe that $\varphi_k(\vec{\boldsymbol{x}}, \boldsymbol{p})$ is the set of optimal consumption choices for agent k given the price system $\boldsymbol{p}$. Similarly, $\varphi_\kappa(\vec{\boldsymbol{x}}, \boldsymbol{p})$ is the set of price systems that maximize the cost of the excess demand $\mathbf{z}$ corresponding to the allocation $\vec{\boldsymbol{x}}$.

Lemma 10.6.7 *$\mathfrak{f}_k(\vec{\boldsymbol{x}})$ is a non-empty, compact, and convex set for every $\vec{\boldsymbol{x}} \in \hat{X}$.*

We define the correspondence $\varphi : \hat{X} \to \hat{X}$ by the condition:

$$\varphi(\vec{\boldsymbol{x}}) = \bigtimes_{k \in K} \varphi_k(\vec{\boldsymbol{x}}).$$

Lemma 10.6.8 *For each $(\vec{\boldsymbol{x}}, \boldsymbol{p}) \in \hat{X} \times P$, set $\mathfrak{f}_k(\vec{\boldsymbol{x}}, \boldsymbol{p})$ is non-empty, convex, and compact.*

Proof The set P is non-empty, convex, and compact by definition. $\mathfrak{f}_k(\vec{\boldsymbol{x}}, \boldsymbol{p})$ is simply $B_k(\boldsymbol{p})$ for $k \in K$. Since there exists $\boldsymbol{x}_k \in \hat{X}_k$ such that

$$\mathbf{0} \leq \boldsymbol{x}_k < \boldsymbol{\omega}_k, \quad \text{and} \quad \boldsymbol{p}\boldsymbol{x}_k \leqslant \boldsymbol{p}\boldsymbol{\omega}_k,$$

it follows that $\boldsymbol{x}_k \in \mathfrak{f}_k(\vec{\boldsymbol{x}}, \boldsymbol{p})$.

The set $\mathfrak{f}_k(\vec{\boldsymbol{x}}, \boldsymbol{p})$ is compact because it is nearstandard and closed.

Let $A_k = \hat{X}_k$ for $k < \kappa$ and $A_\kappa = P$. Then we observe that the previous lemmas imply that

$$\langle K, \{A_k\}_{k\in K}, \{\mathfrak{f}_k\}_{k\in K}, \{u_k\}_{k\in K}\rangle$$

forms a social system. ⊣

By virtue of this, we can establish the following result.

Theorem 10.6.9 *An equilibrium exists in the economy*

$$\mathfrak{E} = \langle (X_k, \succ_k)_{k\in K}, (\boldsymbol{\omega}_k)_{k\in K}, P, \eta\rangle.$$

Proof Let $A = \times_{k\in K} A_k$. Every social system possesses an equilibrium; that is, there exists a profile of actions $(\boldsymbol{a}_k^*) \in A$ that maximizes the utility of each agent. It is easy to see that this equilibrium is a competitive equilibrium. ⊣

10.7 Conclusion

In this way, I conclude this effort to present in a single text a relatively comprehensive view of nonstandard analysis, from its foundations to some of its applications.

Bibliography

Aumann, R. J. (1966). Existence of competitive equilibria in markets with a continuum of traders. *Econometrica, 34*(1), 1–17.

Brown, D. J. (1976). Existence of a competitive equilibrium in a nonstandard exchange economy. *Econometrica, 44*(3), 537–546.

Brown, D. J., & Lewis, L. M. (1981). Existence of equilibrium in a hyperfinite exchange economy: I and II. Cowles Foundation Discussion Papers, 817.

García de la Sienra, A. (2019). *A structuralist theory of economics*. Routledge.

Mas-Colell, A., & Vives, X. (1993). Implementation in economies with a continuum of agents. *Review of Economic Studies, 60*, 613–619.

Rashid, S. (1978). Existence of equilibrium in infinite economies with production. *Econometrica, 46*(5), 1155–1164.

Rashid, S. (1987). *Economies with many agents. An approach using nonstandard analysis*. The Johns Hopkins University Press.

Richter, M. K. (1971). Rational choice. In J. S. Chipman, L. Hurwicz, M. K. Richter, & H. F. Sonnenschein (Eds.), *Preferences, utility, and demand* (pp. 29–58). Harcourt Brace Jovanovich.

Robinson, A., & Brown, D. J. (1975). Non-standard exchange economies. *Econometrica, 43*(1), 41–55.

Scott, D., & Suppes, P. (1958). Foundational aspects of theories of measurement. *Journal of Symbolic Logic, 23*, 113–128.

Skala, H. J. (1975). *Non-Archimedian Utility Theory*. D. Reidel.

Author Index

A. García de la Sienra, *The Logical Foundations of Nonstandard Analysis*, Synthese Library 530, https://doi.org/10.1007/978-3-032-05677-1

Subject Index

A. García de la Sienra, *The Logical Foundations of Nonstandard Analysis*, Synthese Library 530, https://doi.org/10.1007/978-3-032-05677-1

www.ingramcontent.com/pod-product-compliance
Lightning Source LLC
Chambersburg PA
CBHW051347170226
39774CB00009B/24
* 9 7 8 3 0 3 2 0 5 6 7 6 4 *